Biodiversity in HORTICULTURAL CROPS

Biodiversity in HORTICULTURAL CROPS

— Volume 2 —

Editor
Professor K.V. Peter

2023

Daya Publishing House®

A Division of

Astral International Pvt. Ltd.

New Delhi – 110 002

First Published, 2008
Reprinted, 2023

ISBN: 9788170355625 (HB)

Published by : **Daya Publishing House®**
A Division of
Astral International Pvt. Ltd.
– ISO 9001:2015 Certified Company –
4736/23, Ansari Road, Darya Ganj
New Delhi-110 002
Ph. 011-43549197, 23278134
e-mail: info@astralint.com
Website: www.astralint.com

Printed at : **Replika Press Pvt Ltd**

Devotion

Volume II of Biodiversity in Horticultural Crops is devoted to Prof. T.N. Ananthakrishnan, the well known Entomologist and the famous Thrips man and Former Director, Zoological Survey of India (1977-1980) who encourages me to be a good scientist. I also devote this Volume to Dr. A.K. Arora, Former Officiating Director, NBPGR, New Delhi (1987–88) who persuaded me to author books on Bidodiversity.

Acknowledgement

I acknowledge all the eminent scientists who contributed chapters to the book. My tributes are to Prof. S. Kannaiyan, Chairman, National Biodiversity Authority (NBA), Chennai for writing Foreword to the book. I express thanks to Prof.(Dr.) P.I. Peter, Chairman, Health India Laboratories, Adayar, Chennai for support.

NATIONAL BIODIVERSITY AUTHORITY
Govt. of India
475, 9th South Cross Street, Kapaleeswarar Nagar, Neelankarai, Chennai - 600 041.Tamilnadu.
Tel : 044-24490805 Telefax : 044-24492777 e-mail : nba_india1@vsnl.net Website : www.nbaindia.org
सत्यमेव जयते

Prof. Dr. S. KANNAIYAN
M.Sc., (Ag.) Ph.D., C. Biol., M.Biol., (London)
FIBA (Cambridge), FABI (USA), FNAAS, FNABS, FISB, FAMI
CHAIRMAN

Foreword

India is a global player in horticulture and it is one of the world's largest producer of fruits, vegetables, spices, plantation crops and medicinal plants. It is home for a large number of horticultural crops and to name a few, mango, banana, brinjal, black pepper, etc. These horticultural crops, ranging from temperate to tropical; and from arid to humid, are facing the threat of genetic erosion due to the monoculture and introduction of newer varieties. This is a serious concern for all those who are connected with biodiversity conservation.

In the light of the above it is essential to take stock of the available biodiversity in the country. There are many organizations such as National Bureau of Plant Genetic Resources (NBPGR), State Agricultural Universities, array of All India Coordinated Research Projects and the ICAR Institutes which are actively involved in the conservation of biodiversity of various horticultural crops. I compliment Dr. K.V. Peter, formerly Vice-Chancellor, Kerala Agricultural University for having taken the initiative to edit the status of biodiversity of horticultural crops in a series of ten volumes. The present Volume II is a compilation of articles written by 32 eminent Scientists. I hope the book would be of great interest to all those concerned with biodiversity of horticultural crops.

25.01.08
(S. Kannaiyan)

Res: 17C–A1, Sapthaswara Apartment, 3rd Seaward Road-Lane, Valmiki Nagar, Thiruvanmiyur, Chennai – 600 041, Tamil Nadu, Phone: 044-24423857 Mobile: 9884073442 E-mail: skannaiyan@hotmail.com

Contents

Part III: Biodiversity of Spices

Part IV: Biodiversity of Ornamentals

List of Contributors

Chapter 1

N. Muthi Anishetty, Formerly Senior Scientist, FAO, No# 888, Siddhartha Nagar, (Plot No. 91), P.O.Vengalarao Nagar, Hyderabad

S.R. Pandravada & K.S.Varaprasad, NBPGR Regional Station, Rajendranagar, Hyderabad

Z.Abraham, NBPGR Regional Station, Thrissur

Chapter 2

Jayashree Krishnankutty & K.V.Peter, College of Horticulture, Kerala Agricultural University, P O KAU – 680 656

Chapter 3

Savita Bisht, Department of Botany, Government P.G. College, HNB Garhwal University, Kotdwara, Garhwal, Uttarakhand

Chapter 4

V.A. Celine, Department of Oleriiculture, College of Agriculture, Kerala Agricultural University, Vellayani, Trivandrum

Chapter 5

B. Varalakshmi, Division of Vegetable Crops, Indian Institute of Horticultural Research, Hessaraghatta Lake Post, Bangalore – 560 089

Chapter 6

Amish K. Sureja, Department of Vegetable Crops, Central Agricultural University, Pasighat, Arunachal Pradesh – 791 102

E. Sreenivasa Rao, Division of Vegetable Crops, IIHR, Hessarghatta, Bangalore – 560 089

Anilabh D. Munshi, Division of Vegetable Science, IARI, New Delhi – 110 012

Chapter 7

Pritam Kalia, Division of Vegetable Crops, IARI, New Delhi – 110 012

Chapter 8

Rema Menon, R. Keshavachandran & K.V. Peter, Kerala Agricultural University, Vellanikkara, Thrissur – 680 656, Kerala

Chapter 9

G.S. Karibasappa, National Research Centre For Grapes, Manjri Farm, Pune – 412 307

Chapter 10

S. Prasannakumari Amma, V.K. Mallika & E.K. Lalitha Bai, Cadbury- KAU Co-operative Cocoa Research Project, College of Horticulture, Kerala Agricultural University, Vellanikkara, KAU P.O – 680 656, Thrissur, Kerala, India

Chapter 11

D. Singh, V.C. Prahalad, P.S. Chauhan & A. Kavita, College of Horticulture and Forestry, Maharana Pratap University of Agriculture and Technology, Jhalawar, Udaipur

L. Wangchu, College of Horticulture and Forestry, Central Agricultural University, Pasighat, Arunachal Pradesh,

Chapter 12

D. Pandey, G. Pandey, S.K. Shukla & R.K. Pathak, Central Institute for Subtropical Horticulture, Rehmankhera, P.O. Kakori, Lucknow

Chapter 13

Ajay Kr. Singh, Assistant Professor (Hort), KVK, Jakhdhar, Rudraprayag (GBPUA&T, Pantnagar) Uttarakhand, India

T.S. Mehra, Assistant Professor and Head, Department of Forestry, DIBNS Manduwala, Dehradun, Uttarakhand, India

Chapter 14

Jacob Varghese, Department of Botany, S.H.College, Thevara, Cochin – 13

Introduction

Volume II of 'Biodiversity in Horticultural crops' deals Centres of Origin and Diversity of Horticultural crops; convention on biodiversity and sacred groves for biodiversity conservation. Biodiversity (s) in vegetables like amaranth, drumstick, radishes and tropical cauliflower are dealt elaborately. Fruits like jackfruit and grape, commercial crops like cocoa, spices like tamarind, medicinal plants like Indian Gooseberry and orchids of medicinal value and ornamentals like plumeria are discussed in the context of biodiversity.

Knowledge on origin of crops has become important in view of recent developments in Geographical Indications (GI) and Intellectual Property Right (IPR) regimes giving right to communities for the uses of plants and traditional (indigenous) knowledge. In the post Convention on Bio-Diversity (CBD) era, ownership of Genetic variability existing in places of origin shifted from global right to sovereign right of that country. Presence of wild relatives, maximum crop diversity and associated microbial diversity in a location are an essentiality in designating a place as a centre of origin. About 300 plant species are in cultivation throughout the world as major and minor horticultural crops. Vavilov (1951) postulated eight centres where the crops originated leading to the development of agriculture independently and named them as 'Centres of Origin'. Howkas (1988) opined that agriculture had begun not once but several times, more or less simultaneously in different regions of the world. His concept of 'centres of agricultural origin' from where farming spread into one or more regions where he referred as 'Nuclear centres and regions of diversity'. He links nuclear centres with the archeological evidence to provide strong proofs of agricultural origins. Zeen and Zhukovsky (1975) revised boundaries of Vavilov's eight centres by adding Australia, Africa and Siberia and proposed 12 mega gene centres of crop plant diversity. Harlan (1971) recognized only three main centres, each more or less connected but with large diffused non-centres. India is one of the 12 centres/ regions of diversity of crop plants in world. India is a primary centre of diversity for cucurbits (smooth gourd, ridge gourd and pointed gourd), fruits (jack fruit, banana, mango and jamun), spices (large cardamom and black pepper) and several medicinal plants (*Rauwolfia serpentina* and *Saussurea lappa*). It is also a secondary centre of diversity for cluster bean and okra and regional (Asiatic) centre of diversity for Amaranth, buck wheat, cucumber, bittergourd, bottle gourd, snake gourd and Brassicae.

Biodiversity estimates of planet earth range from 5 to over 100 million species, out of which 1.5 million were identified including plants, insects, mammals, birds and other life forms. Unfortunately the earth's biodiversity is fast disappearing. Thousands of species are being eliminated each year due to human activities. Conserving biodiversity is especially crucial in developing countries where people's livelihoods are directly dependent on natural resources like forests, fisheries and wild life. The net economic benefits of biodiversity are estimated at least US$ 3 trillions/year or 11 per cent of the annual world economic output. The Convention on Biological Diversity (CBD) was opened for signature at the Earth Summit in Brazil on 5 July, 1992 and came into force on 29 December, 1993. It contains three national level obligations: to conserve, to sustainably use and to share the benefits of biological diversity. The treaty was signed by 189 nations. Many nations have since ratified the bio- treaty. The treaty/ convention has 42 articles. One of the most controversial issues addressed by the Biodiversity treaty is intellectual property rights(IPR) related to biological and genetic resources. In 2002, the Conference of the Parties of the Convention on Biological Biodiversity (COPCBD) adopted the target to achieve a "significant reduction of the current rate of biodiversity loss at the global, regional and national level as a contribution to poverty alleviation and to the benefit of all life on earth" by 2010. Meeting the targets would require unprecedented efforts.

Globally, there is increased focus on traditional natural resource management and conservation systems in recent times. The people initiated systems for management of natural resources like cultural protected sites and sacred groves are widely acknowledged as an indicator of their sustainable management. Existence and status of sacred groves are indicators of sustainability recognized by various international initiatives as an indicator of good forest management. The tradition of sacred groves in India is one of the best conservation practices. The sacred groves are segments of landscape containing vegetation and other forms of life and geographical features that are delimited and protected by human societies under the belief to keep them in a relatively undisturbed state as an expression of an important relationship of humans with the divine or with nature. The sacred groves are pockets of more or less climax vegetation preserved for religious grounds often found in remote areas. As an ecosystem, the sacred groves play an important role in soil and water conservation. Many sacred groves contain water resources like ponds and streams and the vegetative mass that covers the floor of a grove can absorb water during rainy season and release it during times of drought, hence helping in its conservation. In many parts of Kerala (India), these sacred groves are found in the backyards of numerous traditional houses where a large part of the land is used for growing trees, vegetables and fruits. Traditional folklore is full of such references and these groves evolved as a part of daily life. About 18907 sacred groves are reported in India, Himachal Pradesh alone reporting 5000, followed by Kerala (2000), Maharashtra (1600) and Rajasthan (1036).

Vegetables are food items derived from herbaceous and woody plants which provide micronutrients, antioxidants and are alkaline in reaction on usage. They are non-calorific in nature and provide the much needed fibres to human diet. Based on plant parts used, vegetables are classified as leaf, fruit, stem, tuber and root vegetables. The whole plant is used as vegetable as in amaranth, spinach, celery, parsnip, water leaf, radish, chekkurmanis etc. The present Volume II deals with biodiversity in amaranth, drum stick, tropical cauliflower and radish.

Amaranth is a multipurpose herb grown for leaves, stem and grains.Leaf is rich in vit.A and iron.Grain amaranth is popped and used as nutrient additive.Red pigments are extracted from red types.Amaranth leaf and stem have a lot of fibre very essential for digestion and absorption of nutrients by body.

Drum stick or horse radish tree is a vitamin rich, mineral packed and nutritious vegetable indigenous to North West India. It grows wild in the sub-Himalayan tract from river 'Chinab' eastwards to 'Sarda' and the 'tarai' forests of Uttarakhand in India. There are 13 species of *Moringa* within the monogenic family Moringaceae. The 13 species of Moringa are fitted into three broad categories–bottle trees, slender trees and trees, shrubs and herbs of NE Africa. Drumstick leaf, flowers, pods and stem have culinary and industrial uses.

Tropical cauliflower is a cole vegetable grown globally of which Asia is the leading one followed by Europe and covers small acreages in North America, South America, Africa, Australia and New Zealand. The greatest genetic diversity of cauliflower is in the Mediterranean gene centre including Greece, Syria, Cyprus, Sicily, Italy, Spain and Portugal. Tropical cauliflower is one of the five groups proposed by Crisp (1982)–Asian Group. Asian group is characterized by presence of four maturity groups and is composed of recombinants of European annuals and biennials developed within 250 years and adapted to tropics. The four maturity groups are early, extra early, early (1 a), early (1 b), mid early, mid late and late or snow ball types. The first three groups include typical Indian or Tropical types and the selections made for this purpose perform well, which produce quality curds even during May-June making it possible to grow cauliflower almost all the year round.

Radish is a popular root vegetable in tropical, sub-tropical and temperate regions of the world and is the best recognized for the crisp texture and pungent flavour of its enlarged storage roots and hypocotyls, cooked or salted as a vegetable and also eaten fresh as grated radish, garnish and salad. Radish is one of the most ancient vegetables. The area from Eastern Mediterranean to Caspian sea is likely the origin of European types, but Werth (1937) argued for a more easterly source. Indian group of radishes including the rat-tail types originated in the Near Eastern and Indo-Burma centre of origin though their ancestors have been lost.

Fruits provide vitamins, minerals, anti oxidants and much needed flavour, fragrance and taste to dietary preparations. Among sub-tropical fruits, grapes occupy premier position for its table uses, vines and value added resins and dehydrated berries. Grape being the most important fruit of the world includes more than 10,000 cultivars spread around 100 countries. *Vitis* genus has 70 related species. Efforts are made at National Research Centre for Grapes, Pune, India to utilize indigenous species and varieties in crop improvement programme through biotechnological tools. Grapes are classified based on usage- table and vine. There are seeded, seedless and smooth seeded varieties. Based on colour, there are red, white, yellow, violet, green and purple grapes. Vine grapes yield juice of intrinsic quality. Root stocks of grapes belong to *Vitis vinefera* and related species. Biodiversity of grapes is studied and utilized by many stake holders involved in grape improvement, grape vinery and value added products industry.

Jack fruit is indigenous to the rain forests of the western ghats and is cultivated throughout the tropical low lands in South and South East Asia,parts of Central and Eastern Africa and Brazil.Jackfruit is comparatively a cheap fruit and a fairly good source of vitamins and minerals.Many value added products are made from fruits,seeds,stem and leaves.Property of wood is comparable to teak especially to make wood frames. Ready to serve drink from fruit is a treat of its own.

Use of beverages and confectionaries are indicators of quality food usage. Both provide pleasing and pleasant feeling to food. As appetizer and flavourant, beverages and confectionaries have economic importance in domestic and international trade. Always used after processing, the end product is always value added. There are also stimulatory and energy enhancing properties attributed to beverages. Cocoa is one such beverage crops used to make chocolate, one of the greatest treasures discovered by man. According to legend, cocoa tree was discovered by the Mayan King Quetzalcoatle, 3000 years ago. Central and South America are the centres of origin. Cocoa was introduced to Europe in 16th Century. Most of the genetic diversity is found in the Americas, whereas majority of production comes from West African countries. The genus *Theobroma*, a genus of family Sterculiaceae contains 22 species grouped into 6 sections. The family contains 50 genera and 750 species of trees and shrubs mostly tropical. Cheesman (1944) classified the cultivated and wild cocoas into three groups–the Criollo, Forastero and Trinitario. The major cocoa growing countries are Equador, Brazil, Surinam, Trinidad and Tobago, Venezuala, Colombia, Guyana. The lesser Antilles, St. Lucia, The greater Antilles, Jamaica, Cuba and Mesoamerican countries. and old world countries like Philippines, Indonesia, Sri Lanka, Madagaskar, Thailand and India. The above countries possess notable biodiversity in cocoa. Distibution of cocoa germplasm is through internationally recognized agencies. The International Cacao Gene banks, Trinidad and the collection at CATIE Regional Germplasm Centre. Turrialba, Costa Rica are designated as 'Universal Collection Depositories'. The future of World Cocoa depends mainly on the ready availability of improved planting materials. A lot of improvement is warranted in resistance to major diseases like witches broom, cocoa swollen shoot virus, vascular streak, die back and *Phytophthora* pod rot, bean quality traits, bean size and adaptability to adverse weather conditions.

Spices are the essence of quality food and dietary. As a natural preservative and as antioxidant, spices played historical role in the life and death of Royalty. Mystic aroma and flavour, tentilating colour and majestic look, spices add flavour, fragrance and colour to food items. There are spices of seed, flower, fruit, leaf, stem, bark and root origin. A total of 52 species are listed in the Indian Spices Board Act. Tamarind is a tree spice where pods both mature and immature, leaves, bark and seed have use in food as a spice. Tamarind water is served to all the Guests in Indonesian and Cambodian homes for coolness and freshness. Acidic extracts are taken from leaves and pods which are used in several Ayurvedic drugs. *Tamarindus indica* L. is a member of dicot family Fabaceae, the largest with 700 genera and 17,000 species. The genus *Tamarindus* is monotypic. In the past, a distinction was made between tamarind of the West and East Indies. Tamarind originated in Madagaskar from where it spread to South and East. The tamarind trees are common in S. India particularly in Tamil Nadu and Andhra Pradesh. Wood, root, leaves, inflorescences and flower, fruit and seed are economic plant parts with manifold uses. The genetic variation of tamarind is primarily based on fruit characteristics like length of pod, pod weight, seed number, pod colour, and sweetness of pulp. Rome based Biodiversity International (Former IPGRI) has published a monograph and descriptor on tamarind.

Aonla or Indian Gooseberry is one of the important indigenous fruits of Indian sub-continent. The aonla originated in tropical South east Asia, particularly Central or Southern India, Pakistan, Bangladesh, Sri Lanka, Malaysia, Southern China and Mascarene Islands. Aonla is well distributed in India. Among medicinal trees, aonla occupies a pride of place. Pulp of fruit is highly nutritious and one of the richest sources of Vit. C. Aonla fruits are acidic, acrid, cooling, refrigerant, diuratic and laxative. Normally one aonla fruit contains 20 times vitamin C in terms of antiscorbutic value than two oranges. Aonla is naturally spread due to seed propagation. Grafting on seedlings is also done to propagate elite trees. In fact aonla fruits from wild and forest grown seedling trees are used to prepare 'dathryarishtam', widely recommended to children and women.

Orchids are precious ornamentals belonging to the most highly evolved family of the entire plant kingdom–Orchidaceae. The family comprises over 30,000 individual species and new discoveries are being made every year. About 1220 species are reported from India while 750 species are found in North East India and about 550 species are reported from Arunachal Pradesh alone. About 40 orchid species are used in Ayurvedic formulations like Ashtavarga, Jivinyavarga, Madhuravarga and Chayvanaprash. Orchids also are of ethnobotanical interest linking aboriginal man with plants for medicine. Thirty one lesser known orchids with considerable medicinal value are indicated in the chapter on 'Orchids of Medicinal value'.

Among avenue ornamental trees, plumeria plays a significant role. A member of Apocynaceae family, plumeria or temple tree is native to tropical and sub-tropical America. The ornamental varieties of this tree are grown in tropics in gardens, parks and landspacing around the world. This tree is a common feature of the Hawaii Islands, that many people construe to be indigenous to Hawaii, known as Frangipini or Temple Tree on Champa. It is frequently associated with temples and grave yards. Buddhists, Muslims and Christians alike consider the tree as a symbol of immortality because of its unusual power of producing leaves and flowers even after it has been uprooted from the soil. Plumeria is the loveliest of all tropical American ornamental trees grown in the tropics.

The Vol.II 'Biodiversity in Horticultural Crops' elucidates centres of origin, distribution and variations in characters of selected vegetables, fruits, plantation crops, spices, medicinal plants and ornamentals.

Professor K.V. Peter

Part I

General

Biodiversity in Horticultural Crops Vol. 2 *Pages* **3–52**
Editor: **Professor K.V. Peter**
Published by: **DAYA PUBLISHING HOUSE**

Chapter 1

Centres of Origin and Diversity of Horticultural Crops

N. Muthi Anishetty[1], S.R.Pandravada[2], Z.Abraham[3] and K.S.Varaprasad[2]

[1]Formerly Senior Scientist, FAO, No# 888, Siddhartha Nagar, (Plot No. 91), P.O.Vengalarao Nagar, Hyderabad
[2]NBPGR Regional Station, Rajendranagar, Hyderabad
[3]NBPGR Regional Station, Thrissur

Genetic variation in plants is crucial for human welfare and plays an important role in the evolution and domestication of plants as crops. Genetic resources constitute a unique global heritage and an invaluable asset to meet the ever increasing demand of crop production and productivity. Many plant species originated thousands of years ago in places known as centres of origin basically located between the tropic of Cancer and Capricorn. Ancient tribal groups especially in hills and forests contributed immensely to evolution and accumulation of significant diversity through selection and domestication of crops over a long period of time. Ethnic groups contributed in creating and conserving the diversity based on their wisdom encompassing basically their day to day requirements for food, shelter, medicine or clothing. The potential of landrace diversity and wild relatives in crop improvement programmes cannot be under estimated. There is dire need for conservation of crop genetic resources as their value is immense and indispensable considering diversified new technological options available and human requirements in future.

Diversity in fruits, nuts, vegetables, tubers and medicinal plants contribute greatly both to food security and diversified dietary needs of mankind. Fruits and vegetables come under the protective foods which help in maintenance of health and prevention of diseases. Wild forms of horticultural crops with high content of mineral substances, proteins, vitamins and resistant genes for biotic and abiotic stresses assume significance for their conservation and sustainable utilization from their centres of origin and diversity (Giuseppina *et al.*, 1998).

Knowledge on origin of crops has become vitally important in view of the recent developments in science and property issues. New approaches like Marker Assisted Selection for Crop improvement and scientific capabilities of using genes and gene constructs across the genera and higher taxa surpassing traditional breeding limitations enhanced the importance of variability available in areas of origin. In the post CBD era, ownership of genetic variability existing in places of origin shifted from global rights to sovereign rights of that country.

Presence of wild relatives, maximum crop diversity and associated microbial diversity in a location are an essentiality in designating a place as 'a centre of origin'. About 300 species are in cultivation throughout the world as major and minor horticultural crops. Vegetable cultivation in the new world dates back to 7000 to 5000 BC. Root crops may have been among the earliest to be domesticated in humid tropics preceding the seed crops. Many of the present species have been so altered as a result of selection and breeding that their wild ancestors can no longer be identified with certainty. Only two vegetable crops (Bottle gourd and Sweet potato) are known grown in both the Old and New Worlds, before 1492. Their presence in both Western and Eastern hemispheres suggest possible pre-Columbian agricultural contacts. After the discovery of new world by Columbus, there was rapid exchange of crops among the continents. Almost one third of the world's species originated in South-Eastern Asia and most of the main fruit and vegetable crops did come from the East and West Asiatic Centre and Mediterranean Centre of Origin, while roots and tubers and tropical fruit trees are concentrated in the Central American and Andean Centres. The designated centres of origin and their boundaries were revised and re-revised subsequently by different authors and more theories and concepts were put forth as more detailed studies were taken up on a number of crop species. The view points based on evidences from different sources and applying logic in pioneering efforts put forth by different researchers are given below.

Vavilovian Centres of Origin

Vavilov (1951) based on, differentiating plants into specific and intraspecific taxa on morphological and genetic basis, determining the area of distribution of such species and groups of species, establishing the distribution and magnitude of genetic diversity in geographical centres where it is at it's maximum, the centre where diversity of genetically allied species is concentrated, correlating distribution of diversity with areas of concentration of the nearest wild relatives and supporting/ linking evidences from archaeology and linguistics and history, postulated eight centres where the crops had originated leading to the development of agriculture independently and named them as 'Centres of Origin' of crop plants. These centres were also characterized by the accumulation of dominant genes in the core and the recessive genes in the periphery. The list of crop species that originated in these centres is depicted in Figure 1.1.

Nuclear Centres and Regions of Diversity

Hawkes (1983) opined that agriculture had begun not once but several times, more or less simultaneously in different regions of the world. His concept envisaged centres of agricultural origin from where farming spread into one or more regions which he referred as 'Nuclear centres and regions of diversity'. He linked nuclear centres with the archaeological evidence to provide strong proofs of agricultural origins. There are several regions where crops actually did not originate. In these regions, crops perhaps had spread from the nuclear centres in the past and due to spatial isolation and intensive human selection led to accumulation of genetic diversity. Hawkes further identified 'minor' centres for several crops.

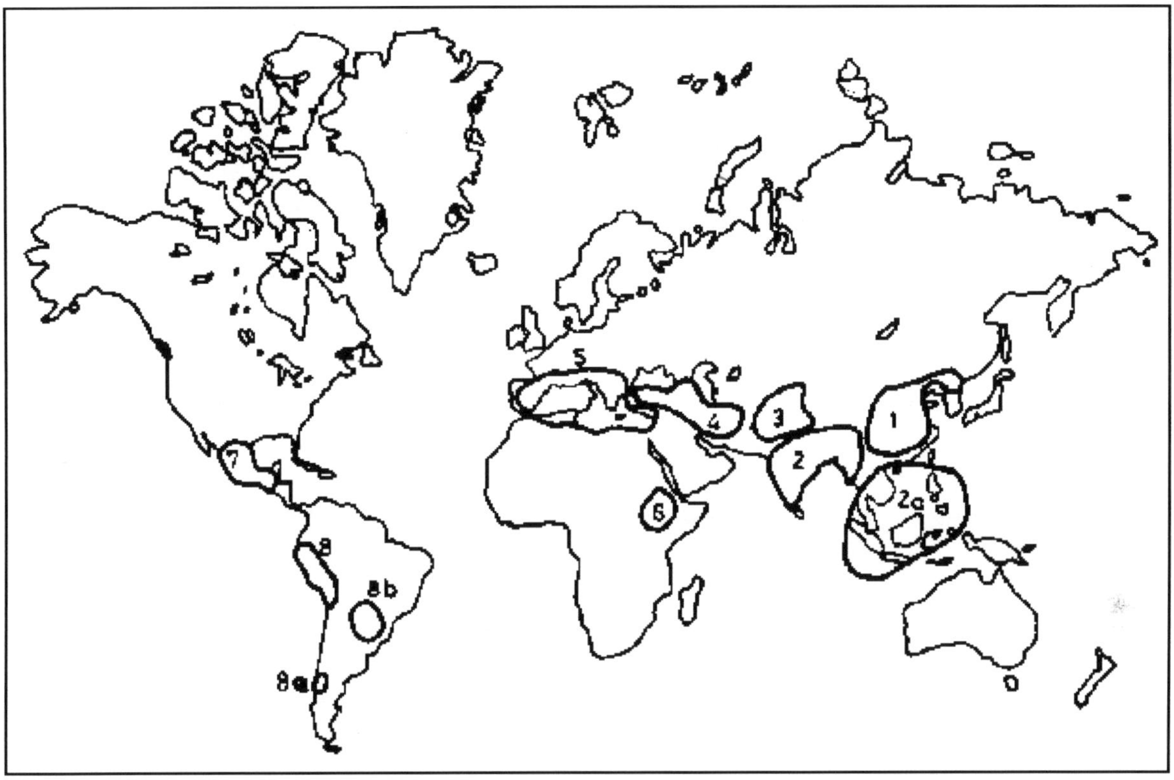

Figure 1.1: The Eight World Centres of Origin of Cultivated Plants Including Horticultural Crops according to N.I. Vavilov (Harlan, 1971)

1. China: Adzuki bean, Common bean, Peach, Orange, Chinese tea; 2. India: Eggplant, Rat's tail radish, Taro, Cucumber, Pepper; 2a. Indo-Malaya: Yams, Pomello, Banana, Coconut; 3. Central Asia: Pea, Carrot, Radish, Pear, Apple, Walnut; 4. The Near East: Pea, Melon, Almond, Fig, Pomegranate, Grape, Apricot, Pistachio; 5. The Mediterranean: Broad bean, Cabbage, Olive, Lettuce; 6. Abyssinia: Pea, Coffee; 7. Southern Mexico and Central America: Common bean, Pepper, Squash/ Pumpkin; 8. South America (Peru, Ecuador and Bolivia): Sweet potato, Potato, Lima bean, Tomato, Papaya; 8a. Chile: Potato; 8b. Brazil and Paraguay: Cassava, Cacao, Pineapple, Purple grandilla/ Passion fruit.

Mega Gene Centres of Crop Plant Diversity

Zeven and Zhukovsky (1975) by adding Australia, Africa and Siberia and revising the boundaries of Vavilov's eight centres proposed 12 mega gene centres of crop-plant diversity and dealt very elaborately on the range and extent of distribution of genetic/varietal/specific diversity. These twelve regions had wider coverage and more acceptability. The prominent crop plants for which rich diversity occurs in these 12 regions are given below.

Centres and Non-centres

Harlan (1971) recognized only three main 'Centres', each more or less connected but with large diffused non-centres. He also recognized smaller areas/pockets of varietal and/or racial diversity within a Vavilovian Centre and termed these as 'Micro centres'. Such small areas, as in Turkey and Africa (Harlan, 1975), contain varietal diversity in several crops either in plains or mountains.

Indian Subcontinent as a Centre of Origin and Diversity

India is one of the twelve centres/regions of diversity of crop plants in the world (Zeven and de Wet, 1982). The antiquity of agriculture and the ethnic diversity in the sub-continent have played a major role in diversification of crop resources in this region. Rich genetic diversity occurs in several crop plants and their wild progenitors. The Indian gene centre possesses about 166 species of agri-horticultural crop plants (Zeven and de Wet, 1982) and 320 species of wild relatives distributed in eight phyto-geographical/agro-ecological zones (Arora and Nayar, 1984). The Indian agriculture is enriched by a continuous stream of introductions of new crops and their cultivars by man since ancient times. The Portuguese, the Spaniards and the British brought several new crops into this country. Prominent horticultural crops introduced are potato, tomato, chillies, coffee, cocoa and cashew nut. Depending on the time, period, degree of variability and the areas of introduction, enormous diversity got built-up within the Indian subcontinent.

Table 1.1: The Twelve Mega Centres of Diversity for Horticultural Crops

Sl.No.	*Regions of Diversity*	*Crops*
1.	Chinese-Japanese Region	Adzuki bean, Leafy mustard, *Orange/Citrus*, Peach, Apricot, Litchi, Tea
2.	Indo-Chinese-Indonesian Region	Winged bean, Cucurbits/Ash gourd, Mango, Banana, Rambutan, Durian, Bread fruit, *Citrus*/Lime, Grape fruit, Nutmeg, Clove, Ginger, Taros and Yams, Betel nut, Coconut
3.	Australian Region	Eucalyptus, Acacia, Macadamia nut
4.	Hindustani Region	Moth bean, *Dolichos* bean, Eggplant, Okra, Cucumber, Leafy mustard, Rat's tail radish, Taros, Yams, *Citrus*, Banana, Mango, Ginger, Turmeric, Cardamom, Arecanut, Black pepper
5.	Central Asian Region	*Allium*/Onion, Garlic, Spinach, Peas, Beet root, Faba bean, Apricot, Plum, Pear, Apple, Walnut, Almond, Pistachio, Melon, Grape
6.	Near Eastern Region	Faba bean, French bean, *Allium, Cucumis,* Melon, Grape, Plum, Pear, Apple, Apricot, Pistachio, Fig, Pomegranate
7.	Mediterranean Region	Lettuce, Beet root, Faba bean, Radish, Olive, *Crocus*, Grape, Fennel, Cumin, Celery
8.	African Region	Bottle gourd, Okra, Yams, Cucumber, Oil palm, Kola, Date palm, Ensete, Melons
9.	European–Siberian Region	Peach, Pear, Plum, Apricot, Apple, Almond, Walnut, Pistachio, Cherry, Mustard (black), Chicory, Lettuce
10.	South American Region	Potato, Sweet potato, *Xanthosoma,* Lima Bean, Amaranth, *Chenopodium, Cucurbita*, Tomato, Papaya, Pineapple, Cassava, Cacao, Passion fruit
11.	Central American and Mexican Region	French bean, Potato, *Cucurbita*, Pepper/Chilli, Amaranth, *Chenopodium*,
12.	North American Region	Jerusalem artichoke, Plum, Raspberry, Strawberry

(Modified from Zeven and Zhukovsky, 1975).

The Indian gene centre is an important region of crop diversity that had strong linkages and contiguity with the Indo-Chinese-Indonesian, Chinese-Japanese and the Central-Asian regions. Further, the influx of germplasm in distant past from the Mediterranean, African and tropical American regions, facilitated build up of enormous locally adapted variability. The different agro-ecological/phyto-geographical regions of India, hold rich diversity in both the cultivated and the wild crop gene pools (Arora, 1991). Based on areas and crops, the information on distribution and extent of accumulated crop-plant diversity is given below.

Geographical contiguity with the Far-East and/or the Indo-Malayan (South/South-East Asian region) belt is largely responsible for accumulation of more regional diversity in sword bean, tomato, citrus, small cardamom, ginger, turmeric, tuber crops, particularly taros and yams.

Table 1.2: Distribution of Horticultural Diversity in the Indian Gene Centre

Diversity Region	*Agro-ecological Region*	*Horticultural Crops*
Western Himalayan region	Western Himalayas, cold arid and Western humid region	Buckwheat, amaranth, french bean, peas, pumpkin, cucumber, *Allium* spp., ginger, leafy Brassicae, pome, stone, soft and nut fruits, medicinal plants
Eastern Himalayan region	Eastern Himalayan humid region	Buckwheat, amaranth, french bean, peas, scarlet bean, pumpkin, cucumber, *Allium* spp. ginger, chayote, tree tomato, pome and stone fruits
North-Eastern region	Assam, Bengal and North-Eastern hills	French bean, dolichos bean, winged bean, pumpkin, cucumber, okra, eggplant, chilli/ *Capsicum* spp., pointed gourd, ash gourd, taros, yams, *Citrus* (lime/lemon/orange/grape fruit), banana, tea, large cardamom, ginger, long pepper
Gangetic plains (includes northern parts of central region)	Northern plains and Central highlands	Okra, eggplant, bittergourd, *Cucumis* spp., *Luffa* spp. jackfruit, mango, lemon/lime, orange, jujube, Indian gooseberry/*Emblica*, *jamun/ Syzygium*, melons
Indus and Western Arid plains	Western plains, Kutch and Kathiawar peninsula	Cluster bean, okra, *Cucumis* spp., jujube, *Khirni/Mimusops*, *Phalsa/Grewia*
Eastern peninsular/ Eastern Ghats region (includes southern parts of central region)	Deccan plateau, Eastern Ghats, Chotanagpur plateau and East coast	*Mucuna*, dolichos bean, taros, yams, elephant-foot yam, banana, mango, lemon/lime, jackfruit, ginger, turmeric, chilli/*Capsicum*, coconut
Western peninsular/ Western Ghats region	Western Ghats and West coast	Dolichos bean, sword bean, okra, eggplant, cucumber, chilli/*Capsicum*, taros, yams, elephant-foot yam, Jackfruit, banana, lime/lemon, orange, *jamun/Syzygium*, black pepper, turmeric, ginger, coconut, arecanut
The islands region	Islands of Andaman and Nicobar and Lakshadweep	Coconut, breadfruit, taros, yams, *Xanthosoma*

(Modified from Arora, 1991).

India is a primary centre of diversity for crops like cucurbits (smooth gourd, ridge gourd and pointed gourd), fruits (jack fruit, banana, mango, *Syzygium cumini/jamun*), spices (large cardamom, black pepper) and several medicinal plants (*Rauwolfia serpentina* and *Saussurea lappa*). It is also a secondary centre of diversity for African crops (cluster bean and okra), tropical American crops (tomato, pumpkin/*Cucurbita* spp., chayote, chillies and *Amaranthus spp.*) and regional (Asiatic) centre of diversity for amaranth, buckwheat, cucumber, bitter gourd, bottle gourd, snake gourd and Brassicae.

Information about the centres of origin and diversity, domestication pattern, movement and introduction of crops and extent, distribution and pattern of diversity for different horticultural crops which had either originated/having a primary or secondary centre of diversity or for crops of recent introductions which have become indispensable in the dietary/food habits of the people and having commercial importance are given below.

Vegetables

The important vegetable crops/crop groups for which good native diversity occurs in the Indian Gene Centre include beans, brinjal, cucurbits (cucumber and other gourds), leaf vegetables, okra, onion and tuber crops (taros and yams) (Rathore *et al.*, 2005).

Brinjal (*Solanum melongena* L.)

The genus *Solanum* originated mainly in Asia and Africa. The economically most important and best-known species, *S. melongena* (n=12) was domesticated from the wild progenitors in the North-Eastern (Indo-Myanmar) region and has spread all over the world and immense diversity in eggplant and its wild relatives exist in this region. Vavilov (1928) regarded the crop as being of Indian origin and opined that a secondary centre of diversity had developed in China. Wild forms of *S. melongena* which are armed perennial herbs/shrubs with bitter fruits are reported from India (Bhaduri, 1951). A total of 22 wild species are reported from the length and breadth of sub-continent (Smartt and Simmonds, 1995). Many *Solanum* species have medicinal values (Bukenya and Carasco, 1999). Two African species, the scarlet egg plant (*S. aethiopicum* L.) and the gboma egg plant (*S. macrocarpum* L.) are extensively cultivated in Africa and at a very limited scale in some other parts of the world. Other species *viz. S. nigrum* L. and *S. nodiflorum* Jacq., are at times cultivated/semi-cultivated in Africa and Asia (Daunay *et al.*, 2000).

It is believed that, Aubergine (brinjal) was introduced to China more than 1,500 years ago where small fruited types got evolved. It was introduced to the countries of Mediterranean region by the traders from Arabia and Persia and to Brazil by the Portuguese. In Koraput region, the *pottangi* variety, a very primitive landrace, presumed as the progenitor of the present day cultivars *of Solanum melongena* occur. Distinct landraces are grown in tribal pockets of India which have specific local preference.

Solanum melongena complex has three species, namely, *S. melongena* L., *S. indicum* L. and *S. insanum* L., very closely related with each other but their taxonomic status is disputed. The wild forms of brinjal, *S. incanum* and *S. melongena* var. *insanum* whose status as the progenitors or derivatives of *S.melongena* are not clear. They are distributed from peninsular region to the foot hills of the Himalayas, but their populations are scanty and rather more sporadic even within the same area/site. Meiotic behaviour of hybrids between these species is regular, suggesting close phylogenetic relationship among the three species (Kalloo *et al.*, 2005).

Okra (*Abelmoschus esculentus* L.)

The genus *Abelmoschus* is of Asiatic origin, but the exact ancestral home of cultivated *A. esculentus* is disputed. Indian origin of *A. esculentus* was advocated by Zeven and Zhukovsky (1975) and Zeven and de Wet (1982). Vredebregt (1990) observed that rich diversity in *A. ficulneus* and that of *A. tuberculatus* in India may be an argument for an Asian origin of *A. esculentus*. The geographical origin of okra based on occurrence of a putative ancestor *A. tuberculatus* seems to be India (Smartt and Simmonds, 1995). *Abelmoschus ficulneus* contributed the second unknown genome to cultivated okra and the primary genome being from *A. tuberculatus*. The ancient Sanskrit literature in India also mentioned *Tindisha* and *Gandhamula* designating *Bhindi* (Okra) (Dhankhar *et al.*, 2005).

South-East Asia as centre of diversity for the genus *Abelmoschus* was considered by van Borssum-Waalkes (1966). It was domesticated in the Ethiopian region (Vavilov, 1936) or in West Africa (Joshi *et al.*, 1974). A late 19th century record shows its occurrence in wild on the White Nile in Sudan (Singh *et al.*, 1975). Joshi and Hardas (1976) suggested that the cultigen, *A. esculentus* originated in Asia or might have present in Africa and Asia as a polytypic species.

A total of ten wild species of *Abelmoschus* namely, *A. angulosus* (2n = 56), *A. tuberculatus* (2n = 58), *A. manihot* (2n = 66), *A. moschatus* and *A. ficulneus* (2n = 72), *A. esculentus* (2n = 130), *A. tetraphyllus* var. *tetraphyllus*, *A. tetraphyllus* var. *pungens* and *A. crinitus* (2n = 138) and *A. caillei* (2n = 196) occur in India. Of these, only *A. esculentus* is cultivated in India. *Abelmoschus moschatus* which occurs wild is also cultivated for its aromatic seeds while the rest of the species are truly wild taxa.

Distribution of cultivated and wild species is in diverse habitats all around the world from Mediterranean to the equatorial areas. An overlapping of cultivated and wild species occurs in South-East Asia, considered as a centre of diversity. The wild okra *Abelmoschus tuberculatus* (related to the cultivated okra, *A. esculentus*) widely occurs in the North-Western plains of India characterized by tall types with tuberculate fruits. Vredebregt (1990) gave an account of world distribution of different *Abelmoschus* species as follows:

1. *Abelmoschus angulosus* in wet and temperate regions of India (Eastern Kerala) to Indonesia (Bali), Sri Lanka, Northern Myanmar and Indonesia (Sumatra) between 750–2000 m amsl. This species tolerates low temperature and little frost.
2. *Abelmoschus crinitus* in Northern Pakistan down to Indonesia (Java) through India, Myanmar and Thailand in open forests and also in dry areas up to 1,400 m amsl.
3. *Abelmoschus ficulneus* in Southern Chad down to Central Australia through Sudan, Ethiopia, Madagaskar, India and Indonesia in low lands up to 600 m amsl with pronounced dry season, in open vegetations and on wastelands.
4. *Abelmoschus manihot* ssp. *tetraphyllus* var. *tetraphyllus* in South Asia (India, Myanmar and Thailand) and South-West Pacific (the Philippines, Moluccas and Papua New Guinea) in dry areas up to 400 m amsl.
5. *Abelmoschus manihot* ssp. *tetraphyllus* var. *pungens* in Indonesia, the Philippines, Christmas Islands, Japan, China, India, Nepal, Indo-Chinese Peninsula and Myanmar in forest glades, hilly slopes between 400 to 3,000 m amsl. This species is tolerant to low temperatures.
6. *Abelmoschus manihot* ssp. *tetraphyllus* var. *megaspermus* having larger seeds is reported only from the Indian States of Maharashtra, Gujarat and Madhya Pradesh primarily occurring on shady hill slopes and foot hills.

The maximum phenotypic variability was observed in cultivated species of *A. esculentus*. The countries with the largest collections of diversity are Turkey and India and to a lesser extent Afghanistan, Iran, Pakistan, Bangladesh and erstwhile Yugoslavia (Dhankhar *et al.*, 2005). Native diversity occurs for plant height, pigmentation and fruit characters. Very tall, robust, purple pigmented, tolerant to cold and late maturing types occur in high altitude tribal areas in the Peninsular plateau. In Okra (*Abelmoschus esculentus*), most of the local landraces were replaced by improved cultivar Pusa Sawani and others in India.

Cucurbits

Cucurbitaceae is divided into two sub-families and eight tribes representing varying degrees of circumscriptive cohesiveness with about 118 genera and 825 species (Jeffrey, 1990). Many cultivated and wild species of Cucurbitaceae were domesticated in prehistoric times in the New and Old worlds and are associated with man's culture. Almost all the cucurbits are of tropical origin (Africa, America and Asia) and these are distributed all over the world in different ecological zones (Sirohi *et al.*, 2005). Among the new world species, Pumpkin (*Cucurbita moschata* (Duch. ex Lam.) Duch. ex Poir.) and Squash (*Cucurbita pepo* L.) are very ancient, recorded in archaeological findings in Mexico dating back to 7000 BC.

The Indian sub-continent is the centre of origin for a number of wild and cultivated cucurbits. Chakravarty (1982) reported that out of 110 genera and 640 species in the world, 36 genera and 100 species are found in India which include 38 endemic species. The important cultivated Cucurbit species for which the Old World especially India is the centre of origin include cucumber (*Cucumis sativus* L.), muskmelon (*Cucumis melo* L.), gherkin (*Cucumis angularis* L.), bitter gourd (*Momordica charantia* L.), Ivy gourd (*Coccinia grandis* (L.Voigt.) and Ridge gourd [*Luffa acutangula* (L.) Roxb.]. For all the above species, the sub-continent is the primary centre of diversity with occurrence of tremendous variability in both cultivated and wild forms.

Among the wild species, *Cucurbita lundelliana* Bailey, *C. martinezii* Bailey and *C. okeechobeensis* Bailey are resistant to powdery mildew. *C. martinezii* also possesses resistance to squash mosaic and water melon mosaic whereas *C. foetidissima* Kunth is resistant to drought. *C. pedatifolia* Bailey is a species intermediate between xerophytic and mesophytic species. Wild species of *Momordica*, *Trichosanthes* and *Cucumis* (*C. hardwickii*) occur predominently in the North-Eastern and peninsular tracts extending to the hills.

The origin and geographical distribution of different cucurbit species are given below.

Cucumis

Cucumis includes about 30 species of annuals and herbaceous perennials of the old world tropics and warm-temperate regions of which cucumber (*Cucumis sativus*) and melon (*C. melo*) are economically important. *C. sativus* originated in India where its close relatives with the same chromosome number ($2n$=14) occur. Among all these species, *C.sativus* and to a limited extent *C. callosus* (Rottle.) Cogn. are cultivated while *C. hardwickii, C. hystrix, C. prophetarum* L. and *C. setosus* Cogn. are wild species (Chakravarty, 1982; Arora and Nayar, 1994). Rich variability is found in foothills of North-West Himalayas and Southern hills for *C.hardwickii*; North-East Himalayas (Meghalaya, Mizoram and Assam) for *C. hystrix*; Rajasthan (Sirohi District) and parts of Gujarat, Maharashtra and Tamil Nadu for *C. prophetarum* and parts of Maharashtra for *C. setosus*. Cucumber is cultivated at least for the past 3,000 years in India and 2,000 years in China. China is a secondary centre of diversity where it occupies the second largest area among vegetables after Chinese cabbage.

The centre of origin for *C. melo* is still not clear although the evidences point to Africa where wild species of *Cucumis* with the same chromosome number ($2n$=24) frequently occur. Domestication might have taken place independently in East Asia and South-East Asia including India. The primary centre of diversity for this extremely polymorphic species is South-West and Central Asia encompassing Turkey, Syria, Iran, Afghanistan, North and Central India, Trans-Caucasian, Turkmenistan, Tajikistan and Uzbekistan and the secondary centre of diversity existing in China and Republic of Korea. Munger and Robinson (1991) proposed seven horticulturally important groups under melons based on fruit characteristics and uses rather than phylogeny.

1. *Cucumis melo* var.*cantaloupensis* (musk melon): It is usually andro-monoecious and has medium-sized fruits with netted, warty or scaly surface. The fruit has usually orange or sometimes green flesh and is aromatic having musky flavour.
2. *C.melo* var.*flexuosus* (snake melon, long melon): It is monoecious and has long slender fruits, consumed immature.
3. *C.melo* var.*momordica* (snap melon): It is monoecious and has ovoid to cylindrical fruits with white to pale orange pulp having low sugar content. The smooth surface of the fruits cracks longitudinally on maturity.
4. *C.melo var.inodorus* (winter melon): It is mostly andro-monoecious and has white or green flesh that lacks musky odour. Fruit shelf-life is longer than musk melon and is indehiscent at maturity.
5. *C.melo var.dudaim* (mango melon, pomegranate melon): It is pickling type and has small, globular fruits having white flesh with acid flavour.
6. *C.melo* var. *conomon* (pickling melon): Cultivated.
7. *C.melo* var. *agrestris* (kachri): It occurs mostly in wild state but is cultivated to a limited extent.

Two other species of some importance in *Cucumis* are *C. anguria* L. (West Indian Gherkin) and *C. metuliferus* E.Mey. (African horned cucumber). *C. anguria* is of American origin, but sexual compatibility and electrophoretic studies give credence that it is a cultigen having descended from a non-bitter mutant of a bitter African species *C. longipes* Hk. f. *C. metuliferus* is also of African origin and has commercial potential (Sirohi *et al.*, 2005).

Cucurbita

Cultivated *Cucurbita* originated in two centres in the new world. While, Central Mexico is the centre of origin for *C. ficifolia*, *C. mixta*, *C. moschata* and *C. pepo*, contiguous area encompassing Southern Peru, Bolivia and Northern Argentina is the centre of origin for *C. maxima*. Areas of diversity are Northern Mexico for *C. pepo*, Central Mexico and the Yucatan peninsula to Costa Rica for *C. mixta*, Central Mexico through the high plateau of Central America along the Andes to Central Chile (1000-2000 m. amsl) for cold tolerant *C. ficifolia*, Southwards from Mexico City through Central America into Northern Colombia and Venezuela for *C. moschata* and Northern Argentina, Bolivia, Southern Peru and Northern Chile for *C. maxima*. Whitaker and Bemis (1975) opined that *C. moschata* is the nearest putative ancestor of the genus. They suggested *C. moschata* as a link between the wild and cultivated species and that all the five cultivated species were independently domesticated. In India, *C. moschata* is more widely distributed than the other species.

According to Decker (1988), archaeological evidences indicate occurrence of wild populations of *C. pepo* in Mexico and Eastern USA as early as 10,000 and 30,000 years ago respectively. Domestication of this species took place independently in these two areas. *C. argyrosperma* was apparently domesticated in Southern Mexico where archaeological evidences of its domestication dating back to 5,200 BC are available. Wild populations of subspecies *C. argyrosperma* ssp.*sororia* that occur in Mexico and Central America gave rise to domesticated subspecies *C. argyrosperma* ssp.*argyrosperma*.

Table 1.3: The Origin and Geographical Distribution of Different Cucurbit Species

Species	*Common Name(s)*	*Origin*	*Geographical Distribution*
Benincasa hispida Thunb. Cogn.	Wax gourd, Ash gourd, White gourd, *Petha*	South-Eastern Asia	Tropical Asia
Citrullus colocynthis L. Schrad.	Colocynth	Africa	Tropical Africa, North-Western India
Coccinia indica Wight and Arn.	Little gourd, Ivy gourd, *Kundru*	Tropical Africa	Tropical Asia and Africa
Cucumeropsis mannii Naud.	Egusi melon	Tropical America	Tropical Africa
Cucumis anguria L.	West Indian Gherkin	Africa	Brazil, West Indies
Cucurbita ficifolia Bouche.	Fig leaf, Malabar gourd	Central Mexico	Central Mexico to Chile
Cucurbita foetidissima Kunth	Buffalo gourd	Mexico, South-Western USA	Arid South-Western USA and Northern to Central Mexico
Cyclanthera pedata (L.) Schrad.	Stuffing cucumber	Tropical America	Andean Countries
Hodgsonia macrocarpia (Bl.) Cogn.	Chinese lard plant	South Asia	South and South-eastern Asia
Lagenaria siceraria (Molina.) Stand.	Bottle gourd, *Lauki*	Africa	All tropical and sub-tropical areas especially in Africa
Luffa acutangula (L.) Roxb.	Ridge gourd	India	India, East Asia
Luffa cylindrica (L.) M. J. Roem.	Sponge gourd, *Chikni torai*	India	Throughout tropics especially in India
Momordica charantia L.	Bitter gourd, *Karela*	Tropical Asia	Throughout tropics especially in South and South-East Asia
Momordica cochinchinensis (Lour.) Spreng.	Cochinchin gourd, sweet gourd	India	South Asia
Momordica dioica Roxb. Ex Willd.	*Kaksa, Kakrol*	Southern tropical Asia	Tropical Asia and Africa
Praecitrullus fistulosus (Stocks) Pang.	Round melon, Indian squash, *Tinda*	India	North-West India
Sechium edule (Jacq.) Swartz	Chayote, Chow-chow, Mirliton	Mexico, Guatemala, Brazil, Peru	Throughout tropics especially Latin America
Sicana odorifera (Vell.) Naud.	Casabanana	Brazil, Peru	Central America, Northern South America
Telfairia occidentalis Hook. *f.*	Fluted pumpkin	West Africa	Tropical Asia
Telfairia pedata (Sims.) Hook.	Oyster nut	Africa	Tropical Africa
Trichosanthes anguina L.	Snake gourd, *Chichinda*	India	South and South East Asia
Trichosanthes dioica Roxb.	Pointed gourd, *Parwal*	India	India

(From Sirohi *et al.,* 2005).

Landraces of subspecies *C. argyrosperma* ssp. *argyrosperma* var.*argyrosperma* were selected first while the other two groups of cultivars, namely, *C.argyrosperma* ssp.*argyrosperma* var.*callicarpa* and *C. argyrosperma* ssp.*argyrosperma* var.*stenosperma* evolved from the Northern and Southern landraces of *C. argyrosperma* ssp.*argyrosperma* var.*argyrosperma*, respectively (Merrick, 1990). *C. filicifolia* was cultivated in the Andes Mountains of Peru from pre-historic times. As the species spread northwards, a large number of perennial xerophytic types got evolved adapted to the desert and plateau regions of Northern Mexico and the South-Western USA. Attempts are made to domesticate another xerophytic species, namely, *C.foetidissima*. This species is perennial and grows wild in the arid deserts of South Western USA and Mexico. Its ability to withstand drought attracted interest in its commercialization. It produces higher yields in seed, rich in protein and oil and edible fleshy tuberous roots. In addition to the xerophytic types, species adapted to humid conditions also migrated to North and one such species *C.okeechobeensis* is found as far North as Florida (Sirohi *et al.*, 2005).

Citrullus

In *Citrullus*, three species *viz. C. lanatus* (Thunb.) Matsum. and Nakai, *C. colocynthis* (L.) Schrad and *C. ecirrhosus* Cogn are recognized. The recently described species *C. rehmii* De Winter represents a valid fourth species (De Winter, 1990). Out of the above four species, *C. lanatus* and *C. colocynthis* are distributed in India. In the species *C. lanatus*, the domesticated water melon is classified as var. *lanatus* and the wild populations as var. *citroides*. De candolle (1882) opined that, there was evidence for a domesticated water melon (*C. lanatus*) being indigenous to tropical Africa specifically that of South Africa. It was probably introduced into cultivation in Egypt and spread to India in very early times. The cultivated forms have two centres of diversity, India and tropical and subtropical Africa. The cultivation of large water melons (*C. lanatus* var. *citroides*) is comparatively of recent origin and the present Soviet cultivars still have the shape of their African ancestor *C. lanatus* var. *catter*. *C. lanatus* ssp. *mucospermus* is a wild form and has small fruits with white bitter flesh. *C. colocynthis*, a potential species for cultivation in the arid lands exhibiting large variation occurs wild in India especially in the North-Western plains and Central and South India and is the ancestor of water melon (Shimotsuma, 1963). *C. ecirrohosus* Cogn. is a perennial species without tendrils and with extremely bitter fruits well adapted to the extreme climatic conditions of desert regions. Wild forms of several *Citrullus* species are found in dry sandy region of Kalahari Desert in South Africa (Sirohi *et al.*, 2005).

Luffa

This genus has a long history of cultivation in tropical Asia and Africa. *Luffa* has nine species out of which six (*L. acutangula* (L.) Roxb., *L. cylindrica* M.Roem., *L. echinata* Roxb. *L. graveolens*, *L. tuberosa* Roxb. *L. umbellata*) occur in India. The probable centre of origin and the primary centre of diversity of *Luffa* are India. *Luffa acutangula* (ridge gourd) and *L. cylindrica* (sponge gourd) are grown throughout India in tropical and subtropical climates. *Luffa acutangula* has two varieties, all wild forms *viz.* var. *amara* confined to peninsular India and var. *forskohlii* (harms.) Heiser and Schilling confined to Yemen. *Luffa echinata* grows in natural habitats in Western Himalayas, Central India and Gangetic plains. *Luffa graveolens* is a wild species distributed in parts of North-Central India and *L. hermaphrodita* (Satputia) that bears fruits in clusters is under cultivation in Eastern Uttar Pradesh and Bihar (Sirohi *et al.*, 2005).

Lagenaria

Lagenaria, previously considered a monotypic genus has now six species which are of African origin. *Lagenaria siceraria* (bottle gourd) is the only cultivated species. It is cultivated throughout the tropical and subtropical regions in Asia, Africa and South America. Based on morphological differences, the cultivars and landraces of African and New World origin are recognized as *L.siceraria* ssp. *siceraria* and those of Asia as *L. siceraria* ssp. *asiatica*. The other three species, namely, *L. abyssinica* Hook.f., *L.breviflora* Robert and *L.sphaerica* (Sond) Naudin are wild perennials (Sirohi *et al.*, 2005).

Trichosanthes

In *Trichosanthes*, a total of 41 species are reported out of which 21 originated in India, parts of tropical Asia and Indo-Malayan region. Among the species, *T. anguina* (snake gourd) and *T.dioica* (pointed gourd) are cultivated throughout India. The major diversity is distributed in North-Central and North-Eastern India for *T. dioica*, whereas for *T. anguina* it is distributed in warm humid areas throughout the country especially in the North-Eastern region, West Bengal, Malabar coast of Karnataka and Kerala and in lower and mid hills of Eastern Ghats where rich diversity is observed (Sirohi *et al.*, 2005).

Momordica

The genus *Momordica* has 45 species, native to tropical Africa. Out of these species, seven are in India and four species (*M. balsamina, M. charantia*, M. *cochinchinensis* and *M. dioica*) are commonly cultivated/semi-domesticated. Balsam apple (*M. balsamina*) occurs in semi-dry North-Western plains, Eastern Ghats and Western Ghats. Bitter gourd (*M. charantia*), is a popular vegetable grown all over tropical and subtropical regions in India. It has large diversity in India, China and South-East Asia. It is the most likely that its cultivation started in India, although a number of related forms exist in Africa. Sweet gourd (*M. cochinchinensis*) is popular in Tripura, Assam and West Bengal whereas, kakrol (*M. dioica*) is cultivated in West Bengal, Assam, parts of Bihar and adjoining areas. There are several other *Momordica* species of minor importance which include *M. foetida* Schum. and Thonn, a traditional medicinal plant of West Africa and *M. grosvenori* Sw. for extraction of sweet glycoside component (Sirohi *et al.*, 2005).

Coccinia

The genus *Coccinia* has nearly 30 species, confined to tropical Africa except *C.grandis* (ivy gourd) which occurs in cultivated and wild forms in South and South-East Asia and Africa. In India, it is cultivated in warm humid areas (Sirohi *et al.*, 2005) and is also grown in Latin America, Australia and other tropical areas.

Cyclanthera

Cyclanthera has about 30 species and all are native to tropics of South America. *Cyclanthera pedata* var. *edulis* (stuffing cucumber) is cultivated in tropical America and Asia. It is a potential crop for wide cultivation in high altitude areas. It's closely related species *C.brachystachia* (Ser.) Cogn. is also cultivated for its edible fruits in Nepal and South Taiwan (Sirohi *et al.*, 2005).

Benincasa hispida

Ash gourd (*Benincasa hispida*) a monotypic genus is a native of Indonesia (Java) and Japan. It was domesticated in India during prehistoric times. It is widely grown all over the country in tropical and subtropical regions and possesses large variability for morphological and quality characters. Ash gourd fruits have long storability and the creeper is a potential crop of tropical Africa and America (Sirohi *et al.*, 2005).

Sechium edule

Chow-chow (*S.edule*) is a native of Mexico and Guatemala and is under cultivation for centuries. *Sechium* formerly, a monotypic genus with a total of eight species are now recognized. The wild forms of this species and it's close relative *S. compositum* are found in South American region (Sirohi *et al.*, 2005).

Praecitrullus fistulosus

Praecitrullus fistulosus (round melon), earlier included in the genus *Citrullus* is a closely related species. It originated in India. In India, its variability is rather limited suggesting its partial or recent domestication (Sirohi *et al.*, 2005).

Tomato

Peru and Central Mexico are the centres of origin and domestication for cultivated tomato. Evidences suggest that even though, *Lycopersicon* spp. originated in the Peru-Ecuador-Bolivian area of South America, the domestication took place in Central Mexico and Coastal Peru. There are about 12 species in genus *Lycopersicon*. On the basis of fruit colour, all these species are classified into two subgenera, namely, *Eulycopersicon* (characterised by red fruits with carotenoid pigmentation and annual growth habit) and *Eriopersicon* (characterised by green fruits with anthocyanin pigmentation). The cultivated species of tomato (*L. esculentum*, 2n=24) and wild species (*L. pimpinellifolium* (Jusl) Mill) belong to *Eulycopersicon* and these two species are classified into five botanical varieties. One of these varieties, namely, *L. esculentum* var.*cerasiforme* (cherry tomato) is the immediate ancestor of the cultivated tomato (Kalloo, 1985). A characteristic feature of the evolutionary process is that, while cultivated tomato is highly self-pollinated, there is considerable natural cross-pollination in the cherry tomato (Kalloo *et al.*, 2005). Origin and evoluation related issues of tomato are dealt extensively and elaborately by many workers mostly coming to a consensus about it's South American (Peru/Ecuador/Bolivia) origin.

Alliums (Onions, Garlic, Shallots, Leeks and Chives)

The genus *Allium* was earlier classified under the family Liliaceae and retained for a long time, later under family Amaryllidaceae and presently placed under the family Alliaceae (Purseglove, 1972). *Allium* is a very complex genus accounting for over 700 species of bulbous herbs. According to Bujisen (1990), the genus comprises about 1000 species, distributed throughout the world, mostly in the Northern hemisphere with the greatest number in the former USSR/CIS.

The common names, ploidy and areas of cultivation of different *Allium* species in India are given below. The shallot is indigenous to Palestine and owes its name to the town of Ascalon (Giles, 1943). It is treated as a separate species by some authorities. In India, it is grown in parts of Himalayas and also in Peninsular India. Welsh onion is native to Central and North-Western China and cultivated to some extent in India, particularly in North-Eastern region and Western Himalayas. The chive has a Mediterranean origin but the Chinese chive is domesticated in the Orient and it changed little from the wild. Leek has its centre of origin in the Eastern Mediterranean from where it spread to Europe by the Middle Ages. In India, it grows well in the Western Himalayas. The other domesticated species occur in Turkey, Iran, Iraq, Pakistan and the erstwhile USSR. The remaining species occurring in wild, were classified primarily, based on leaves or floral morphology under different sections and groups, although, their taxonomic classification, correct nomenclature and synonymy continue to be vague, complex and unresolved (Stearn, 1946; Pandey *et al.*, 2005).

Table 1.4: Common Name, Ploidy and Areas of Cultivation of Different *Allium* Species in India

Botanical Name	Common Name	Chromosome Number (2n)	Areas of Cultivation
Allium cepa var. *cepa* L.	Common onion	16	Throughout India
A. cepa L. var. *aggregatum G.Don*	Shallot	16	Himalayas, Peninsular India
A. fistulosum L.	Welsh onion	16	Western Himalayas, North-East region
A. tuberosum Rottl. ex Spreng.	Chinese chives	32	Western Himalayas North-East region
A. sativum L.	Garlic	16	Throughout India
A. ampeloprasum L. var. *porrum* (L.) J. Gay	Leek	32	Western Himalayas
A. schoenoprasum L.	Chives	16/24/32	Western Himalayas

The regions of high species diversity exist in Commonwealth of Independent States (CIS), Turkey, Iran, North Iraq, Afghanistan, Pakistan and hilly tracts of North and North-Eastern India. According to Bujisen (1990), all taxa of the genus *Allium* cultivated in South-East Asia are introduced either from China, Central Asia or Europe. Besides, wild species are generally restricted to the Northern Hemisphere and are totally absent in tropical part of South-East Asia. The species cultivated in South-East Asia which have their origin and secondary centre of diversity in Sino-Japanese centre include Welsh onion, Rakkyo, Chinese chives etc. The above species are also found in Indian gene centre, a few of them still in the wild state.

Jones and Mann (1963) discussed many aspects of onion and its allies. The domesticated *Allium* species came from the Near East, Central or Eastern Asia. In India, the genus *Allium* is widely distributed in the temperate and alpine regions. About 30 species occur in Indian gene centre (Babu, 1977). Majority of the wild species are new records. Negi and Pant (1992) reported occurrence of lesser-known wild species in mountainous regions and discussed information on their habitat, local uses etc. The prominent wild species included in the study were *A. carolinianum* DC. (syn. *A. blandum* Wall.), *A. chinense* G. Don (syn. *A. bakeri* Regel.), *A. consanguineum* Kunth, *A. humile* Kunth (syn. *A. govanianum* Wall. ex Baker), *A. przewalskianum* Regel. (syn. *A. jacquemontii* Regel., *A. stolczkii* Regel.), *A. stracheyi* Baker, *A. victorialis* L. and *A. wallichii* Kunth. According to Hooker (1973), the most widely cultivated *Allium* species in the Indian sub-continent include onion, garlic, shallot and leek. Genetic diversity in cultivated alliums is reported from different parts of the country. The prominent areas for genetic diversity are Jammu and Kashmir, Himachal Pradesh, Uttarakhand, North-East region, Rajasthan, Gujarat, Maharashtra, Andhra Pradesh, Karnataka and Tamil Nadu (Pandey *et al.*, 2005).

Onion, *Allium cepa* var. *cepa* L. does not occur in wild state. It originated in an area encompassing the present Iran, West Pakistan and the mountainous countries to the North (Purseglove, 1972). Onion has the primary centre of origin in Central Asia and the secondary centre in the Near East (McCollum, 1976). It was domesticated at several places and occupied a vast area in Western Asia extending from Palestine to India. Most of the Botanists believe that it has it's origin in Iran, Pakistan, Afghanistan, Kazakhistan, Russia and adjoining regions. Onions are under cultivation since the ancient times in the Middle East and India. Onion was introduced to the New World shortly after its discovery and was under cultivation as early as 1629 and now spread to most parts of the world.

Cole Crops

All cole crops originated in the Mediterranean region (Thompson, 1976) mainly by mutations and introgression from wild species during evolution or human selection. There are several perennial wild species of the *B. oleracea* cytodeme distributed in the Mediterranean and along the North Atlantic Coast. All these wild species are entomophilous. The main centre of genetic diversity of the *B. oleracea* is in Sicily, where several wild species are found, namely, *B. incana* Ten., *B. rupestris* Rafin., *B. macrocarpa* Guss. and *B. villosa* Biv. The other wild species occurring in the Mediterranean are *B. montana* Pourr in Southern Europe, North-East Spain and Southern Italy, *B. insularis* Moris in Corsica and Sardinia, *B. hilarionis* Post. and *B. cretica* Lam. and its subspecies *laconica, nivea* and *cretica* in Crete (Greece), Cyprus, Syria and Lebanon.

The greatest diversity of cole crops, particularly cauliflower, broccoli and kale are in the Mediterranean gene centre which includes Greece, Syria, Cyprus, Sicily, Italy, Spain and Portugal. Wild species or subspecies and improved heterogeneous cultivars were the primary genetic resources for various characters in cole vegetables. Introgression from wild species in primitive form of *B. oleracea* in the Mediterranean area and North-Western Europe resulted in accumulation of great genetic diversity in cauliflower, broccoli, Brussels sprouts, kohlrabi and kale (Crisp and Ives, 1982). Several biennial types of kale and broccoli exist in UK. The annual types of cauliflower and broccoli and their landraces occur in Italy and adjoining regions. Introduction of the improved open pollinated cultivars and F_1 hybrids in Europe, developed in Japan and the USA, replaced many of the landraces, primitive cultivars and traditional varieties. There is extensive genetic erosion but fortunately many genetic resources are still available in the Mediterranean region. Rich genetic diversity of landraces of Kale and cabbage including savoy cabbage occur in Portugal. In addition to landraces in Italy, genetic diversity in cauliflower occurs in France, UK, Sweden, Denmark and the Netherlands. Many types of cauliflower were developed in Europe. Originals or Italians developed in Italy in 16th Century, are very diverse and include both annuals and biennials with different kinds of curd conformations and colours. The common Italian types include Jazi, Naples, Romanesco and Florablanca. The original Italian types gave rise to all the European types including Snowball, Old English, Walcheren, Roscoff and Angers.

The North European annuals are also of Italian or East Mediterranean origin. These include Alpha, Erfurt, Danish, Le Cerf and Mechelse. The two other groups of cauliflower suggested by Crisp and Tapsell (1993) are Asian and Australian origin. The Random Amplified Polymorphic DNA (RAPD) analysis showed that populations within overlap and are indistinguishable when the variation within and among population for each regional type and among population for all regions were studied. Although, cauliflower and broccoli are very closely related and indistinguishable in many cases, they have now become distinct vegetables as a result of selection, isolation and adaptation over a number of years. Genetic diversity in cabbage is mainly found in North-West Europe. The wild perennial forms of *B. oleracea* occur in South-Western Britain, Western France, Northern Spain and the Mediterranean. F_1 hybrids and improved cultivars of cabbage are widely cultivated in Europe, the USA, Japan, China and South Korea and these have replaced the traditional types (Chiang *et al.*, 1993 and Swarup and Brahmi, 2005).

Kale (*Brassica oleracea* L. var. *acephala* (DC.) Alef. var. *sabellica* L.)

Probably, the kale was the first crop to be selected and adapted by man. Use of kale as human food was mentioned by Theophrastus as early as 350 BC. The Greeks cultivated it as early as 600 BC. From the leafy kale plants, the other cole crops like collard, broccoli, cauliflower and cabbage got evolved. The Restriction Fragment Length Polymorphism (RFLP) studies indicated a monophyletic origin of cultivated morphotypes in *B. oleracea* and that the thousand headed kale is an ancient form of *B. oleracea*. The wild diploid species *B. oleracea* var. *sylvestris* (L.) DC. a leafy kale-like plant which occurs in the Northern shore of the Mediterranean sea and on lime cliffs in North-Eastern England and along the North sea, is the progenitor of the cultivated kale and leafy non-heading cabbage. The narrow-stem kale or the stem kale originated from *rupestris-incana* complex in Adriatic area of Southern Italy. The branching stem kale descended from *B. cretica* in Greece. Similarly, from the wild leafy-kale, *B. olderaca* var. *costata* DC., the Portuguese kale got developed.

Broccoli (*Brassica oleracea* L. var. *botrytis* (L.) Alef. var. *italica* Plenck)

The ancestral broccoli (*B. cretica* ssp. *nivea* Lam.) is a wild annual with cymose white flowers. This is the primitive form of broccoli from which the cultivated broccoli evolved several thousand years ago (Gray, 1982; 1993). The ancestral form of broccoli after moving to Southern China in the East, gave rise to the leafy kale types or the Chinese kale, *B. alboglabra* (Bailey) Sun. Natural crossings between ancestral broccoli and the biennial wild types/species resulted in many morpho and cultivated types of *B. oleracea* like cultivated broccoli including sprouting broccoli and its various forms and also cauliflower. Sprouting broccoli was probably used first during the Christian era or even earlier in Southern Europe but not documented till 1660.

Cauliflower (*Brassica oleracea* L. var. *botrytris* Prain)

Cauliflower is comparatively of a later origin, may be about 500 years ago. It probably originated by introgression within broccoli gene pool. Syria is the place of origin for cauliflower. It was probably introduced into Italy by the Genoese from Levant or Cyprus around 1490. Italy is a centre of genetic diversity for cauliflower and broccoli where several Italian landraces of these two crops occur even today. It's first description was published in 1576 as *B. florida botrytis* and in 1583 as *B. cauliflora* Gars. After introduction of cauliflower in Italy, its cultivation began in England in 1619 but commercial cultivation started later during 1860. In France, Germany and Holland, cultivation of cauliflower started in the beginning of 17th century. As all flower buds open in sprouting broccoli compared to pre-floral fleshy apical meristems in cauliflower, it is thought by many that broccoli must have preceded cauliflower in the evolution of species. Probably, the Romans and Celts introduced the ancient leafy kale, leafy cabbage, broccoli and cauliflower in the chalky coasts of England and North-Western France from Mediterranean region. The northwards migration of these primitive forms of cole crops, accompanied by introgression from wild species, *B. oleracea* var. *sylvestris* occurring there, gave rise to biennial types with winter hardiness and cold vernalization requirement for flowering. The cultivated species selected by man had biennial habit with yellow flowered racemose inflorescence unlike the ancestral kale or broccoli of the Mediterranean, which had cymose inflorescence with white flowers.

Cabbage [*Brassica oleracea* L. var. *capitata* (L.) var. *capitata* L.]

The hard-headed white cabbage selected in North Europe descended from wild *B. oleracea*, the non-heading type cabbage originated in Eastern Mediterranean region and Asia Minor, which was probably brought into Western Europe by the Celts. The earliest record of cabbage was during 2000-2500 BC. Perhaps modern day white headed cabbage was selected in Germany around 1150 AD and moved to England during 14th century. Savoy cabbage [var. *sabauda* (L.) O. Schwarz] which evolved in Italy might have probably originated from the Portuguese kale (var. *costata*) (Silva-Dias *et al.*, 1990).

Kohlrabi (*Brassica oleracea* L. var. *caulocarpa* (DC.) Alef. var. *gongylodes* L.)

Kohlrabi was evolved from a narrow-stem non-heading type of kale, perhaps var. *costata*, the Portuguese kale or the var. *medullosa* Thell. in Central Mediterranean region. It was first described in Europe in 1485 and grown in Ireland in 1724 and in England in 1837. It may have come from Southern Europe as indicated by the names of the two most popular varieties, White Vienna and Purple Vienna.

Brussels Sprout (*Brassica oleracea* L. var. *gemmifera* Zenker.)

Brussels sprouts originated as a vegetable later, perhaps during 1587 or 1696, though it was described earlier in 1213 in Belgium. It became wide-spread in Europe in the late 18th and early 19th centuries. It's origin is obscure and probably, it descended from various forms of kale (thousand headed kale) or Savoy cabbage or both. The other view is that Brussels sprout (*B. oleracea* var. *gemmifera*) was evolved from *B. oleracea* var. *ramose* that originated from *B. oleracea* var. *sylvestris* (Ockendon and Smith, 1993 and Swarup and Brahmi, 2005).

Cluster Bean [*Cyamopsis tetragonoloba* (L.) Taub.]

Cluster bean is probably indigenous to India although it is not observed in a wild state (Purseglove, 1972).

Leaf Vegetables

Variability occurs in amaranth, spinach, *Rumex, Basella, Trigonella* etc. throughout the country. In Amaranth, about four-five species are under cultivation which are robust annuals and domesticated prehistorically in the highlands of tropical and subtropical Americas (Smartt and Simmonds, 1995). Local diversity with plant types, green, completely/partially purple and spiny generally occurs.

Root and Tuber Crops

The important tuber crops for which rich diversity occurs in the Indian Sub-continent are cassava (*Manihot esculenta* Crantz), sweet potato (*Ipomoea batatas* [L.] Lam.), greater yam (*Dioscorea alata* L.), white yam (*D.rotundata* Poir.), lesser yam (*D.esculenta* [Lour.] Burk.), potato yam (*D.bulbifera* L. var.*sativa*), taro (*Colocasia esculenta* [L.]Schott.), tannia (*Xanthosoma sagittifolium* [L.] Schott.), elephant foot yam (*Amorphophallus paeoniifolius* [Dennst.] Nicolson), Chinese potato (*Coleus parviflorus* Benth.), yam bean (*Pachyrrhizus erosus* [L.] Urban), moghania (*Moghania vestita* Benth. ex Bak.), winged bean (*Psophocarpus tetragonolobus* [L.] DC.), arrowroot (*Maranta arundinacea* L.), Queensland arrowroot (*Canna edulis* Ker-Gawler) and giant taro (*Alocasia macrorrhiza* [L.] G. Don) (Edison *et al.*, 2005).The important tuber crops for which rich diversity occurs in the Indian sub-continent include Taro (*Colocasia esculenta* (L.) Schott var. *esculenta*), Tannia/Cocoyam (*Xanthosoma sagittifolium* (L.) Schott), Greater Yam (*Dioscorea alata* L.) and Elephant foot (*Amorphophallus campanulatus*)

Potato

The two main centres of diversity of tuber-bearing species of *Solanum* are Central America and the Andean region of North-Western Argentina, Peru and Southern Bolivia. The main cultivated potato species *S. tuberosum* L., a tetraploid (4*x*=48) originated in the Andes of Peru and Bolvia in South America, more specifically the lake Titicaca basin in Peru-Bolvian region (Ross, 1986). It has two subspecies, namely subspecies *tuberosum* adapted for long days and subspecies *andigena* adapted to short days. *Solanum tuberosum* ssp. *andigena* Hawkes is widely distributed in the Andean regions of Venezuela and Northern Argentina whereas *S. tuberosum* ssp. *tuberosum* occurs naturally in Southern Chile only. Among the primitive cultivated species, *S. phureja* Juz. et Buk. (2*x*=24) is distributed from Venezuela to Northern Bolivia between 2,000 to 3,500m amsl, *S. chaucha* Juz. et Buk. (2*x*=36), *S. stenotomum* Juz. et Buk. (2*x*=24) and *S. goniocalyx* Juz.et Buk. (2*x*=24) in the Northern Peru and Central Bolivia between 3,000 to 3,900 m amsl and *S. curtilobum* Juz. et Buk. (2*x*=60), *S. juzepczukii* Buk. (2*x*=36) and *S. ajanhuiri* Juz. et Buk. (2*x*=24) in Central Peru and Northern Argentina at altitudes ranging from 3,800 to 4,200m amsl. The wild species exhibit a great deal of phenotypic plasticity and genotypic variability and are adapted to a wide range of habitats. They are found from sea level to 4,000m amsl in the hills and from equator to 40^{0}S.

Potato is introduced to the Old World by the Spaniards. It was probably introduced to Spain first during 1570 from the Northern parts of South America. It probably belonged to subspecies *andigena*, the primitive putative progenitor of subspecies *tuberosum*. The initial introductions from the tropics were day neutral or short day adapted and from these genotypes, the types suited to the temperate long day conditions of Europe were developed. In India, potato is generally thought to have introduced in late 16th century possibly by the Portuguese or the British missionaries. It was chronicled by Terry (1655) that he saw during 1617-18 in the northern most parts everywhere good roots as carrots, potatoes and other like them". Evidences suggest that potato might have been known in India even prior to arrival of Europeans into India in modern times (Vasco de Gama landed in Calicut, Kerala in 1498). Further, botanical evidences also show that the early Indian potato varieties were more closely related to the subspecies *andigena*. By the end of 19th century, potato cultivation spread throughout Northern India (Pushkarnath, 1976; Shekhawat *et al.*, 2005).

Yams (*Dioscorea* spp.)

Yams contribute towards food security to about 300 million underprivileged people in the low-income food-deficit countries in tropical Asia, Africa and Americas. Yams originated in Africa, South-East Asia and South America.

Out of all the yams, greater yam (*Dioscorea alata* L.) is the most important edible species in India, probably originated in Myanmar and Thailand. The crop would have developed from putative parental species *D.hamiltonii* or *D. persimilis* which grow wild in the South-East Asia especially in the Assam-Burma region (Alexander and Coursey, 1969). Yams have very large diversity for ploidy status and their geographic distribution. The allogamy and the formation of winged seeds facilitate gene flow to a great extent.

The genus contains about 600 spp., which occur mainly in the tropics and subtropics throughout the world. A total of 11 species are grown for their stem tubers, domesticated independently in four centers. These include, three species (white yam (*D. rotundata*), yellow yam (*D. cayennensis* Lam.), trifoliate yam (*D. dumetorum* (Kunth) Pax) indigenous to West Africa, three species (greater yam (*D. alata*), lesser yam (*D. esculenta*), *D. opposita* Thunb.) native to Asia, potato yam (*D. bulbifera* L.) found in both Africa and Asia and cush-cush yam (*D. trifida* L.) native to America (Edison *et al.*, 2005).

Out of the old world origin species, *D.alata* is very important which is polymorphic and not known to occur in a wild state. The probable centre of origin is in the South-East Asia. New Guinea and Polynesia region are the secondary centres of diversity of this crop which spread to this area probably before 100 BC. In the first millennium AD. it was introduced into Madagaskar. During the sixteenth century, the Portuguese and Spanish traders carried this to West Africa and the New World and is now pantropical in distribution.

Aroids

Aroids mainly constitute tuber bearing crop species belonging to family Araceae namely, taro (*Colocasia esculenta*), elephant foot yam (*Amorphophallus paeoniifolius*), tannia (*Xanthosoma sagittifolium*) and giant taro (*Alocasia macrorrhiza*).

Taro [*Colocasia esculenta* (L.) Schott var. *esculenta*]

Taro is the most widely cultivated aroid worldwide. Taro originated in South-East Asia and North-East India is the centre of origin (Watt, 1889). About seven species are reported, native to the Indo-Malaysia and the Pacific region. *C. esculenta* occurs wild in South-Eastern Asia (Purseglove, 1972). It spread by introduction to different places which included China, Japan, Egypt, Arabia, Mediterranean region, Africa, Pacific, Polynesia, Hawaii and New Zealand. Taro consists of two distinct morpho types, the big corm and the small corm or cormel. Plucknett (1985) differentiated *C.esculenta* var.*esculenta* as *darsheen* type (big corm) and *C.esculenta* var.*antiquorm* as *eddoe* type (cormel). In India, *darsheen* is mostly found in the Northern region and *eddoe* and mixed type in the Southern region.

Elephant Foot Yam [*Amorphophallus paeoniifolious* (Dennst.) Nicolson]

Elephant foot yam is a common vegetable and it's low acrid morpho types are cultivated. In India, variety *karunakizhangu* is commonly cultivated in Tamil Nadu which is characterized by formation of distinct cormels, unlike the larger circular corms, formed in other local forms. In elephant foot yam maximum diversity is reported from the Western Ghats where several wild species occur in the forests and adjoining areas (Edison *et al.*, 2005).

Tannia/Cocoyam [*Xanthosoma sagittifolium* (L.) Schott]

About 40 species are natives to tropical America. *Xanthosoma* was domesticated in the new world and was under cultivation in tropical America and the West Indies in the pre-colombian times. South America is the centre of origin for tannia.

Sweet Potato

Sweet potato (*Ipomoea batatas*) is of American origin. *Ipomoea batatas* has a total of 13 closely related wild species (Austin and Huaman, 1996). All the species have x=15 as the basic chromosome number. While the cultivated sweet potato is a hexaploid (having 90 chromosomes), most of the wild species are either diploid and tetraploid. There are some species *viz. I.trifida* which has 2x, 3x, 4x and 6x cytotypes (Edison *et al.*, 2005).

Cassava

Cassava (*Manihot esculenta*) is the most important root crop in the tropics and ranks fourth after rice, sugarcane and maize as the source of food for human consumption. The genus *Manihot* comprises about 98 species. Cassava originated in South America most probably Brazil where wild species are found in the natural habitats. None of the existing wild species can clearly be identified as an ancestor of cassava. It is more likely that all the species of the genus, including *M.esculenta* might have derived from some of the extinct ancestral species. Genetic erosion of wild species occurs at a faster rate in Brazil (Nassar, 1979, Edison *et al.*, 2005).

Minor Tuber Crops

In addition to major root and tuber crops, there are several tuber species of minor importance namely, arrowroot (*Maranta arundinacea*), Chinese potato (*Solenostemon rotundifolius*), Queensland arrowroot (*Canna edulis*), winged bean (*Psophocarpus tetragonolobus*) and yam bean (*Pachyrrhizus erosus*) which are important sources of food.

The arrowroot originated in South America and the Lesser Antilles. The genus *Maranta* includes 20 species of perennial herbs. It is widely distributed throughout the tropical countries like India, Sri

Lanka, Indonesia, the Philippines, West Indies and in Australia. The Chinese potato is an important minor tuber crop cultivated in India, Sri Lanka, South-East Asia and parts of Africa. In India, it is cultivated in Tamil Nadu and Kerala. Chinese potato originated in the Andean region or Peruvian coast from Venezuela to Northern Chile. The genus *Canna* includes about 25 species of herbs, widely distributed throughout the tropics and subtropics. Queensland arrowroot is grown as a fodder crop in Japan. It is cultivated in Australia for production of starch. In India, it is grown for edible and tuberous rhizome. The winged bean is exceptional among all the food crops as practically all parts of the plant are edible. It is a perennial climber with tuberous root and is cultivated in Asia including India, Myanmar and Africa. The genus *Psophocarpus* has five species which are distributed throughout the tropics in Africa, Asia and Madagascar. Tropical Asia might be it's probable centre of origin where it is primarily domesticated. Burkill (1966) considered that it originated on the African side of the Indian Ocean, possibly in Madagaskar or Mauritius. It is a native of either India or Mauritius or African region where all the wild species of this genus are found. The yam bean is a high yielding leguminous root crop of good nutritional value. It is cultivated in India, Mexico, China, Singapore, the Philippines, Hawaii and Indonesia. *Moghania* is a less known leguminous root crop of the Khasi and Jaintia hills of Assam. *Moghania tuberosa*, a wild species bearing succulent roots, occurs in the coastal belt of Konkan region in Maharashtra (Edison *et al.*, 2005).

Tropical Fruits

Mango (*Mangifera indica* L.)

The genus *Mangifera* belonging to family Anacardiaceae originated in the South-East Asia at an early date. Evidences based on phytogeographical, morphological, anatomical, palynological and cytological studies indicate that the genus had it's origin in the continental region of Myammar, Thailand, Indo-China and the Malay Peninsula (Mukherjee, 1951). The natural spread of the genus is limited to the Indo-Malaysian region, stretching from India to the Philippines and New Guinea in the East (Mukherjee, 1958). Records suggest that it has been in cultivation in the Indian subcontinent for well over 4,000 years now (De Candolle, 1904). India is the most important centre of diversity for Mango having a large number of cultivated varieties as well as wild types. The Sunda Ilands (Java, Sumatra and Borneo), the Philippines and the Celebes-Timor form the secondary centre of diversity.

The genus *Mangifera* has 41 species with the highest concentration being in the Malay Peninsula (19), Sunda Ilands (16) and the Eastern Peninsula (14) with many of the species being common between them (Mukherjee, 1949). Yadav and Rajan (1993) reported six important zones of variability of wild mangoes. They are (1) Humid subtropical region comprising Manipur, Tripura, Mizoram and South Assam (2a) Chhota Nagpur Plateau and tri-junction of Madhya Pradesh, Bihar and Orissa (2b) Santhal Parganas (Rajmahal Hills) in Jharkhand (3), South Madhya Pradesh (tribal areas) and adjoining Orissa and Andhra Pradesh (4), Dhar Plateau of Madhya Pradesh and adjoining South Rajasthan and Gujarat (5) Humid tropical southern peninsular India and (6) Andaman and Nicobar Islands. Other *Mangifera* species namely, *M. andamanica* King, *M. khasiana* Pierre, *M. sylvatica* Roxb. and *M. camptosperma* Pierre are also reported from India (Mukherjee, 1985; Yadav and Rajan, 1993). *Mangifera andamanica* is widespread in tropical, wet evergreen forests of Andaman Islands (Karihaloo *et al.*, 2005). *Mangifera sylvatica* occurs up to 1,300m amsl in the Eastern Himalayas in Sikkim, Darjeeling district of West Bengal, Khasi Hills of Meghalaya, Upper Assam and Surma valley (Assam) and in the Andaman Islands (Mukherjee, 1949; Agharkar and Roy, 1951). *Mangifera khasiana* is an endemic but insufficiently known species from Khasi hills and is distinguished from *M. sylvatica* in having small and narrow leaves, small flowers, inflorescence with fasciculate divergent branching, petals ovate to lanceolate and the disc being glandular or with distinct lobes (Mukherjee, 1985). Truly wild mango trees are recorded in North-Eastern India, sub-Himalayan tracts, deep gorges of the Bahraich and Gonda hills in Uttar Pradesh and in outer hills in Kumaon and Garhwal in Uttarakhand (Brandis, 1874; Kanjilal *et al.*, 1934). Semi wild trees are present in the forests of almost all parts of the Indian subcontinent.

Recent taxonomic and molecular evidences suggest that mango probably evolved within a large area comprising North-Western Myanmar, Bangladesh and North-Eastern India (Mukherjee, 1997). Based on cytological studies, Mukherjee (1950) indicated that *M. indica* ($2n$=40) is an ancient allopolyploid with the basic chromosome number x=8. Cultivar differentiation has taken place primarily through mutation and inter-varietal hybridization with the resulting varieties, maintained through vegetative propagation (Mukherjee, 1951; Karihaloo *et al.*, 2005). Occurrence of wild forms of *M.indica*, allied species of *M.sylvatica*, *M.caloneura*, *M.pentandra* (leaf fossil impressions) and presence of numerous cultivated and wild varieties in India give evidence for it's origin in the Indo-Burma region through allopolyploidy, possibly through amphidiploidy (Mukherjee, 1958). The tract from Eastern India to Myanmar and possibly further in the Malay region, is suggested as the native home of mango (Popenoe, 1920; Vavilov, 1951).

Bananas (*Musa* spp.)

References to plantain in ancient Indian literature in Valmiki's Ramayana (2029 BC), Kautilya's Arthashastra (250 to 300 BC) and the Tamil Classic Silappadikaram (500 to 600 AD) suggest very early domestication of the plant in India (Krishnamurthy and Seshadri, 1958; Karihaloo *et al.*, 2005).

About 40 species are reported under the genus mainly distributed in the South-Eastern Asia and the Pacific with the centre of origin and diversity in the Assam-Burma-Thailand area. All the edible bananas are derived from *M. acuminata* (A genome) and *M. bulbisiana* (B genome). The centre of origin and diversity for *M.acuminata* is in the humid tropics and wild seeded diploid forms are abundant in the Malaysian area. *M. bulbisiana* is widely distributed in India, Philippines and New Guinea. The wild bananas are distributed in South-East Asia and the Pacific (Simmonds, 1962). The cultivated bananas originated from *M. acuminata* Colla ($2n$ =22, *AA* genome) and *M. balbisiana* Colla ($2n$=22), *BB* genome). The wild types are seed bearing with poor edibility and *M. acuminata* was the first species to possess edible traits. Probably, human selection favoured parthenocarpy, structural heterozygosity and seed sterility leading to the production of edible seedless fruits (Simmonds, 1994). Most cultivated bananas are triploid comprising three major groups, AAA, AAB and ABB. Diploids (AA, AB) and tetraploids (AAAB, AABB, ABBB) are less abundant (Karihaloo *et al.*, 2005). *Ensete glaucum*, *Musa balbisiana* and *M. ornata* are distributed in the Eastern and Western ghats in a diffused fashion and also in the North-Eastern India to Philippines and the New Guinea.

India possesses rich diversity in wild *Musa* spp. in the North-Eastern region whereas variability among cultivars, especially belonging to genomic groups *AA*, *AB*, *AAB* and *ABB* exists in the Southern region. Maximum genetic variability of *M. acuminata* occurs in South-East India. *Musa flaviflora* Simmonds is localized to Manipur and Meghalaya. Several wild species are also recorded from Southern and Western regions of India. These include *M. acuminata* ssp. *burmannica*, *M. balbisiana*, *M. nagensium* Prain, *M. sikkimensis* Kurz, *M. cheemanii* Simmonds, *M. ornata* Roxb., *M. laterita* Chees., *M. velutina* Wendl. Ex Drude and *M. sanguinea* Hook. Banana cultivation is predominantly practiced in Tamil Nadu, Karnataka, Kerala, Andhra Pradesh, Maharashtra, Bihar, West Bengal and North-Eastern States. The variability in cultivars decreases as one proceeds from extreme South. It is also interesting to note that all the Indian *AA* and *AB* cultivars occur only in the extreme South. The most common commercial varieties grown in different states are: Poovan (*AAB*) in Bihar; Pachabale (*AA*) in Karnataka; Nendran (AAB) in Kerala; Karpuravalli (ABB), Monthan (AAB), Morris (AAA), Mysore Poovan (AAB), Nendran (AAB), Pachanadan (AAB), Rasthali (AAB) and Robusta (AAA) in Tamil Nadu; and Dwarf Cavendish (AAA), Poovan (AAB) and Rasthali/Morthaman (AAB) in West Bengal (Karihaloo *et al.*, 2005). Good local varieties occur both in table and vegetable types. Diversity occurs in plant height, maturity, bunch size, fruit shape, size, aroma, bearing and maturity. The *Virupakshi* (Hill banana) landrace occurs in the Palni hills (1,600 m). The traditionally popular landraces (*amruthapani* and *chakrakeli*) are alarmingly replaced by the improved varieties.

Citrus (*Citrus* spp.)

South-East Asia, Australia, the intervening islands between Asia and Australia and Central Africa are recognized as centres of diversity of important *Citrus* and related genera (Swingle and Reece, 1967). The *Citrus* related genera are found in Asia, Australia and Africa. *Citrus reticulate, C. sinensis* and *C. aurantifolia*, the commercially important species grown in India, originated in India and China (McPhee, 1967). The cultivated species of Citrus originated in the South-East Asia probably in Eastern India, Burma and Southern China (Tanaka, 1977). Recent surveys show that considerable genetic diversity exists in the Yunnan province to consider that area as one of the centres of origin and diversity (Gmitter and Hu, 1990).

In India, North-Eastern Himalayan region and foothills of the Central and Western Himalayan tracts are rich sources of citrus diversity. As many as 17 *Citrus* species, their 52 cultivars and seven probable natural hybrids originated in the North-Eastern region (Bhattacharya and Dutta, 1956). Native diversity occurs in the *Citrus* spp. in *C. aurantifolia, C. limon, C. paradisi, C. reticulata* and *C. sinensis*. Variability occurs mainly for fruit size, quality and juiciness. In mandarin orange, the important landraces include *Coorg, Desi, Khasi, Nagpur and Sikkim* etc. Primitive forms of *C. indica* occur in the Tura range of Garo hills, Meghalaya in the North-Eastern region where several local types are distributed which bear very small sub sweetish fruits and a Gene Sanctuary for Citrus was established there (Singh, 1981). The indigenous variability among the cultivated species is significant and a number of commercial varieties are developed from the native germplasm through selection and crop improvement (Karihaloo *et al.*, 2005). Singh and Chadha (1993) listed 30 *Citrus* species and botanical varieties, indigenous to India and under cultivation for a very long time.

Among the *Citrus* species in South India, sweet orange and acid lime are very important and *Sathgudi, Bathayee* and *Narinja* are the popular local landrace cultivars in Andhra Pradesh which are adaptable in arid/semi-arid/low rainfall situations.

Jack fruit (*Artocarpus heterophyllus* Lam.)

Artocarpus heterophyllus originated in Western Ghats of India and later distributed to South-East Asia and other tropics. It has high degree of variability in South-East Asia. Being protandrous and mainly cross-pollinated, the plants show great variability in many characters including texture, odour and flavour of pulp (Karihaloo *et al.*, 2005). In jack, the important local types include *kharja panasa* and *tene panasa* in the Eastern Ghats. Good local diversity for fruit size and shape occur in the coastal areas.

Sapota (*Achras sapota* L.)

Significant local diversity exists especially in fruit size (small/medium/big) and shape (round/oval), fibrousness, aroma and taste. The popular endemic landraces *include Bangalora, Dwarapudi, Jonnavalasa, Kiratabarati* and *Vavilivalasa* in South India and *Chatri* and *Kalipatti* in Western India.

Litchi [*Litchi chinensis* (Gaertn.) Sonn.]

Litchi is a native of China. It is bought to India towards the end of 18th century. The only other species of the genus, *L. philippinensis* Radlk. is a wild plant grown largely in the Philippines and is of little commercial importance except for using it as a rootstock. The cultivated types of litchi form a complex evidently derived from more than one wild progenitor. The better types have larger fruits with small, frequently abortive seeds and heavy bearing. Various cultivars are classified on the basis of growth and flowering pattern (Arora *et al.*, 1996; Singh, 1998). For large scale cultivation, cultivars namely, Bedana Dehradun, Calcuttia, China Rose, China Sahi, Dehra Rose, Desi, Early Bedana, Late Bedana, Purbi and Kasba are recommended. Among these, the most popular cultivars are Bedana, Calcuttia, China Sahi and Rose Scented. Based on vegetative and fruit characteristics, litchi cultivars are classified into five groups (Singh, 1998). Group 1: Ajhauli, Dehradun, Deshi, Rose Scented and Shahi; Group 2: China, Calcutta, Durbi and Manroji; Group 3: Kasba; Group 4: Late and Early Bedana, Seedless and Swarna Rupa; and Group 5: Longia (Karihaloo *et al.*, 2005).

Guava (*Psidium guajava* L.)

Guava is indigenous to tropical America where it occurs in wild state. The Portuguese introduced this species into the Indian sub-continent and it was naturalized in some areas where significant variability exists (Purseglove, 1981). Being a partially cross-pollinated species, substantial variability got accumulated in seedling populations in different agro-climatic zones of guava cultivation. This enabled selection of several promising cultivars in different parts of the country. In the Indo-Gangetic plains, a wide range of variability is available in Allahabad, Mirzapur, Kanpur, Unnao and Fatehpur districts and adjoining areas in Uttar Pradesh. Over 100 cultivars of guava are grown in the country and the important ones are Lucknow49, Allahabad Safeda, Red Fleshed, Chithidar, Harijha, Banarasi, Seedless, Supreme, Kerala and Baruipur (Mitra and Bose, 1999; Karihaloo *et al.*, 2005).

Pineapple [*Ananas comosus* (L.) Merr.]

Five species are reported under this genus, native in the Eastern South America. The centre of origin is probably in the Parana-Paraguay river drainage area, where related seedy species *A. bracteatus*, *A. ananassoides* and *A. erectifolius* occur in wild state. It was presumed that the Tupi-Guarani Indians first selected and domesticated *A. comosus* at it's centre of origin and later it spread to different parts of the new and old world. The Hawaii is the second centre of diversity because of early introduction and somatic mutations. In pineapple, *Simhachalam* is a popular landrace adapted to South Indian hill slopes.

Temperate Fruits

Most temperate fruits originated in three broad centres suggested by Vavilov (1951). These are (*i*) Chinese Centre comprising mountainous regions of North-Eastern, Central and Western China as far as Kansu Province and the North-Eastern Tibet, (*ii*) Central Asiatic Centre representing mountainous area extending from Tien-Shen South, including the Hindu Kush to Kashmir and (*iii*) Near-Eastern Centre including the mountainous contiguous area that extends from North-Eastern Iran to Caucasus and Central Turkey. A few species of the temperate fruit genera like *Malus*, *Pyrus*, *Prunus*, *Rubus* etc. escaped to the Himalayas and over a period of a few thousand years, not only survived but got established in the hills, leading to evolution of new species and varieties. Randhawa (1987) gave a comprehensive account of wild temperate fruit species native to Himalayas including their taxonomic status, distribution, occurrence and distinguishing features. Subsequently, several workers reviewed the status of biodiversity in temperate fruits in the North-Western Himalayas (Kishore and Randhawa,

1993; Singh *et al.*, 2000; Rana *et al.*, 2002; 2003). As such, there exists rich genetic diversity in temperate fruits, primarily in the North-Western Himalayas and to a relatively low extent in the North-Eastern hilly region (Sharma, *et al.*, 2005).

Apple, (*Malus pumila* Mill.) is the most important fruit of the temperate region, originated in the Caucasus mountains of Western Asia where wild apple exists in the forest areas. *Malus baccata*, a wild relative of apple (*Malus domestica*) is a good root-stock.

Arid Fruits

Many arid fruits like *aonla, bael, ber, jamun, ker, khejri, lasoda, phalsa, pilu* and wood apple are indigenous. The cactus pear, custard apple, date palm, mulberry, pomegranate and tamarind though exotic, possess fairly good diversity in several pockets of India in view of introduction and acclimatization over a long period of time (Vashishtha, *et al.*, 2005).

Table 1.5: Common Name, Family, Ploidy and Origin of Different Arid Fruit Species

Botanical Name	Common Name (s)*	Family	Chromosome No. (2n)	Origin
Aegle marmelos L. Correa	Bael fruit, *Bael*	Rutaceae	36	Northern India
Opuntia ficus-indica L. Mill.	Cactus pear, Prickly pear	Cactaceae	22-88	Mexico
Capparis decidua (Forsk.) Edgew	Capparis, *Ker, Teet*	Capparidaceae	14,16 (2x)	Arid region of India
Buchanania lanzan L.	Almondette tree, *Chironji*	Anacardiaceae	-	Western peninsula
Cordia myxa Roxb.	Cordia, *Lasoda, Gonda*	Boraginaceae	-	India
Annona squamosa L.	Custard apple, *Sarifa*	Annonaceae	14, 16	Tropical America
Phoenix dactylifera L.	Date palm, *Khajur*	Palmaeae	36	Around, Persian Gulf, Iraq
Ficus carica L.	Fig, *Anjir*	Moraceae	26	South-East Asia
Emblica officinalis Gaertn.	Indian gooseberry, *Aonla*	Euphorbiaceae	18, 28	Peninsular India
Ziziphus mauritiana Lamk.	Indian jujube, *Ber*	Rhamnaceae	24	Indian sub-continent
Syzygium cumini (L.) Skeels	Java plum, *Jamun*	Myrtaceae	40	Indian sub-continent
Carissa carandas L.	Karanda, *Karonda*	Apocynaceae	12	India
Prosopis cineraria (L.) Druce	*Khejri, Jand, Sami*	Leguminosae	-	Thar desert of India
Morus alba L.	Mulberry, *Tut*	Moraceae	308	Northern China/Japan
Grewia subinaequalis DC.	Grewia, *Phalsa*	Tiliaceae	36	India
Salvadora oleides Decne	Pilu	Salvadoraceae	24	India/Tropical Africa
Punica granatum L.	Pomegranate, *Anar*	Punicaceae	18	Iran, Afghanistan to Himalayas
Tamarindus indica (L.) Gaertn.	Tamarind, *imali*	Leguminosae	24	Tropical Africa
Feronia limonia (L.) Swingle	Wood apple, *Kainth*	Rutaceae	18	India, Sri Lanka

*: Vernacular name is given in italics.

Bael (*Aegle marmelos* Correa ex Roxb.)

Bael fruit is native to India and grows wild in plains and sub-mountainous regions of South-East Asian countries. In India, it is found in sub-Himalayan tract and plains, particularly in Indo-Gangetic plains, Central India and North-Eastern region. The tree is deciduous and hardy and withstands arid conditions and salinity. The fruit is rich in nutritive value and contains fairly high amount of vitamin A (186 IU/100 g) and riboflavin (1191 mg/100 g). The fruit, when ripe, is eaten fresh. The pulp is used to make a delicious drink. The tender green fruit is used for making preserve, jam, squash and nectar. The importance of *bael* fruit also lies in its curative properties, which make it one of the most useful medicinal plants of India (Kirtikar and Basu, 1935). It's fruit is stomachic, demulcent and anti-gonorrhoeic; pulp is laxative and root bark is used as anti-diarrhoeic. The genus has three species, of which only *Aegle marmelos* is cultivated. The main bael growing states are Uttar Pradesh (specifically Bahraich district), Bihar, West Bengal, Madhya Pradesh and Orissa. There are hardly any standard cultivars. Some popular local types are named after the region of their occurrence. Local cultivars, namely, Mirzapuri, Darogaji, Ojha, Rampuri, Azamati, Khamaria, Kagzi Gonda, Kagzi Etawah, Sewan Large and Deoria are described from important *bael* growing areas. Srivastava *et al.* (1998) gave an account of the inherent genetic diversity, available in Bael.

Capparis (*Capparis decidua* Edgew.)

Capparis is a large genus comprising of more than 300 species distributed throughout warm region of the world (George Ushar, 1974). Out of about 40 species occurring in India, a few are economically important and the most important being *C. decidua*, useful as a fruit. Capparis is a succulent xerophytic shrub found extensively in the arid regions. The red ripe berries are eaten afresh. It is also used as a medicine, the top shoot is used in toothache and boils, bark is laxative and used to cure cough, asthma and inflammation whereas the fruit is used to cure cardiac problems. The unripe fruit is used as a vegetable and both fresh and dehydrated unripe fruits are also pickled. *Capparis decidua* is distributed in India, Pakistan, Saudi Arabia, Egypt and Tropical Africa. In India, it is found mainly in Rajasthan, Punjab, Haryana and Gujarat. Being an open-pollinated plant, large variability occurs in seedlings for plant vigour, morphology, productivity and quality parameters. Big fruit size and prolific bearing should be given emphasis while collecting germplasm (Vashishtha, *et al.*, 2005).

Custard Apple (*Annona squamosa* L.)

The genus *Annona* includes about 120 species indigenous to the American tropics and subtropics. The custard apple, even though of South American origin got introduced to the country during the 14 century long before the discovery of America by the Portuguese. It got naturalized in semi-wild state especially in the semi-arid and sub-humid climates throughout the country (Singh, 1969)

The whole group including the hybrids is loosely called custard apple. Many *Annona* species bear edible fruits but the four most important species are *A. cherimola* Mill. (cherimoya, laxman phal), *A. muricata* L. (sour sop), *A. reticulata* L. (Bullock's heart or *ramphal*) and *A. squamosa* (sweet cop, sugar apple, *sitaphal*). *Annona squamosa* is the most favourite fruit and is cultivated in India, China, the Philippines, Cuba, Egypt and Central Africa. In India, it occurs in naturalized state in Andhra Pradesh, Tamil Nadu, Madhya Pradesh, Gujarat, Maharashtra and Assam and has accumulated good diversity for habit, plant type, fruit and reticulation size, seed size and number. *Annona reticulata* is also spread throughout the country (Singh, 1969) and has a fewer seeds with the pulp being of poor quality. *Annona cherimola* is found in the Nilgiri and Nandi hills in Southern India, Uttarakhand (Mussorie), Madhya Pradesh, Maharashtra, Andhra Pradesh and North-Eastern region and can better withstand frost than custard apple. *Annona glabra* syn. *A. palustris* (pond apple) is used as a root stock and is tolerant to flooding and excess moisture stress. *Atemoya*, a hybrid of *A. cherimola* x *A. squamosa* is better adapted to colder climates and alkali soils.

Date Palm (*Phoenix dactylifera* L.)

Date palm is the oldest cultivated fruit tree. It is indigenous to countries around Persian Gulf. Date palm is dioecious and its fruit is a rich source of sugars and minerals. Three species of *Phoenix* which occur in India are *P. dactylifera*, *P. sylvestris* (L.) Roxb. and *P. canariensis* Chaubad. *P. sylvestris* found throughout the drier tracts up to an altitude of 1500 m. in India is underutilized and has a lot of potential. The commercial date growing countries of the world are Iraq, Saudi Arabia, Algeria, Egypt, Morocco, Tunisia, Pakistan and Sudan besides many other Middle East and North African countries. Date palm was introduced to deserts of South America, Kalahari Desert of South Africa and Great Central Desert of Australia. In the Indus Valley, date palm was introduced during 4th century BC by the soldiers of Alexander, the Great and during the beginning of the 8th century by the Muslims. Date palm cultivation in India is limited to Gujarat (Kachchh region) where wild grooves of seedling date palm are found (Vashishtha *et al.*, 2005).

Fig (*Ficus carica* L.)

The edible fig is a native of Southern Arabian Peninsula, Italy, Balkan Peninsula and Russia. The genus *Ficus* has many other species with edible fruits that grow wild in India in addition to *F. carica*. Others include, *Ficus benghalensis* L., *F. glomerata* Roxb., *F. hispida* L. and *F. lucescens* Blume throughout India; *F. auriculata* Lour. (timla), a tall tree in sub-Himalayan region, North-Eastern hills and parts of Eastern India; *F. heterophylla* L. a creeping shrub in hot parts of India mainly near water bodies; *F. hirta* Vahl. a small tree in the tropical Himalayas; *F. hispida* L. and *F. hookeriana* Corner in Eastern Himalayas and Meghalaya; *F. lanceolata* Buch-Ham, a branched shrub in Khasi hills; *F. lucescens* and *F. nemoralis* in Himalayas; *F. palmata* Forsk in Himalayas and in hills of peninsular tract; *F. rumphii* Blume in sub-Himalayan tract and in the peninsular region; *F. scandens* Roxb. a climbing shrub throughout Himalayas; *F. semicordata* Buch.-Ham. Ex J.E. Smith in sub-Himalayan tract, Western and North-Eastern India; *F. subincisa* Buch.-Ham. in Himalayas and Meghalaya; and *F. tsiela* Roxb., a large tree in the Deccan Peninsula. Singh (1969) viewed that since the fig varieties grown in India have unlobed or faintly lobed leaves, they are probably hybrids between the common fig and the Indian wild species (Vashishtha *et al.*, 2005).

Indian Gooseberry (*Emblica officinalis* Gaertn.)

Indian gooseberry or aonla (*Emblica officinalis* Gaertn.) is native to peninsular India. It is commonly found in the mixed deciduous forests from tropical hill slopes up to an elevation of about 1200 m. The related species *E. acidus* Skeel. syn. *P. distichus* Muella, popularly known as star gooseberry or country gooseberry is mostly grown for the small edible fruits, whereas the wild species *E. fischerry* Gamble, found in the forests of South India, bears fruits suitable for preparation of pickles (Vashishtha *et al.*, 2005).

Indian Jujube (*Ziziphus mauritiana* Lam.)

The genus *Ziziphus* has more than 600 species (Baiely, 1947) distributed mainly in India, China, Korea, Russia and Thailand of which 18 to 20 species are native to India (Watt, 1893). Nine edible species found in different parts of India are listed by Arora and Pandey (1996). They include *Ziziphus apetala* Hook. f. Fl., *Z. funiculosa* Buch-Ham. and *Z. incurve* Roxb which occur in North-Eastern hills. *Z. mauritiana* and *Z. nummularia* Wight and Arn. are found all over the drier tracts particularly in North-West India and Uttar Pradesh; *Z. oenoplia* (L.) Mill. and *Z. rugosa* Lamk. occur throughout India particularly in Central and Eastern India; *Z. jujube* Mill. and *Z. oxyphylla* Edgew. grow naturally in the North-Western Himalayas. In addition to the above mentioned nine species, *Z. rupicola* T. Andern is found in Central and Eastern India and *Z. xylocarpus* Willd. is found in Madhya Pradesh and Peninsular region (Pareek, 1988). The other important species are *Z. xylocarpus*, *Z. lotus* Lamk., *Z. spina-christi* (L.) Willd., *Z. zoaziroo* Mart., *Z. fumiculosa* Buch-Ham., *Z. glabrata* Heyne, *Z. oxyphylla* and *Z. mistol* Griseb. (Bal and Uppal, 1992; Vashishtha *et al.*, 2005).

Among these, *Z. mauritiana* (Indian jujube or ber) and *Z. jujube* (Chinese jujube) are cultivated and are distinct from each other with respect to morpho-physiological characters (Chatterjee and Randhawa, 1959). Maheswari and Singh (1965) recognized six economically important species in India. They are *Z. nummularia* (*jharber*, used as rootstock), *Z. mauritiana*, *Z. oenoplia* (*makoh*), *Z. rugosa* (*suran ber*), *Z. sativa* Gaertn. (*Z. vulgaris* Lam. Kandiari) and *Z. xylopyrus* Hochst.ex A. Rich. Tent. (*kat ber*).

Java Plum [*Syzygium cuminii* (L.) Skeels]

Java plum or jamun is in cultivation in Malayan Archipelago and India. The genus has many species that bear edible fruits and some of these are of ornamental value. *Syzygium jambos* (L.) Alston (rose apple or gulab jamun) found in South India is a popular tree. Its fruits are yellow and the tree is of ornamental value. *Syzygium zeylanicum* (L.) DC., a small tree with edible fruits, is found in the Western Ghats. *Syzygium malaccense* (L.) Merr. and L.M. Perry (Malay rose-apple) and *S. uniflora* L. (*Surinam cherry* or *Pitanga cherry*) are also found in South India. *Syzygium javanicum* Miq. (water apple) is distributed in South India and West Bengal. One wild species, *S. fruticosum* DC., with small, edible, dark purple fruits is grown as a wind break and *S. densiflora* (Blume) DC., resistant to termites is used as a rootstock. Other species which grow wild are *S. arnottianum* (Wight) Walp., *S. calophyllifolium* (Wt.) Walp. *S. caryophylletum* (L.) Alston., *S. heyneanum* Mill., *S. operculatum* Gamble, *S. zeylanicum* (L.) DC. and *S. aqueum* Alston. Seedling strains with significant variability for fruit shape, size, pulp colour and content, TSS, acidity and earliness occur in Uttar Pradesh and Maharashtra providing good scope for selection of better genotypes. Though, no standard jamun variety exists, the type grown in North India is purple-red, ripens early and produces relatively bigger and oblong fruits. The other type is late ripening and bears round, blackish and small fruits (Vashishtha *et al.*, 2005).

Karonda (*Carissa congesta* Wight)

Karonda is an evergreen, densely branched and spiny shrub. It is widely grown in India in plains and foot hills for it's fruits and is also an excellent bio-fencing material. The genus *Carissa* has about 20 species, of which five are of Indian origin. Among these, *C. congesta* is the most important and grows wild in Bihar, West Bengal, South-Eastern Chhattisgarh, parts of Orissa and Southern parts of India. Diversity in *Carissa* also occurs in the North-Western India, particularly, in Mount Abu in Rajasthan and Khandala area in Maharashtra (Mehra and Arora, 1982). Species *C. inermis* Vahl. grows wild in Deccan peninsula and West Coast and *C. spinarum* L. in Punjab, Jammu and Kashmir and Western Uttar Pradesh. *Carissa paucinervia*, an evergreen shrub is found in drier tracts throughout India (Vashishtha *et al.*, 2005).

Phalsa (*Grewia* spp.)

Grewia, commonly known as *phalsa*, is one of the oldest indigenous fruits of the country. Grewia has 140 species, out of which 40 occur in India. Among them, *G. subinaequalis* is the only species of commercial importance. The other edible species are *G. damine* Graerth., a small tree, found in Rajasthan, Bihar and drier parts of South India; *G. elastica* Royle found in the sub-Himalayan tract, Central India and Western Ghats; *G. flavescens* Juss. in upper Gangetic plains, Bihar and Central and Southern India; G. *hirsuta* Vahl. in warm parts of India ascending to sub-Himalayan tract; *G. rothi* DC. in Peninsular region; *G. sapida* Roxb. in North-Western India Eastwards to Assam hills and also in the

Eastern Ghats; *G. sclerophylla* Roxb. in tropical Himalayas from Kumaon to Eastern hills of Assam; *G. tenax* Forsk. an evergreen shrub in Rajasthan and arid regions of Central, Eastern and Western India; *G. tiliaefolia* Vahl. throughout India and *G. villosa* Willd. found in hot parts of North-Western, Central and Western India (Vashishtha, *et al.*, 2005).

Pomegranate (*Punica* spp.)

Pomegranate is one of the favourite table fruits in tropical and subtropical regions. Genus *Punica* has two species namely, *P. granatum* and *P. protopunica* and further has two subspecies, namely *chlorocarpa* and *prophyrocarpa* (Vashishtha, *et al.*, 2005). It has a very narrow genetic variability. Pomegranate is extensively cultivated in Mediterranean countries (Spain, Morocco and Egypt), Afghanistan and to some extent in Myanmar, China, Japan and the USA.

Mulberry (*Morus* spp.)

Vavilov (1926) considered 'China-Japan' region which includes East China, Korea and Japan as the centre of origin of genus Morus. There are about 70 species recognised in the genus, of which 24 species are in China, 19 in Japan, 6 in Korea, 4 each in Taiwan and India, 3 each in Myanmar and Indonesia, 2 each in Thailand and Afghanistan and 1 each in Saudi Arabia, Oman and Muscat. Further, 14 species are in North America and 7 in Central and South America (Sanjappa, 1989). The four species in India are *M. indica, M. alba* L., *M. serrata* Roxb. and *M. laevigata* Wall. ex Brandis (Hooker, 1894). Mulberry grows all over India from temperate to tropical zones. It is distributed naturally in the Sub-Himalayan regions up to an altitude of 2100 m extending to Indus and Brahmaputra rivers. *Morus rubra* L. and *M. nigra* L. are cultivated for fruits while *M. alba* is cultivated for rearing silk worm. *Morus laevigata* is from Andaman and Nicobar Islands (Parkinson 1923, Vashishtha *et al.*, 2005). Mulberry is an introduction to India from Persia.

Wood Apple [*Feronia limonia* (L.) Swingle]

Wood apple is native to India and distributed all over the plains and foothills, particularly in arid and semi-arid regions (Vashishtha *et al.*, 2005).

Cordia (*Cordia myxa* Roxb.)

Cordia also known as *lasoda*, a native of India, grows as stray plantations in semi-arid and arid regions of Rajasthan and Haryana. In addition to *C. myxa*, the other species in the genus, *C. gharaf* (Forsk.) Ehrenb and Asch. is found in the sub-Himalayan tract and Central India while *C. vestita* Hook. F. and Thomas grows wild in tropical Himalayas. There are no defined cultivars in this crop. Great variation exists in fruitsize and their pulp content. Large-fruited cultivar in Gujarat is recognized as *Paras*. This drupaceous fruit has edible mucilaginous pulp, which constitutes about 70 per cent of the fruit, rich in sugars but is free of acid. Its unripe fruit is used as vegetable and pickle, while ripe is eaten fresh and also used for making country liquor. The desirable characters in germplasm include coloured fruit skin, less stickiness, high TSS, small seed and pleasant flavour (Vashishtha *et al.*, 2005).

Pilu (*Salvadora* spp.)

Salvadora or *pilu*, is an evergreen shrub native to tropical Africa. It is commonly found in warmer parts of India and grows luxuriously in Thar desert in Rajasthan and Gujarat. Two species of *Salvadora* are found in India (Chundawat, 1990). *Salvadora oleoides* is distributed in arid regions of Punjab, Rajasthan and Gujarat and the other species *S. persica* L. is more common in other drier tracts of the country. The fruit of *S. persica* is also eaten though it is less pulpy than *S. oleoides*. No established varieties are available.Natural variability is prevalent which should be surveyed and exploited for commercial purpose. The important areas for germplasm collection include North-West Rajasthan (Sekhawati region) and adjoining arid regions of Haryana and Punjab (Vashishtha *et al.*, 2005).

Khejri (*Prosopis cineraria* (L.) Druce)

The genus *Prosopis* has 44 species of trees and shrubs. Important species in *Prosopis* are *P. cineraria*, *P. juliflora* (SW.) DC., *P. alba* Griseb, *P. nigra* (Griseb) Hieron, *P. chilensis* (Molina) Stuntz., *P. glandulosa* Torr., *P. pallida* (Humb and Bonpl. ex Willd.) Kunth. and *P. tamarugo* F. Pjil. *Prospis cineraria*, commonly known as *khejri*, is well adapted to desert ecosystem due to it's capacity to tolerate climatic stresses and multiple utility. Indian arid region is the home of *P. cineraria* and even today, it is the main species in arid ecosystem. As per Indian scriptures, khejri is mentioned in Vedas, Ramayana and Mahabharata. Pollen stratigraphic studies (Singh, 1970) revealed abundance of *khejri* in Western Rajasthan 10,000 years back when it also enjoyed high ecological status (Vishnu Mittre, 1975). The natural occurrence of khejri is reported from Afghanistan, Iran, Pakistan and India (Puri *et al.*, 1964) (Vashishtha *et al.*, 2005).

Cactus Pear [*Opuntia ficus-indica* (L.) Mill.]

Cactus pear, aso known as prickly pear, is a succulent, spiny/spineless xerophytic plant with multiple utilities. Besides ecological restoration, ornamental horticulture and biofencing, it produces edible fruits for human consumption and as fodder for livestock. Fruits are rich source of sugar, vitamins and minerals. Fruits are also used for preparation of various processed products like jam, jelly, marmalade, syrup, squash and wine. The tender vegetable–cladodes are used as salad. Castus pear has several medicinal uses (Hegwood, 1990). The edible cactus pear is native to the Western hemisphere occurring naturally in Alberta, Canada to 40^{0}S latitude in Argentina (Felker *et al.*, 1997). Over 300 species of genus *Opuntia* are known. In Mexico only, 104 species and varieties are recorded. Schweinvar (1985) described 11 subgenera of *Opuntia*. *Opuntia albicarpa* spp. nov., *O. ficus-indica O. robusta* Wendl. and *O. cochenillifera* (L.) Mill. are important fruit type species while *O. cochenillifera* (L.) Mill. is an important vegetable type. The germplasm needs to be enriched with accessions, less spiny, bearing large fruits and having good keeping quality and of vegetable type from Argentina, the USA, Mexico, Brazil and Chile (Vashishtha *et al.*, 2005).

Almondette Tree (*Buchanania lanzan* Spreng.)

It is a native of India and grows wild in drier parts of Western Peninsula. The kernel has a pleasant sub-acidic flavour, is nutritious and eaten raw or roasted. It yields light yellow sweet oil (35.4-47.2 per cent) with a mild, pleasant aroma, used as common substitute for almond oil in flavouring sweetmeats, in confectionery and betel nut powder. Tribals use it as brain tonic. The kernel is rich in protein (19.0 per cent), fat (59.1 per cent) and minerals (3.0 per cent). Diversity in panicle, fruit size and quality of kernel occur mainly in Madhya Pradesh (Panna, Umaria, Mandla, Shivni, Chhindwara, Betul, Balaghat), Chhattisgarh (Bastar), Uttar Pradesh (Lalitpur) and Maharashtra (Nagpur, Akola, Amravati). Efforts are made to collect and conserve the germplasm and so far 60 accessions are collected and cryopreserved at the National Bureau of Plant Genetic Resources (NBPGR), New Delhi (Vashishtha *et al.*, 2005).

Spices

India has the unique distinction of contributing both the King and the Queen of spices, namely black pepper (*Piper nigrum* L.) and cardamom (*Elettaria cardamomum* L.) both having originated in the tropical evergreen forests of Western Ghats of South India where their wild forms still exist (Ravindran *et al.*, 2005). The crops of importance under spices include chillies, turmeric, ginger, Indian long pepper etc.

Chillies (*Capsicum annuum* L.)

Chilli is the most important spice crop grown all over the world except in the colder parts. The centre of diversity for *Capsicum* was South-Central South America with majority of the species having some range in Brazil and/or Bolivia (Gonzalez and Bosland, 1991 and WWF and IUCN, 1997). The genus *Capsicum* (2n=24) contains 25 wild and five domesticated species, *viz.,C. annuum* L., *C. baccatum* L., *C. chinense* Jacq., *C. frutescens* L and *C. pubescens* Ruiz and Pavon (Bosland and Votava, 2000). The primary centre of origin for domesticated chillies (*Capsicum annuum* L.) was semi-tropical region of Mexico (Whitmore and Turner, 2002). The other four domesticated species originated in South America (Eshbaugh *et al.*, 1983; Denevan, 2001). Amazonia is the centre for *C. chinense* (Toquica *et al.*, 2003) and Western Amazonia as the centre for *C. frutescens*, which is more variable in Central America (Heiser, 1985 and Hernandez-Verdugo *et al.*, 1999). Bolivia is the centre of origin and domestication for *C. baccatum* and the mid-elevation Andes for *C. pubescens* (Eshbaugh, 1993). Zeven and Zhukovsky (1975) and Zeven and de Wet (1982) considered the Central American and Mexican region as the region of diversity of Pepper/Chilli where the greatest variability exists in. Based on gene flow through conventional breeding and crossability, *Capsicum* species are grouped in to three species complexes, namely, *C.annuum* complex, *C.baccatum* complex and *C.pubescens* complex. Among the five cultivated species, *C. annuum* L., *C. frutescens* L. and *C. chinense* Jacq. belong to *annuum* complex, while *C. baccatum* and *C. pubescens* Ruiz and Pav. belong to *baccatum* and *pubescens* complex, respectively.

Among the five cultivated species (*C. annuum, C. baccatum var.pendulum, C. fruitescens, C. chinense* and *C. pubescens*), the most widely grown species is *C. annuum* (chilli and sweet pepper). *C.annuum* was domesticated in highlands of Mexico and includes most of the Mexican chilli peppers, most of the hot peppers of Asia and Africa, and various cultivars of sweet peppers of temperate countries. Due to non-adaptability of *C. annuum* to lowland tropics of Latin America, *C. frutescens* and *C. chinense* replaced its cultivation. *C. frutescens* is also cultivated in Africa and India for either intact fruits or for oleoresin extraction. The cultivation of *C. baccatum* and *C. pubescens* is mostly restricted to Latin America (Pickersgill, 1997; Kalloo *et al.*, 2005). It was introduced into the Indian sub continent towards the end of 15th Century by the Portuguese (Heiser, 1976). Chillies with long history of cultivation, natural selection pressures and adaptability to different environments led to the origin of a great diversity of forms or cultivars differing in their growth habit, size, colour, shape, flavour and pungency. India is the secondary centre of diversity (Mehra and Arora, 1982).

In chillies locally adapted landrace variability occurs throughout the country especially in the Eastern and South Eastern uplands, Southern peninsula, Northern plains, Central highlands, Western ghats, North-East and the Himalayan regions. Good variability occurs for plant type, fruit colour, bearing, shape and size and pungency. Chillies with small round cherry types, small oblong stout and small conical stout, extra long broad and tapering and cultivars with yellow fruits are found in the North coastal areas. Indigenous endemic non-pungent/sweet paprika landraces, the fruits of which are bright red and having more colouring principle occur in the country. The landraces Warangal (tomato chilli) of Andhra Pradesh, Bydagi in the Dharwad area of Karnataka and Degchi landrace of Kashmir are well known. *C. frutescens* (Bird's pepper) mainly occurs in the hilly terrain often under semi-domestication by tribal communities. *C. baccatum* is generally found in the Northern hills under cultivation along with *C. annuum*. *C. chinense* is mainly restricted to Kerala and Western ghats under cultivation either for fruits or for ornamental purposes. Many varieties of chillies are recorded in India, varying in shape, colour, size and pungency. As such there is large scope for selection and breeding from *Capsicum* genetic resources for developing genotypes suitable to hot wet tropics and different climates in India.

Turmeric (*Curcuma longa* L.)

Turmeric is one of the ancient spices used in India. It is a part of the Indian culture and also the socio-religious practices of most of tribal groups of the country. The processed and dried rhizomes of *C.longa* constitute the spice of commerce.

The centre of origin for Turmeric is South East Asia extending from India to Indo-China, East Indies and parts of China (Rao and Rao, 1994). *Curcuma* has about 100 species mostly naturalized in the Indo-Malayan region especially in the Western Ghats, North Eastern region and Eastern Ghats of India and Java (Ravindran *et al.*, 2005). Very little is known about its ancestors and phylogenetic relationship. Holttum (1950) reported that at least two other species, *C.colorata* Val. and *C.xanthorhiza* Roxb. have equally deep yellow rhizome colour but they never gained any prominence either as spice or as a source of colour.

The South Asian region is the centre of origin. Domestication of turmeric occurred in India where it was under cultivation from ancient times (Ravindran, 1999). Turmeric is a triploid with $2n$ $(3x) = 63$. It is propagated by rhizomes but it occasionally sets seeds, used in breeding programmes. India is the largest producer and exporter of turmeric. Native and wild types occur in the Western and Eastern ghats and North-East in India. In turmeric, landraces are still popular because of their adaptability and significant levels of tolerance/resistance to biotic/abiotic stresses. Good variability exists for days to maturity, rhizome shape and size, inside colour, surface colour, aroma and yield in the endemic landraces. The cultivars are mainly known after the place where the particular landrace is traditionally grown.

Ginger (*Zingiber officinale* Rosc.)

Ginger was under cultivation in India from time immemorial. *Zingiber* is included in the tribe Hedychieae along with other genera, namely, *Curcuma, Hedychium* and *Kaempferia* (Holttum, 1950) and in the series Zingiberae which contains only the genus *Zingiber*. The genus is divided into four sections by Baker (1886) and ginger is included in section Lampuzium. Baker (1886) described 24 species from the Indo-Malayan region. Sabu (1991) described eight species from Western Ghats and adjacent areas. A botanically distinct variety *Z.officinale* var.*rubrum* having pink outer skin of rhizome is reported in India by IISR,Calicut which is under cultivation in Malaysia. The wild species (*Z. roseum*) also occurs in both the Western and Eastern ghats.

Ginger (*Zingiber officinale* rosc.) being a tropical species originated in the South-East region of Asia, probably India or China (Bailey, 1949; Purseglove, 1972). The South Asian region is the centre of origin and it is under cultivation from ancient times (Ravindran, 1999). There is no conclusive evidence with regard to origin of ginger. It is probably a sterile hybrid between two distinct species that got thriving because of the successful vegetative propagation by rhizomes. Sato (1960) proposed that the proto karyotype of *Zingiber* is a species with four chromosomes and the present basic number is derived from such a species. The basic chromosome number of ginger is 22 and all the Indian species, studied so far have this number. Ginger flowers under certain climatic conditions but does not set seed (Ravindran *et al.*, 2005). It was introduced to the Mediterranean region from India during first century

A.D. (Burkill, 1966). From the sub-continent, Ginger also travelled to East Africa through Arabs during the thirteenth century and to West Africa through Portuguese during the fifteenth century.

In India, the important areas of diversity for cultivated ginger exist mainly in the North-Eastern region, Western and Eastern ghats where native types possess variability for rhizome shape, size, in side and surface colour, aroma and resistance/tolerance to rhizome rot and other diseases. It is grown in most of the states in India from Himachal Pradesh to Kerala including the North-Eastern region which are the major producing areas of ginger.

Garlic (*Allium sativum* L.)

Garlic is the second most important *Allium* species widely used and cultivated other than onion. For garlic, Central Asia is the primary centre and Mediterranean region is the secondary centre of origin (McCollum, 1976). *A. longicuspis* Regel ,endemic to Central Asia is the wild ancestor of Garlic. It is well known to Egyptians, Greeks and Romans before 3,000 BC and spread to the Mediterranean region in ancient times. It was in cultivation for a long time in India and China and Garlic was introduced to the Western Hemisphere by the Spanish, Portuguese and the French.

Black Pepper (*Piper nigrum* L.)

Black pepper is the dried mature fruit of the perennial climber and is the most important and widely used spice. Pepper originated in the tropical evergreen forests of the Western Ghats where wild Pepper is still common in many pockets. A native of humid tropical forests of the Western ghats from where it spread throughout the tropics (Ravindran, 1999). Genus *Piper* has over 1,000 species of which about 110 are of Indian origin (Ravindran *et al.*, 2000 and 2005). About 16 species are from the forests of Western Ghats, while most of the other species are from North-Eastern region of India (Ravindran and Babu, 1994). The earliest record of description of *Piper* species was by Van Rheede (1678) in his '*Hortus Indicus Malabaricus*' in which he described five types of wild peppers including black pepper and long pepper. Some species of *Piper* are affected by deforestation and many of the taxa are now confined to a few niches and may soon be extinct if not collected and conserved. *Piper barberi*, *P. hapnium*, *P. silentvalleyensis*, *P. wightii*, and *P. schmidtii* are extremely rare and are probably on the verge of extinction.

In India, the North-Eastern and the South-Western (Western Ghats) regions are recognized as two independent centres of diversity. The germplasm diversity in black pepper comprises of cultivars, wild relatives and genotypes characterized by variability for length of inflorescence, bearing habit, fruit size, colour and quality aspects. Endemic diversity is very rich in Kerala and Karnataka and over one hundred landraces exist. Unlike the wild forms of *P. nigrum* which are mostly dioecious, cultivated ones are bisexual. Local cultivar diversity of black pepper in India includes Aimpiriyan, Arakkulammunda, Arimulaku, Balankotta, Cheppukulamundi, Cheriyakanikkadan, Cholamundi, Chumala, Doddigae, Jeerakamundi, Kalluvally, Karimkotta, Karimunda, Karimundi, Karivilanchi, Kottanadan, Kurimalai, Kuriyalmundi, Kuthiravally, Malamundi, Maligesara, Mundi, Narayakodi, Nedumchola, Neelamundi, Perambramundi, Perumkodi, Poonjaran munda, Thulamundi, Hommankodi, Uddaghere, Uthirankotta, Vadakkan, Valiyakaniakkadan, Vattamundi, Vellanamban and Velliyaranmunda (Ravindran *et al.*, 2005). Inter and intra cultivar variability were reported by Ratnambal *et al.* (1985) and variability for morphological and bio-chemical characters by Ravindran and Nirmabl Babu (1994). Variability for major quality parameters like essential oils, oleoresin and piperin are reported by Raju *et al.* (1983) and Gopalam and Ravindran (1986).

Cardamom (*Elettaria cardamomum* Maton)

Cardamom of commerce is the dried fruit of rhizomatous and herbaceous perennial species. The origin and phylogenetic relationship of cardamom are not clearly known. It is probably of an amphidiploid origin and parents might have become extinct. Holttum (1950) felt that *E.cardamomum* and the Malaysian *Elettaria* represent parallel development from different points of origin in the *Alpinia* stock. It's closest relative is the wild cardamom *E.ensal* (Abheyv) (syn *E.major* Thw.) from Sri Lanka. Earlier workers recognized two botanical varieties in *E.cardamomum*, namely, var. *major* and var. *minor* (or var.*cardamomum*). Now var. major is treated as a separate species *E.ensal*. This is the native cardamom of Sri Lanka. Cardamom has chromosome number $2n$=48 (Sato, 1960; Ramachandran, 1969).

In India, only one species (*E.cardamomum*) occurs in tropical evergreen forests of Western Ghats and in the Silent Valley forest area where wild Cardamom is still common. It has three morphologically distinct eco- types namely, Malabar, Mysore and Vazhukka. Efforts of domestication of cardamom in India are relatively recent starting only in the 19th century with the European settlers beginning large scale clearance of forest areas for establishment of plantations of tea, coffee and also cardamom. The wild populations of cardamom gradually declined as a result of the large scale destruction of forest habitats (Ravindran *et al.*, 2005). The centre of diversity for genus *Elettaria* is the Malaysian-Indonesian region from where eight species are reported (Sakai and Nagamasu, 2000). Variability mainly occurs for earliness, bearing habit, fruit shape, size and quality parameters.

Seed Spices

The major seed spices grown in India are coriander, cumin, fennel, fenugreek, celery, bishops weed (ajowan), dill, caraway, black caraway, black cumin, white mustard, black mustard and Indian mustard.

Coriander (*Coriandrum sativum*) is the most important crop under seed spices. It is native to Southern Europe, Asia Minor and Caucasus where it grows in wild. It was introduced to India from the Mediterranean region possibly by the army of Alexander, the Great. Cumin (*Cuminum cyminum*) is a native of Turkistan or Upper Egypt and it's use is known from the biblical times. It is under cultivation in Mediterranean region for a very long time. It is cultivated in India (especially in Gujarat and Rajasthan), Ukraine, Morocco, Algeria, Syria, Malta and Cyprus. Cuminaldehyde is the major component in cumin oil.Fennel (*Foeniculum vulgare*) has two subspecies namely, *capillaceum* (garden fennel) and *piperitura* (wild fennel). Sub-species *capillaceum* comprises var.*vulgare* (bitter fennel), var.*dulce* (sweet fennel or French sweet fennel or Roman fennel) and var.*panmorium* (Indian fennel). Bitter fennel is commercially cultivated in India, Romania, Russia, Hungary, Czechoslovakia, Germany, France and Italy. It grows wild in France, Spain, Morocco and Algeria. Sweet fennel is cultivated in France, Italy and Greece and has the best odour and flavour quality. *Foeniculum vulgare* ssp. *capillaceum* var.*panmorium* is grown in India. Fenugreek (*Trigonella foenum-graecum*) has its original home in the region bordering the Eastern shores of Mediterranean extending to Central Asia. In addition, Ethiopia is an independent centre of origin. It is grown in India, Argentina, Egypt, France, Morocco, Lebanon etc.Celery (*Apium graveolens*) is a native of Eastern Europe where it grows in wild. It is highly tolerant to salinity. Celery is grown in temperate zones as a garden crop and its bleached leaf stalk is relished as vegetable. The main celery producing countries are India and France. In India, celery is grown mostly for seed. It is also cultivated to some extent in subtropical America. Bishop's weed/*Ajowan* (*Trachyspermum ammi*) is cultivated in North Indian states. It has two morphological types: a larger type, grown in most *ajowan*-growing regions and a small form, grown in Madhya Pradesh (Indore and adjoining regions). The former is used as a medicine and in the preparation of curry powder which is mainly exported. Other *ajwan*-growing countries are Egypt and Iran. Dill/Garden dill (*Anethum graveolens*) is a native of Mediterranean region and grows wild in parts of Europe, Africa and Asia. Anise/Aniseed (*Pimpinella anisum*) is yet another seed spice grown in India, Egypt, Syria, Tunisia, Morocco, China, Chile, Mexico and the USA. In India, anise is grown in North Indian states (Ravindran *et al.*, 2005).

Tree Spices

There are many tree spices in India, out of which a few are native, others introduced and a few are under commercial cultivation. Of these, nutmeg, clove and cinnamon all of which are of recent introductions are widely grown in Kerala, Karnataka and Tamil Nadu in the high altitude areas. Allspice, a recent introduction is less popular. Tamarind, curry leaf and *Garcinia* are indigenous and are grown widely in India (Ravindran *et al.*, 2005).

Nutmeg (*Myristica fragrans* Houtt.)

Nutmeg is the only tree that produces two separate spices, namely nutmeg (kernel or seed) and mace (aril covering the seed). The species originated in the Moluccas Islands of Indonesia. Nutmeg was introduced to Kerala coast by the Portuguese from Indonesia during the 16th century. The wild species of nutmeg *viz.*, *M.amygdalina* Wall., *M.andamanica* Hook., *M.attenuata* Wall., *M.dactyloides* Gaertn. (syn *M.laurifolia* Hook. f.), *M.beddomeii* King., *M.gibbosa* Hook., *M.glabra* Blume, *M.glaucescens* Hook., *M.irya* Gaertn., *M.kingii* Hook., *M.longifolia* Wall. and *M.magnifica* Bedd. are endemic to the Western Ghat forest areas and also to the Indo-Malayan region. The mace and seeds of *M.dactyloides* are used in the tribal and indigenous medicical systems (Ravindran *et al.*, 2005). Indonesia, New Guinea and Grenada are the major producers of nutmeg.

Cinnamon and Cassia (*Cinnamomum* spp.)

True cinnamon is obtained from *C.verum*, indigenous to Sri Lanka and South Western Ghats in India. The earliest description of cinnamon is in van Rheede's *'Hortus Malabaricus'* in which two species were described under the local name *karuva* and *kattukaruva* which are *C. verum* and *C.malabathrum* Bercht. and Presl. respectively (Ravindran *et al.*, 2005). *Cassia cinnamon* is obtained from three main sources, the most important being *C.cassia* Bercht. and Presl. (Chinese cassia). The other cassia cinnamons are: Indonesian (Javan) cassia (*C.burmanii* C.G. and Th. Nees) and the Indian cassia (*C.tamala* Nees, *C.bejholghota* and *C.impressinervium*).

Clove (*Eugenia caryophyllata* Thunb.)

The dried, mature and unopened flower buds are consumed as a spice (Ravindran *et al.*, 2005). Clove is a native of Moluccas Islands in Indonesia from where it spread to many tropical countries. Clove was brought to India by the British planters who introduced the first trees in Courtallam and Nilgiri areas of Tamil Nadu. Under the natural conditions in Moluccas, both the clove producing (aromatic) and non-clove producing (non-aromatic) trees co-exist. It was the selective multiplication and culture of the aromatic clove producing trees, which led to all clove plantations in various countries.

Garcinia spp.

The genus *Garcinia* is distributed in the tropics of the old world, chiefly Asia, with approximately 200 species. Malaysia and Africa with large number of endemic species appear to be the two main centres of diversity. There are two major centres of diversity in the Indian sub-continent: the North-Eastern region from Assam extending up to Mayanmar and the other in the Western Ghats. In India, three important species namely, *Garcinia cambogia* (Malabar tamarind), *G.indica* (Kokum) and *G.mangostana* (Mangosteen) are cultivated on a small scale (Ravindran *et al.*, 2005).

Tamarind (*Tamarindus indica* L.)

Tamarind, also known as Indian date is a common spice used in South India and other parts of the country. It originated in tropical Asia most probably in India, but also occurs in Central Africa and dry Savannas of tropical Africa. Being a crop native to India, there is accumulation of indigenous genetic variability which needs to be collected and conserved (Ravindran *et al.*, 2005). Tamarind. being highly cross-pollinated and owning to its wide geographical distribution and adaptability to different agro-climatic regions, a large genetic diversity exists in the seedling populations of tamarind all over the country particularly, in Maharashtra, Chhattisgarh (eastern part) and Tamil Nadu. In Tamarind, individual plants having distinct phenotypic diversity with good yield potential are naturally available. Variability was observed for tree height, canopy, bearing, duration, fruit shape and size, pulp content, taste and yield. Promising selections of tamarind are PKM1, Pratisthan and Yogeswari (Vashishtha *et al.*, 2005).

Curry Leaf [*Murraya koenigii* (L.) Spreng.]

Curry leaf is an essential leafy spice in South Indian food preparations. It originated in India and cultivated extensively as a backyard plant in most households. It is grown as a commercial crop in Tamil Nadu, Karnataka and Andhra Pradesh (Ravindran *et al.*, 2005).

Vanilla (*Vanilla planifolia* Andr.)

Vanilla is a commercially important spice, originated in Mexico and grown in many Pacific Ocean Islands, Indonesia and many African countries. It was introduced into India probably in the beginning of the 20th century (Ravindran *et al.*, 2005).

Plantation Crops

The important plantation crops for which good native diversity accumulated in the Indian Gene Centre include coconut, cashew nut and betel palm arecanut. The oil palm, cocoa and rubber are the introductions made in recent times and these crops have become indispensable and very crucial in the overall agricultulral commodity requirements and economy.

Coconut (*Cocos nucifera* L)

Cocos is a monotypic genus. The centre of origin of Coconut where most of the primitive types are found is in the North-Western South America (Purseglove, 1968). Evidences in favour of it's origin somewhere in South-East Asia are rather speculative to further specify the area of origin. From the centre of origin, it moved eastwards to Pacific and further to Americas and towards the West to India and Madagascar over the sea (Harries, 1978). The species now is very widely cultivated and almost pantropical in distribution. Over the years, tremendous variability got generated in the crop due to cross-pollination. Two distinct varieties are generally recognized. The talls (typica) occupy most of the areas (Nampoothiri *et al.*, 2005). The Indian collections exhibit variation in height, bearing and nut quality.

Cashew nut (*Anacardium occidentale* L)

It is a native of tropical Americas from Mexico to Peru and Brazil and also of the West Indies. It is naturalized in many tropical countries especially in the coastal zone. It was introduced to the Indian sub-continent from Brazil by the Portuguese in sixteenth century to control soil erosion in coastal areas. Cashew was one of the first fruit tree species from New World to have widely introduced in the tropics by early Portuguese and Spanish adventurers (Johnson 1973; Puseglove 1988). The local Tupi Indians probably played a major role in dispersion of the species in their temporary migrations towards the coast of North-Eastern Brazil, where a considerable intra-specific variation was recorded (Ascenso 1986). From its origin in North-Eastern Brazil, cashew spread into South and Central America (Van Eijnatten 1991). The Portuguese discovered cashew in Brazil and from there, it first spread to Mozambique (Africa) and later to India between 1563 and 1578. It was first tested in gardens in Cochin on the Malabar Coast of Kerala. Cochin served as a dispersal point for cashew in India and perhaps for South-East Asia as well (Johnson, 1973; Nampoothiri *et al.*, 2005).

Oil Palm (*Elaeis guineensis* Jacq.)

Oil palm originated in West Africa. The American oil palm (*Elaeis oleifera* Cortes) formerly known as *E.melanococca* Range and *Corozo oleifera* (Kunth) Cortes, is native to Central America. There is no detailed information on another related species, *E.odora* (formerly known as *Barcella odora*). Oil palm was introduced to Brazil in 15th century by the Portuguese. During 1848, four seedlings of oil palm were planted in Bogor, Indonesia, the progenies of which were planted in Malaysia to be later known as Deli dura. It was with its commercial plantings from 1917 that the crop really flourished in Malaysia which still leads the world in oil palm industry. Apart from Malaysia, it is cultivated in India, Indonesia, Nigeria, Republic of Zaire, Costa Rica, Ivory Coast and Papua New Guinea (Nampoothiri *et al.*, 2005).

Betel Palm (Arecanut) (*Areca catechu* L.)

The genus *Areca* has about 76 species and among them, *A. catechu* L. is the only cultivated species, even though nuts of a few other species *viz. A.triandra* Roxb. are also used for masticatory purposes (Nampoothiri *et al.*, 1999). The *Areca* species mostly of monoecious palms are spread from Indo-Malaysia to the Solomon Islands and Northern Australia. Various centres of origin are suggested which include India, China, Malayan peninsula, East Indies, the Sunda Islands, the Philippines and Malaysia (Burkill, 1966). Prevalence of germplasm diversity supports East Indies group of Islands as the centre of origin (Nampoothiri *et al.*, 2005). It is certainly of very ancient cultivation in Malaysia from where it might have introduced to the Indian sub-continent very long time ago. The germplasm is comparatively less diverse and productive. Diversity occurs for plant height, earliness and yield attributing characters. Five botanical varieties of *A.catechu*, namely, A.*communis, A.silvatica, A.batanensis, A.deliciosa* and *A.longicarpa* are reported based on size and shape of fruits and kernels. In cytological investigations, mostly bivalents were observed with a rare quadrivalent in *A.catechu* and only bivalents in *A.triandra* and a secondary allotetraploid origin for *A.catechu* were suggested.

Cocoa (*Theobroma cacao* L.)

Cocoa is a native of South America and was introduced to India in the first half of 20th century. It is well adapted to Indian conditions and is cultivated in Kerala, Karnataka, Andhra Pradesh and Tamil Nadu. It is popular as an intercrop in coconut and arecanut gardens (Nampoothiri *et al.*, 2005).

Rubber [*Hevea brasiliensis* (H.B. and K.) Muell.- Arg.]

Currently, a total of 10 species are recognized in genus *Hevea* which occupies the whole of Amazon river basin in Brazil. Genetic base of the crop in the East is narrow, limited to a few seedlings originally collected from a miniscule of genetic range in Brazil, referred to as 'Wikham base' (Simmonds, 1989). These were collected from Boim, near the Tapajos river in Brazil. The base has futher narrowed down through unidirectional selection for yield, a cyclical generation-wise assertive breeding pattern and a wider adoption of clonal propagation by budding. The parentage of popular clones bred in various rubber growing countries can be traced to a handful of parental clones (Tan, 1987; Nampoothiri *et al.*, 2005).

Beverages

Among the beverage crops, tea occupies a pivotal place in importance in having significant indigenous diversity, covered in almost 60 per cent of tea growing area in the world.

Tea [*Camellia sinensis* (L.) Kuntze]

Various authors traced origin of tea to Central China, somewhere in the Mangolian plateau (Weatherstone, 1992). From this primary centre, it spread towards South to the secondary centre, near the source of the Irrawady river (Barua, 1989). From this secondary centre, further dispersal took place in three different directions along the sources of three great rivers, Yang-tse, Mekong and Brahmaputra (Bezbaruah, 1999). The present day commercial tea populations belong to *C.sinensis* (L.) O.Kuntze (China type), *C.assamica* (Masters) Wight (Assam type) and *C.assamica* ssp. *lasiocalyx* (Planch. ex Watt) Wight (Cambod type). These three caffeine producing taxa, due to their out-breeding character underwent natural hybridisation with the non-tea producing taxon *C.irrawadiensis* Barua and other closely related species which resulted in high heterogeneity in the present day commercial seedling populations (Satyanarayana and Sharma, 1993). The genetic diversity of tea consists of cultivated species, wild genera and species, weedy relatives, old seedling populations, biclonal and polyclonal seed stocks and clones. While most of the genera of family Theceae and different species of *Camellia* are found in Yunnan province in South-West China, nevertheless, North-East India bordering Myanmar also has rich diversity (Weatherstone, 1992). The native variability is conspicuous for it's adaptability to colder climate and having narrow/broad leaves.

Medicinal and Aromatic Plants

India is known to be a rich repository of medicinal and aromatic plants from time immemorial (Pareek *et al.*, 2005). Some of the cultivated medicinal plants in India and their origin are given below.

Psyllium (*Plantago ovata* Forsk.)

Psyllium or *isabgol* belongs to family Plantaginaceae and is a native of Persia.

Opium poppy (*Papaver somniferum* L.)

Opium poppy belongs to family Papaveraceae. It is historically the most important medicinal plant and is the main source of commercial opium. It is the oldest and the best known pain killer drug of modern medical system. The centre of origin of opium poppy is Western Mediterranean region of Europe from where it spread through the Balkan Peninsula to Asia Minor (Mortan, 1977).

Senna (*Cassia angustifolia* Vahl.)

Senna is a native of Saudi Arabia. It was introduced to Tamil Nadu during the mid 18th century.

Periwinkle (*Catharanthus roseus* (L.) G.Don)

Periwinkle is a pantropical species occurring mainly in West Indies and Madagascar. It is well adapted to diverse agro-climate of India and is commercially grown in Tamil Nadu, Karnataka, Gujarat, Madhya Pradesh and Assam (Virmani *et al.*1978).

Dioscorea (*Dioscorea* spp.)

There are four species of yam as sources of steroids, namely, *Dioscorea deltoidea* Wall. and *D.prazeri* Prain and Burkil of the old world's temperate region and *D.composita* L. and *D.floribunda* L. of the new world's tropical region. *D. deltoidea* is indigenous to India and is distributed throughout the Himalayas and in West Bengal. *D.prazeri* occurs in the North-Eastern region of India. *D.composita* and *D.floribunda* were introduced from Central America.

Solanum (*Solanum viarum* Dunal)

A number of *Solanum* spp. possess high concentrations of steroid solasodine in the foliage, growing stems and berries. In India, *S.viarum* is commercially exploited for the steroids whereas *S.laciniatum* L. is the main source of herbal drugs in Soviet Union, Hungary, Poland, Czech Republic, Slovakia and Newzealand (Telek *et al.*, 1977).

Belladonna (*Atropa* spp.)

Indian belladonna (*A.acuminata* Royle ex Lindl.) contains atropine and its isomer hyoscyamine, used as pre-anaesthetic agents to check secretion in the throat and respiratory tracts. It is native to the temperate Western Himalayas. *A.belladonna*, the main source of the drug is a native of Central and Eastern Europe.

Ashwagandha (*Withania somnifera* Dunal)

Ashwagandha is a native of India. Another species, *W.coagulans* Dunal is also reported from India, found in the wild state in Mediterranean regions in North Africa. The alkaloids present in the roots are the main source of the drug. Ashwagandha is used as a tonic and commercially cultivated in North-Western Madhya Pradesh for its roots, used extensively in the indigenous systems of medicine in India (Nigam, 1984).

Sarpagandha (*Rauvolfia serpentina* Benth. Ex Kurz)

Sarpagandha is native to India and is found in the foot-hills of Himalayas and in Andaman and Nicobar Islands and the Western Ghats. The roots yield different alkaloids of which reserpine and some related alkaloids are hypotensive and also used to control certain forms of insanity. It is cultivated in Central India but largely exploited from the wild for use in the indigenous systems of medicine.

Woolly foxglove (*Digitalis purpurea* L. and *D.lanata* Ehrh.)

Woolly foxglove or Tilpushpi is native to West-Central Europe and was introduced to India after the Second World War. They yield cardiotonic glycosides like digoxin, digitalin and lanatocides, used in regulating heart beat and related heart diseases.

Aloe (*Aloe barbadensis* Mill.)

Aloe is indigenous to African continent and Mediterranean regions. In India, it is naturalised and occurs in semi-wild state in the drier tracts. Its cultivation is attempted in Haryana, Rajasthan, Gujarat and Andhra Pradesh.

Babchi (*Psoralea corylifolia* L.)

Psoralea or babchi is widely distributed in tropical and subtropical regions of the world. In India it occurs in parts of Central and Northern India.

Giloe (*Tinospora cordifolia* Miers)

Giloe is indigenous to India and distributed throughout the country. The natural population from the wild is exploited for use in indigenous system of medicine.

Henbane (*Hyoscyamus* spp.)

Genus *Hyoscyamus* is a native of Europe, Central Asia, India and Tropical Africa. *Hyoscyamus niger* L. and *H.muticus* L. are the main drug yielding species. In India, these species occur in North-Western Himalayas.

Kalmegh [*Andrographis paniculata* (Burm.) Nees]

Kalmegh is widely distributed in moist shady soils of Himalayan foothills and in large areas of Central and Southern India. It is valued for its alkaloid andrographalide and is also used in treating hepatitis.

Kewanch [*Mucuna pruriens* (L.) DC.]

Kewanch has a sedative property and used in the treatment of Parkinson's disease.

Pyrethrum (*Chrysanthemum cinerarifolium* Vis.)

The flowers contain pyrethrins for preparation of safe household insecticides. The crop is grown in temperate hills of Jammu and Kashmir and Kodaikanal tract of Tamil Nadu.

Indian Bdellium/Guggal [*Commiphora mukul* (Hook. ex Stocks) Engl.]

Africa and Asia are the centres of origin of *Commiphora* species. Its cultivation is promoted in Rajasthan.

Safed musli (*Chlorophytum* spp.)

This originated in tropical and subtropical Africa (Hooker, 1894). Of the 13 species occurring in India, eight are endemic to the Indian subcontinent (Nair, 1974). All these species are grouped under one trade name safed musli, extensively used in the preparation of ayurvedic medicines as a tonic and aphrodisiac. Safed musli is distributed in Southern Rajasthan, Western Madhya Pradesh and Northern Gujarat and farmers took up commercial cultivation of this crop in many parts of the country.

Datura (*Datura* spp.)

There are three species of *Datura* with medicinal values namely *D.metel* L., *D.stramonium* L. and *D.innoxia* Mill. They are rich in hyoscyamine and related alkaloids, used as pre-anaesthetic in surgery and childbirth, in ophthalmology and prevention of motion sickness. It is widely distributed in semi-arid and arid regions of Punjab, Haryana, Rajasthan, Gujarat and Tamil Nadu.

Aromatic Plants

India is well known for a wide spectrum of aromatic species diversity spread through the sub-continent.

Mentha (*Mentha* spp.)

Important species used as sources of essential oils are *Mentha arvensis* L., *M.piperita* L. and *M.citrata* L. These are mostly distributed in temperate and sub-temperate regions of the world and cultivated in India, the USA, Europe, South-East Asia and Latin America.

Lemongrass [*Cymbopogon flexuosus* (Nees ex Steud.) Wats.]

Lemongrass is a native of India and cultivated in Southern India, particularly Kerala. India is the main producer and exporter of lemongrass oil. The lemongrass oil, obtained by hydro-distillation of leaves is the richest source of citral, used as an industrial raw material for production of alpha- and beta-ionones and Vitamin A.

Java citronella (*Cymbopogon winterianus* Jowitt)

Java citronella contains citronella (Java) oil in the leaves. It was introduced from Indonesia and the crop is now commercially established in Assam, Meghalaya and Karnataka.

Palmarosa [*Cymbopogon martini* var. *motia* (Roxb.) Wats.]

Leaves and inflorescence yield the palmorosa oil, which has rose like aroma. It is cultivated in Northern and Southern states of India.

Vetiver [*Vetiveria zizanoides* (L.) Nash]

Vetiver or khus is highly priced for its oil in perfume industry. In India, commercial cultivation of vetiver is done in wet and damp environments and marshy areas in South-Eastern Rajasthan. The South Indian vetiver cultivation in the river banks has shrunk over the years.

Basil (*Ocimum* spp.)

Species of *Ocimum* are distributed in the tropical, subtropical and warmer parts of temperate regions of the world. The maximum diversity at species level is observed in the tropical rain forests of Africa. Certain species are under cultivation in a very small scale.

Rose/Damask Rose (*Rosa damascena* Mill.)

Damask rose flower is the source of rose oil (otto of rose), highly priced and used in high grade perfumes. It is grown in a small scale in Western Uttar Pradesh and Rajasthan.

Rose geranium (*Pelargonium graveolens* L'Herit.)

Rose geranium is a South African plant introduced to Tamil Nadu in early 20th century. Its small scale cultivation is done in Karnataka (Bangalore region) and Tamil Nadu (Anamalai, Nilgiris and Kodaikanal hills). It is now popular in the lower hills of Uttarakhand also.

Anise (*Pimpinella anisum* L.)

Anise is widely cultivated in most countries of Central and South-Eastern Europe. In India, anise is grown in many North Indian states and has high potential for introduction in regular agriculture.

Conclusion

There are implications on exchange of horticultural germplasm in view of international treaties. In any country, among the crops that are under cultivation a few may be of native and some may be of non-native origin. For food security, dependence on several crops is the order of the day and exclusive dependence on native crops may not be possible. The enforcement of the CBD from 1993 and provisions under TRIPS led to the apprehension that exchange of germplasm would get restricted. The international treaty on plant genetic resources for food and agriculture (ITPGRFA) facilitated establishment of a multilateral system of access and benefit-sharing which applies to a list of crops in Annexure I (35 food crop species and 29 forage crop species). Under this treaty, germplasm exchange is ensured with an SMTA for food security and interdependence irrespective of origin of crops. In the post CBD era, on analysis it was revealed that there was a decline by 14.5 per cent in access and introduction of germplasm (Anonymous, 2007). To increase food production at global level on sustainable basis, dependence on crop genetic resources originated from different geographical locations through introduction and exchange is inevitable. This holds good more in underutilized crop species in whose cases the centres of origin itself are the centres of diversity as the domestication efforts having just started and still in the evolutionary process.

Studies on confirmation of centres of origin of different crops and conservation of germplasm variability from the centres of origin assume importance as a valuable source of genes for utilization especially for sources of resistance as the pests and pathogens also co-evolve with the germplasm in those centers for improvement of the existing crop species and development of improved varieties.

References

Agharkar, S.P and Roy, N.1951. On the origin and distribution of cultivated mangoes. Indian J. Genet., 11:48.

Alexander, J. and Coursey, D.G.1969. The origin of Yam cultivation. (In): *The domestication and exploitation of plants and animals.* P.J.Ucho and G.W.Dimbleby (eds.), London, Duckworth. pp 405-25.

Anonymous 2007. State of plant genetic resources for food and agriculture in India 1996-2006. NBPGR, New Delhi. pp 70.

Arora, R.K. 1991. Plant diversity in the Indian gene centre. *In* R.S.Paroda and R.K.Arora (eds.), *Plant genetic resources conservation and management,* IBPGR, New Delhi, India. pp 25–54.

Arora, R.K and Nayar, E.R. 1984. *Wild relatives of crop plants in India*. NBPGR Sci. Mongr. No.7, NBPGR, New Delhi. Pp 90.

Arora, R.K. and Pandey, A. 1996. *Wild Edible Plants of India: Diversity, Conservation and Use*. Indian Council of Agricultural Research, NBPGR, New Delhi, India, 294 p.

Arora, R.K., Rao, V.R. and Rao, A.N. 1996. Status Report on Genetic Resources of Litchi in China. IPGRI Regional Office for Asia, the Pacific and Oceania, Singapore, pp1-18.

Ascenso, J.C. 1986. Potential of the cashew crop-1. *Agric. Int*. 28:324-327.

Austin and Huaman, Z. 1996. A synopsis of *Ipomoea* (Convolvulaceae) in the Americas. *Taxon.*, 45:3-38.

Babu,C.R. 1977. *Herbaceous Flora of Dehra Dun*. Publication and Information Department, CSIR, New Delhi, India, 721 p.

Baiely LH 1947. *The Standard Encyclopaedia of Horticulture*. MacMillan and Co, New Delhi, India.

Bailey, L.H. 1949. Manual of cultivated plants (2nd Edition). Macmillan Company, New York.

Baker, J.G. 1886. Scitamineae. (In) J.D. Hooker (ed.), *Flora of British India* 6: 198-264.

Bal, J.S. and Uppal, D.K. 1992. *Ber Varieties*. Associated Publishing Co., New Delhi, India.

Barua, D.N. 1989. Science and practice in tea culture. Tea Research Association, Calcutta, India, 509 p.

Bezbaruah, H.P. 1999. Origin and history of development of tea. (In) *Global Advances in Tea Science*, pp 383-392.

Bhaduri, P.N. 1951. Inter-relationship of non-tuberiferous species of *Solanum* with some consideration on the origin of brinjal. Indian J. Gent. Pl. Br., 2: 75–86.

Bhattacharya, S.C. and Dutta, S. 1956. *Classification of Citrus fruits of Assam*. Indian Council of Agricultural Research. Scient. Monogr. 20:1-110.

Bosland, P.W and Votava, E.J. 2000. Peppers: Vegetable and Spice Capsicums. Crop Production Science in Horticulture 12. CAB International Publishing, Wallingford, England, UK, 204 p.

Brandis, D.D. 1874. Forest Flora of North West and Central India. Allen, London, UK.

Bujisen, J.R.M. 1990. Taxonomic Survey of Allium Species Cultivated in South-East Asia. Rijks Herbarium, Rapenburg, Netherlands, 47p.

Bukenya,Z.R. and Carasco, J.F. 1999. Ethnobotanical aspects of *Solanum* L. (Solanaceae) in Uganda. In: M. Nee, D.E. Symon, R.N. Lester and J.P. Jessop (eds.), Solanaceae IV: Advances in Biology and Utilization. Royal Botanic Gardens, Kew, UK.

Burkill, I.H. 1966. A dictionary of the economic products of the Malay Peninsula (2 vols.), Ministry of Agriculture and Cooperative, Malaysia, pp 2444.

Chakravarty,H.L.1982. *Fascicles of Flora of India, Fascicle II, Cucurbitaceae*. Botanical Survey of India, Howrah, India.

Chatterjee, D. and Randhawa, G.S. 1959. Standardized names of cultivated plants in India, I Fruits. *Indian J. Hort*. 9:24-36.

Chiang, M.S., Chong, C., Landry, B.S. and Crete, R. 1993. Cabbage, *Brassica* oleracea subsp. *capitata* L. (In) G. Kalloo and B.O. Berg. (eds.), *Genetic Improvement of Vegetable Crops*. Pergamon Press Ltd., Oxford, England, UK, pp.113-155.

Chundawat, B.S. 1990. *Arid Fruit Culture*. Oxford and IBH Co., Ltd., New Delhi, India.

Crisp, P. and Ives, S.J. 1982. Genetic Conservation of Crucifers. *Rep. Nat. Veg. Res. Stn.*, Wellesbourne, UK, 58 p.

Crisp, P. and Tapsell, C.R. 1993. Cauliflower, *Brassica oleracea* L. (In) G. Kalloo and B.O.Bergh. (Ed.), *Genetic Improvement of Vegetable Crops*. Pergamon Press Ltd., Oxford, England, 58p.

Daunay,M.C., Lester, R.N., Hennart, J.W. and Duranton, C. 2000. Eggplants: Present and Future. *Capsicum Eggplant Newsl*. 19: 11-18.

De Candolle, A. 1882. *Origin des Plantes Cultivies*. Germes Bailliere, Paris, France, 377p.

De Candolle., A. 1904. Origin of cultivated plants. Kegan Paul. London.

De Winter, B. 1990. A new species of *Citrullus* (Benincaceae) from the Namib Desert, Namibia. *Bothalia* 20:209-211.

Decker, D.S. 1988. Origin, evolution and systematics of *Cucurbita pepo* (Cucurbitaceae), Econ. Bot., 42: 4-15.

Denevan, W.M. 2001. Cultivated Landscapes of Native Amazonia and the Andes: Triumph Over the Soil. Oxford University Press, Oxford, UK and New York, USA, 432 p.

Dhankhar, B.S., Mishra, J.P. and Bisht, I.S. 2005. Okra. In: B.S. Dhillon, R.K.Tyagi, S.Saxena and G.J. Randhawa. (eds.). Plant Genetic Resources: Horticultural Crops. Narosa Publishing House, New Delhi, pp 59-74.

Edison, S., Velayudhan, K.C., Easwari Amma, C.S., Santha V Pillai., Mandal, B.B., Sheela, M.N., Vimala, B., Unnikrishnan, M. and Zakir Hussain. 2005. Tropical Root and Tuber Crops. In: B.S.Dhillon, R.K.Tyagi, S.Saxena, G.J.Randhawa. (Ed). *Plant Genetic Resources: Horticultural Crops*. Narosa Publishing House, New Delhi, pp 228-250.

Eshbaugh, W.H. 1993. Peppers: History and exploitation of a serendipitous new crop discovery. In: J. Janick and J.E. Simon. (Ed.). New Crops. John Wiley and Sons, New York, pp 132-139.

Eshbaugh, W.H., Guttman, S.I. and. McLeod, M.J.1983. The origin and evolution of domesticated *Capsicum* species. *Journal of Ethnobiology*, 3: 49-54.

Felker, P., Singh, G. and Pareek, O.P. 1997. Opportunities for development of cactus (*Opuntia* spp.) in arid and semi-arid regions. *Ann. Arid Zone*, 36:267-278.

George Ushar.1974. *A Dictionary of Plants Used by Man.* Constable and Co. Ltd., London, UK, 121 p.

Giles, W.F. 1943. Onion and other edible alliums. *J. Roy. Hort. Soc.*, 36:193-200.

Giuseppina, Morico., Fabrizio, Grassi. and Carlo, Fideghelli. 1998. Horticultural genetic diversity: Conservation and sustainable utilization and related international agreements. World Conference on Horticultural Research. ISHS. 17–20 June, 1998, Rome, Italy.

Gmitter, F.G. Jr., and Hu, X. 1990. The possible role of Yunnan, China in the origin of contemporary Citrus species (Rutaceae). Econ. Bot., 44: 257-77.

Gonzalez, M. and Bosland, P. 1991. Strategies for stemming genetic erosion of *Capsicum* germplasm in the Americas. *Diversity*, 7 (1&2): 52-53.

Gopalam, A. and Ravindran, P.N. 1986. Indexing quality parameters in black pepper cultivars. Indian Spices. 22–23: 8–11.

Gray, A.R. 1982. Taxonomy and evolution of broccoli (*Brassica oleracea* var. *italica*). Econ. Bot.,36:397.

Gray, A.R. 1993. Broccoli [*Brassica oleracea* L. (Italica group)]: (In) G.Kalloo and B.O. Berg (eds.). *Genetic Improvement of Vegetable Crops*. Pergamon Press Ltd., Oxford, England, pp.61-86.

Harlan, J.R. 1971. Agricultural origins: centres and non-centres. Science, 174: 468-474.

Harlan, J.R. 1975. Crops and man. American Society of Agronomy, Madison, Wisconsin. 295 p.

Harries. 1978. The evolution, dissemination and classification of *Cocos nucifera* L. *Bot. Rev.*, 44:265-319.

Hawkes, J.G. 1983. The diversity of crop plants. Harvard University Press, Cambridge, Mass., 184 p.

Hegwood, D.A. 1990. Human health discoveries with *Opuntia species* (Prickly pear). *Hort. Sci,*. 25:1115-1516.

Heiser, C.B.1976. Pepper *Capsicum* (Solanaceae). (In) *Evolution of crop plants.* N.W.Simmonds. (Ed.). Longman, London, pp 265-268.

Heiser Jr., C.B. 1985. How many kinds of peppers are there? In: *Of Plants and People.* C.B.Heiser Jr. (ed.). University of Oklahoma Press, Norman and London. Pp. 142-154

Hernández-Verdugo, S., Dávila Aranda, P. and. Oyama. K.1999. Síntesis del conocimiento taxonómico, origen y domesticación del género. *Capsicum. Boletín de la Sociedad Botánica de México,* 64: 65-84.

Holttum, R.E. 1950. The Zingiberaceae of the Malay-Peninsula. *Gard. Bull.* 13:1-249.

Hooker, J.D. 1894. *The Flora of British India, Vol. 6.* L. Reeve and Co., Ltd., The Oast Book House, Ashford, Kent, England.

Hooker, J.D. 1973. Liliaceae *Allium* L. Flora of British India. Vol. VI, reprinted by Singh and Singh, New Connaught Place, Dehra Dun and Ms. Periodical Experts, Delhi, India, pp 337-345.

Jeffrey, C. 1990. Systematics of the Cucurbitaceae: An overview. (In) *Biology and Utilization of Cucurbitaceae.* D.M. Bates, R.W. Robinson and C.Jeffrey. (Ed.). Cornell University Press, Ithaca, New York, USA, pp 3-9.

Johnson, D. 1973. The botany, origin ad spread of the cashew (*Anacardium occidentale* L.). *J. Plantation Crops,* 1:1-7

Jones, H.A and L.K.Mann. 1963. *Onions and their Allies.* London and New York, USA, 286 p.

Joshi, A.B. and Hardas, M.W. 1976. Okra. In: *Evolution of Crop Plants.* N.W.Simmonds. (Ed.). Longmans, London, UK, pp 194-195.

Joshi, A.B, Gadwal, B.R. and Hardas, M.W. 1974. Okra. (In) *Evolutionary Studies in World Crops, Diversity and Change in the Indian Sub-continent.* J.B.Hutchinson. (Ed.).Cambridge University press, pp 99-105.

Kalloo,G. 1985. *Tomato.* Allied Publishers Pvt. Ltd., New Delhi, India.

Kalloo, G., Umesh Srivastava., Major Singh. and Sanjeet Kumar. 2005. Solanaceous Vegetables. (In) *Plant Genetic Resources: Horticultural Crops.* B.S.Dhillon, R.K.Tyagi, S.Saxena and G.J.Randhawa. (Ed.). Narosa Publishing House, New Delhi, pp 19-33.

Kanjilal, U.N., Kanjilal, P.C. and Das, A. 1934. *Mangifera indica* L., Flora of Assam, 1, pp 335-336.

Karihaloo, J.L., Malik, S.K., Rajan, S., Pathak, R.K. and Gangopadhyay, K.K. 2005. Tropical Fruits. In: *Plant Genetic Resources: Horticultural Crops.* B.S.Dhillon, R.K.Tyagi, S.Saxena and G.J.Randhawa. (eds.). Narosa Publishing House, New Delhi, pp 121-145.

Kirtikar, K.R. and Basu, B.D. 1935. *Indian Medicinal Plants Vol.1.* LM Basu, Allahabad, India.

Kishore, D.K. and Randhawa, S.S. 1993. Wild germplasm of temperate fruits. (In) *Advances in Horticulture–Fruit Crops. Vol.1.* K.L.Chadha and O.P. Pareek (eds.). Malhotra Publishing House, New Delhi, India, pp 227-241.

Krishnamurthy, S. and Seshadri, V.S. 1958. Origin and evolution of cultivated banana. Indian J. Hort., 15:135-145.

Maheswari P and U Singh 1965. *Dictionary of Economic Plants in India.* Indian Council of Agricultural Research, New Delhi, India.

McCollum GD 1976. Onion and allies. In: NW Simmonds (Ed.), *Evolution of Crop Plants,* Longman Group Limited., London, UK, 339p.

McPhee, J. 1967. *Oranges.* Farar, Strans and Giroux, New York, USA.

Mehra, K.L. and Arora, R.K. 1982. *Plant Genetic Resources of India–Their Diversity and Conservation.* National Bureau of Plant Genetic Resources, New Delhi, India, Sci. Mongr. No.4, Pp 60.

Merrick, L.C., 1990. Systematics and evolution of a domesticated squash. *Cucurbita argyrosperma* and its wild and weedy relatives. In: DM Bates, RW Robinson and C Jeffrey (Eds.), *Biology and utilisation of the Cucurbitaceae.* Cornell University Press, Ithaca, New York, pp 77-95.

Mitra, S.K. and Bose, T.K. 1999. Guava. In: TK Bose, SK Mitra, AA Farooqi and MK Sadhu (Eds.), *Tropical Horticulture*. Vol.I. Naya Prokash, Calcutta, India, pp 297-318

Mortan, J.F. 1977. *Major Medicinal Plants*. Botany, Culture and Uses. CC Thomas Publishers, USA.

Mukherjee, S.K. 1948a. Lloydia.12: 73-136.

Mukherjee, S.K. 1949. A monograph on the genus *Mangifera* L. Lloydia 12:73-136.

Mukherjee, S.K. 1950. Mango. Its allotetraploid nature. *Nature* 151:196-197.

Mukherjee, S.K. 1951. Origin of mango. *Indian J. Genet. Pl. Breed*. 11:49-56.

Mukherjee, S.K. 1958. Indian journal of Horticulture. 16: 129-34.

Mukherjee, S.K. 1985. Systematic and Ecogeographic Studies on Crop Genepools 1. Mangifera L. IBPGR, Rome, Italy.

Mukherjee, S.K. 1997. Introduction: Botany and importance. (In) R. Litz. (Ed.), *The Mango*. Commonwealth Agricultural Bureaux International, UK.

Munger, H.M. and Robinson, R.W. 1991. Nomenclature of *Cucumis melo* L. *Cucurbit Genetic Cooperative Report* 14:43.

Nair, V.N. 1974. Cytological studies in *Chlorophytum bruchae* Ansari. *Curr. Sci.* 13:161.

Nampoothiri, K.U.K., Ananda, K.S. and Parthasarathy, V.A. 1999. Biodiversity in arecanut. *The Botanica,* 49:127-135.

Nampoothiri, K.U.K., Kumaran, P.M., Swamy, K.R.M., Annamma Varghese, Y., Ilango, R.V.J. and Rekha Chadhury. 2005. Plantation Crops. (In) B.S.Dhillon, R.K. Tyagi, S.Saxena, G.J. Randhawa. (Ed.). *Plant Genetic Resources: Horticultural Crops*. Narosa Publishing House, New Delhi, pp 251-278.

Nassar, N.M.A. 1979. The need for germplasm conservation in wild cassavas. *Indian J. Genet.,* 39:465-470.

Negi, K.S. and Pant, K.C. 1992. Less known endemic wild species of *Allium* (Amaryllidaceae) *Econ. Bot.* 46:112-116.

Nigam, K.B. 1984. Aswagandha cultivation. *Indian Hort.* 28:39-41.

Ockendon, D.J. and Smith, B.M. 1993. Brussels sprouts. *Brassica oleracea* var. *gemmifera* DC. In: G Kalloo and BO Bergh (Ed.), *Genetic Improvement of Vegetable Crops*. Pergamon Press Ltd., Oxford, England, pp. 87-112.

Pandey, U.B., Ashok Kumar, Ruchira Pandey and Kamala Venkateshwaran, 2005. Bulbous crops. (*In*) Dhillon, BS, RK Tyagi, S Saxena, GJ Randhawa (Ed). Plant Genetic Resources: Horticultural Crops. Narosa Publishing House, New Delhi. Pp.108-120.

Pareek, O.P. 1988. Present status and future needs for genetic resources in arid zone fruits. In: RS Paroda, RK Arora and KPS Chandel (eds.), *Plant Genetic Resources–An Indian Perspective.* National Bureau of Plant Genetic Resources, New Delhi, India, pp.320-334.

Pareek, S.K., Veena Gupta, K.C. Bhatt, K.S. Negi and Neelam Sharma, 2005. Medicinal and Aromatic Plants. (*In*) Dhillon, BS, RK Tyagi, S Saxena, GJ Randhawa (Ed). *Plant Genetic Resources: Horticultural Crops*. Narosa Publishing House, New Delhi, pp.279-308.

Parkinson, C.E. 1923. *A Forest Flora of Andaman Islands*. Bishan Singh Mahendra Pal Singh, Dehra Dun, India.

Pickersgill, B. 1997. Genetic Resources and Breeding of *Capsicum* spp. *Euphytica* 96:129-133.

Plucknett, D.L. 1985. Taxonomy of the genus *Colocasia*. (In) J.K. Wang and S. Higam. (eds.). *Taro: A Review of Colocasia esculenta and its Potential*. Univ. Hawaii Press, Honolulu, USA, pp 14-19.

Popenoe, W. 1920. Manual of Tropical and Subtropical Fruits. Macmillan, New York, USA.

Puri, G.S., Jain, S.K., Mukherjee, S.K. Shanti Swarup and Kotwal, N.N. 1964. Flora of Rajasthan. *Rec. Bot. Survey India*, 19:1-159.

Purseglove, J.W. 1968. The origin and distribution of the coconut. Trop. Sci., 10: 191-199.

Purseglove, J.W. 1972. *Alliaceae*. (In) JW Purseglove (Ed.), Tropical Crops–Monocotyledons. Longman Group Limited, Harlow, UK, 607p.

Purseglove, J.W. 1981. *Tropical Crops: Dicotyledons*. The English Language Book Society and Longman, UK.

Puseglove, J.W. 1988. *Tropical Crops–Dicotyledons*. English Language Book Society, Longman, London, UK, pp 18-32.

Pushkarnath, 1976. *Potato in Sub-tropics*. Orient Longman, New Delhi, India.

Raju, K., Ravindran, P.N and Nair, M.K. 1983. Quality evaluation of black pepper cultivars. Indian Spices. 20: 3–5.

Rama Rao, M. and Rao, D.V.R. 1994. Genetic resources of Turmeric. (In) Chada, K.L and Rethinam, P. (Ed.) Advances in Horticulture. Vol. 9–Plantation and Spice Crops Part 1. Malhotra Publishing House, New Delhi-110 064, India. p 131–149.

Ramachandran, K., 1969. Chromosome numbers in Zingiberaceae. *Cytologia* 34:213-221.

Rana, J.C., Sharma, B.D. and Gautam, P.L. 2002. Conservation and Sustainable Utilization of Agri-Horticultural Plants Diversity in Himachal Pradesh. In: AC Kapoor (Ed.), *Sustainability of Hill Agricullture: Emerging Threats and Possible Solutions. Proc. Int. Symp. Sustainable Agriculture in Hill Areas*. Chawdhary Sarwan Kumar Himachal Pradesh Krishi Viswa Vidyalaya, Palampur, Himachal Pradesh, India, pp.59-79.

Rana, J.C., Verma, V.D., Yadav, S.K. and Pradheep, K. 2003. Genetic diversity of wild relatives and minor fruits in the Indian Himalayas. *Proc. 7th Int. Symp. Temperate Zone Fruits in the Tropics and Subtropics*. Dr. YS Parmar University of Horticulture and Forestry, Nauni, Solan, Himachal Pradesh, India, pp.62-63.

Randhawa, S.S., 1987. *Wild Germplasm of Pome and Stone Fruits*. Indian Agricultural Research Institute, Regional Station, Shimla, Himachal Pradesh, India, 51p.

Rathore, D.S., Srivastava, Umesh and Dhillon, B.S. 2005. Management of genetic resources of horticultural crops: Issues and strategies. (*In*) Dhillon, B.S., Tyagi, R.K., Saxena, S and Randhawa, G.J. (Ed.). Plant Genetic Resources: Horticultural crops. Narosa Publishing House, New Delhi, India. Pp 1-18.

Ratnambal, M.J., Ravindran, P.N and Nair, M.K. 1985. Variability in the black pepper cultivar Karimunda. J. Plant. Crops. 13:154–157.

Ravindran, P.N. and Nirmal Babu, K. 1994a. Genetic resources of black pepper. (In) KL Chadha and P Rethinam (Ed.), *Advances in Horticulture Vo. 9* (1). Plantation Crops and Spices, Malhotra Publishing House, New Delhi, India, pp. 99-120.

Ravindran, P.N., Nirmal Babu, K., Sasikumar, B. and Krishnamoorthy, K.S. 2000. Botany and crop improvement of black pepper. (In) PN Ravindran (Ed.), *Black pepper, Piper nigrum*. Harwood Academic Publishers, Amsterdam, Netherlands, pp. 23-142.

Ravindran, P.N., Nirmal Babu, K., Peter, K.V., Abraham, Z. and Tyagi, R.K. 2005. Spices. (*In*) Dhillon, B.S., R.K. Tyagi, S. Saxena, G.J. Randhawa (Ed). Plant Genetic Resources: Horticultural Crops. Narosa Publishing House, New Delhi. Pp.190-227.

Ravindran, P.N. 1999. Genetic diversity of major spices and their conservation in India. (*In*) Sasikumar, B., Krishnamoorthy, B., Rema, J., Ravindran, P.N. and Peter, K.V. (Ed.) . Biodiversity, Conservation and Utilization of Spices, Medicinal and Aromatic Plants. IISR, Calicut, Kerala, pp 219-228.

Ross, H. 1986. *Potato Breeding–Problems and Perspectives*, Springer-Verlag, Berlin, Germany, 132p.

Sabu, M. 1991. *A Taxonomic and Phylogenetic Study of South Indian Zingiberaceae*. Ph. D. Thesis, University of Calicut, Kerala, India.

Sakai, S. and Nagamasu, H. 2000. Systematic studies of Bornean Zingiberaceae: II *Elettaria* of Sarawak. *Edinb. J. Bot.* 57:227-243.

Sanjappa, M. 1989. Geographical distribution and exploration of the genus *Morus*. (In) Sengupta and S.B. Dandin (Ed.), *Genetic Resources of Mulberry and Utilization*. Central Silk Research and Training Institute Publication, Mysore, India.

Sato, D. 1960. Karyotype analysis in Zingiberales with special reference to the protokaryotype and stable karyotype. *Sci. Paper Coll. Genet. Education. Univ.Tokyo Biol.* 10:225-243.

Satyanarayana, N. and Sharma, V.S. 1993. An overview of tea plant improvement in south India. (In) M.J. Mulky and V.S. Sharma. (Ed.). *Tea culture, processing and marketing*. Oxford and IBH Publishing Company Ltd., New Delhi, India, pp 36-44.

Schweinvar, L. 1985. Cactaceae. In: J Rzedowski and G Rzedowski (eds.), *Flora Fanerogamica del valle de Mexico*. Vo. 2. Esc. Nac. Cienc. Biol. and Institute de Ecologia, Mexico DF.

Sharma, S.D., Kumar, K., Gupta, S., Rana, J.C., Sharma, B.D. and Rathore, D.S. 2005. Temperate Fruits. (*In*) Dhillon, B.S., R.K. Tyagi, S. Saxena, G.J. Randhawa (Ed). Plant Genetic Resources: Horticultural Crops. Narosa Publishing House, New Delhi. Pp.146-167.

Shekhawat, G.S., Gopal, J., Pandey, S.K., Kang, G.S. and Randhawa, G.J. 2005. Potato. (*In*) Dhillon, B.S., R.K. Tyagi, S. Saxena, G.J. Randhawa (Ed). Plant Genetic Resources: Horticultural Crops. Narosa Publishing House, New Delhi. Pp.89-107.

Shimotsuma, 1963. Cytogenetics and evolutionary studies in genus *Citrullus*, *Seiken Zin Report. Kihara Inst. Biol.* R.S.15:24-34.

Silva-Dias, J.C., Barnard, J., Phippen, W.B. and Kresovich, S. 1990. Genetic diversity of landraces of Portughese cabbages and kales, *Brassica oleracea* L. In: J.R. McFerson, S. Kresovich and S.G. Dwyer (Ed.), *Proc. 6th Crucifer Genetics Workshop*, Cornell University, New York, USA, p.35.

Simmonds, N.W. 1962. *The Evolution of the Bananas*. Longman, London, UK.

Simmonds, N.W. 1989. Rubber breeding. (In) C.C. Webster and W.J. Baulkwill. (Ed.). *Rubber*. John Wiley and Sons, New York, USA, pp 97-99.

Simmonds, N.W. 1994. Bananas. (In) J Smartt and NW Simmonds (Ed.), Evolution of Crop Plants, London, UK.

Singh, B. 1981. Establishment of first gene sanctuary in India for Citrus in Garo hills. Concept Publ. Co., New Delhi.

Singh, D., Kumar, K. and Chandel, R.S. 2000. Biodiversity of wild fruits in the Western Himalayas. *Asian Agri-History* 4: 233-240.

Singh, G. 1970. Palaeobotanical features of Thar Desert. *Ann. Arid Zone* 8:188-195.

Singh, H.P. 1998. Genetic diversity, breeding and utilization of the genepool of litchi. In: R.K. Arora and V. Ramanatha Rao (Ed.), Tropical Fruits in Asia: Conservation and Use. Proc. IPGRI-ICAR-UTFANET Regional Training Course of the Conservation and Use of Germplasm of Tropical Fruits in Asia. Indian Institute of Horticultural Research, Bangalore and International Plant Genetic Resources Institute, Rome, Italy.

Singh, H.P. and Chadha, K.L. 1993. Genetic resources of Citrus: In: KL Chadha and OP Pareek (eds,.), *Advances in Horticulture*, Vol.1. Malhotra Publishing House, New Delhi, India.

Singh, H.P., Swarup, B. and Singh, B. 1975. *Three Decades of Vegetable Research in India*. Indian Council of Agricultural Research Technical Bulletin, New Delhi, India.

Singh, R. 1969. *Fruits*. National Book Trust, New Delhi, India.

Sirohi, P.S., Munshi, A.D., Kumar, Gunjeet and Behera, T.K., 2005. Cucurbits. (*In*) Dhillon, B.S., R.K. Tyagi, S. Saxena, G.J. Randhawa (Ed). Plant Genetic Resources: Horticultural Crops. Narosa Publishing House, New Delhi. Pp.34-58.

Smartt, J., and Simmonds, N.W. 1995. Evolution of crop plants (Second Edition). Longman Scientific and Technical, England, pp 1–531.

Srivastava, A.K., Singh, H.K. and Singh, I.S. 1998. Genetic diversity in bael (*Aegle marmelos* Carrea.) *Indian Hort.* 42 (4).

Stearn, W.T. 1946. Nomenclature and Synonymy of *Allium odorum* and *A.tuberosum*. *Herbertea* 11:227-245.

Swingle, W.T. and Reece, P.C. 1967. The botany of Citrus and its wild relatives. (In) W. Reuther, L.D. Batchelor and H.J. Webber (Ed.), *The Citrus Industry*. University of California Press, Berkeley, USA, pp.190-340.

Tan, H. 1987. Strategies in rubber tree breeding. (In) A.U. Campbel, A.J. Abbot and R.K. Atkin. (Ed.). *Improving Vegetatively Propagated Crops*. Academic Press, London, UK, pp 27-62.

Tanaka, T.1977. Fundamental discussion of Citrus classification, Studia Citrologia, 14:1-146.

Telek, L., Delphin, H. and Cabanillas, E. 1977. *Solanum mammosum* as a source of solasodine in the low land tropics. *Econ. Bot*. 31:20-128.

Terry, E. 1655. *A Voyage to East India*. (In) J.Wilkie (Ed), London, Reprinted (1777), 547p.

Thompson, K.F. 1976. Cabbages, kales etc. *Brassica oleracea* (Cruciferae). (In) N.W. Simmond (Ed.), *Evolution of Crop Plants*, Longman, London, UK.

Toquica, S.P., Rodríguez, F., Martínez, E., Duque, M.C. and Tohme, J. 2003. Molecular characterization by AFLPs of *Capsicum* germplasm from the Amazon Department in Colombia, *Genetic Resources and Crop Evolution* 50: 639-647.

van Borssum-Waalkes, J. 1966. Malaysian Malvaceae Revised. *Blumea* 14:1-251.

Van Eijnatten, C.L.M. 1991. *Anacardium occidentale* L. (In) E.W.M. Verheij and R.E. Coronel. (Ed.). *Plant Resources of South-East Asia–No.2. Edible Fruits and Nuts*. PUDOC-DLO, Wageningen, Netherlands, 446 p.

Van Rheede, H. 1678. *Hortus Indicus Malabaricus*, Vol. 7. Amsterdomi, pp.23-32.

Vashishtha, B.B., Saroj, P.L. Kumar, Gunjeet and Awasthi, O.P. 2005. Arid Fruits. (*In*) Dhillon, B.S., R.K. Tyagi, S. Saxena, G.J. Randhawa (Ed). *Plant Genetic Resources: Horticultural Crops.* Narosa Publishing House, New Delhi. Pp.168-189.

Vavilov, N.J. 1926. Studies on the origin of cultivated plants. *Trudy Byuro Prikl Bot.*, 16: 139-248.

Vavilov, N.I. 1928. Geographical centres of our cultivated plants. Proc. V. Intnl. Congr. Genet. New York, pp 342–369.

Vavilov, N.I. 1936. Studies on the origin of cultivated plants. *Bull. Appl. Bot.* 26:1-248.

Vavilov, N.I. 1951. Phytogeographical basis of plant breeding. The origin, variation, immunity and breeding of cultivated plants (K. J. Choster, translator). Chronica Botanica. 13: 366.

Virmani, O.P., Srivastava, G.N. and Singh, Pratap 1978. *Catharanthus roseus,* the tropical periwinkle. *Indian Drugs* 15:231-252.

Vishnu Mittre, 1975. Problem and prospects of palaeobotanical approach towards the investigation of the history of Rajasthan Desert. (In) *Proc. Problem of Desert in India* (*Mimeo*), Geological Survey of India, Jaipur, India.

Vishnu Swarup and Pratibha Brahmi, 2005. Cole Crops. (*In*) Dhillon, B.S., Tyagi, R.K., Saxena, S., Randhawa, G.J. (Ed). Plant Genetic Resources: Horticultural Crops. Narosa Publishing House, New Delhi. Pp.75-88.

Vredebregt, J.H. 1990. Taxonomic and Ecological Observations on Species of *Abelmoschus* Medik. In: *Report of an International Workshop on Okra Genetic Resources*, NBPGR, New Delhi, India, pp. 69-76.

Watt, G. 1893. *A Dictionary of Economic Products of India* Vol. VI, Part IV. Cosmos Publication, Delhi, India, pp. 366-374.

Watt, G. 1889. *Dictionary of the Economic Plants of India*. Goot Printing, Calcutta, 2: 509-513.

Weatherstone, J. 1992. Historical introduction. (In) K.C. Willson and M.N. Clifford. (Ed.). *Tea: Cultivation to Consumption*. Chapman and Hall, London, UK, 23 p.

Whitaker, T.W. and Bemis, W.P., 1975. origin and evolution of the cultivated Cucurbita. Bulletin Torrey Botanical Club 102:362-368.

Whitmore, T.M., and Turner II, B.L. 2002. Cultivated Landscapes of Middle America on the Eve of Conquest. Oxford University Press, New York. 338 pp.

WWF and IUCN. 1997. Centres of Plant Diversity: Vol. 3, The Americas. IUCN Publications Unit, Cambridge, England, UK. 562 pp.

Yadav, I.S. and Rajan, S. 1993. Genetic resources of Mangifera. (In) K.L. Chadha and Pareek O.P. (Ed.), *Advances in Horticulture*, Vol.1. Part-1. Malhotra Publishing House, New Delhi, India, pp.77-93.

Zeven, A.C. and De Wet, J.M.J. 1982. Dictionary of cultivated plants and their regions of diversity. Centre of Agricultural Publicity and Documentation, Wageningen, The Natherlands. Wageningen. 259 p.

Zeven, A.C. and Zhukovsky, P.M. 1975. *Dictionary of Cultivated Plants and their Centres of Diversity*. Centre of Agricultural Publishing and Documentation, Wageningen, Netherlands, 219 p.

Biodiversity in Horticultural Crops Vol. 2 *Pages* **53–62**
Editor: **Professor K.V. Peter**
Published by: **DAYA PUBLISHING HOUSE**

Chapter 2

Convention on Biodiversity

Jayashree Krishnankutty and K.V. Peter
College of Horticulture, Kerala Agricultural University, P O KAU – 680 656

Biodiversity is the variety of life on Earth. This includes all of the plants and animals that live and grow on the Earth, all of the habitats that they call home, and all of the natural processes of which they are a part. The earth supports an incredible array of biodiversity with plants and animals of every kind. This fantastic variety of life is found in diverse habitats ranging through the extremes of climates. Biodiversity is essential to every aspect of the way that humans live around the world. Plants and animals provide people with food and medicine. Trees play an important role in absorbing greenhouse gases and cleaning the air we breathe, and rivers and watersheds provide the clean water that we drink. Forty per cent of all prescriptions written today for our ailments are composed from the natural compounds found in different plant species. Approximately 80,000 edible plants are found in the world, and one in every three mouthfuls of the food you swallow is prepared from plants pollinated by wild insects and animals.

Biodiversity estimates of Planet earth range from 5 to over 100 million species. Some 1.5 million were identified or studied, including plants, insects, mammals, birds and other life forms; the great bulk of these are insects. Unfortunately, the earth's biodiversity is fast disappearing. Scientists now believe that thousands of species are being eliminated each year due to human activities. Dr. E. O. Wilson of Harvard University estimates that about 17,500 species are lost per year–a rate about 1,000 to 10,000 times the rate before modern human interventions (USAID). Extinction of each additional species results in the irreversible loss of unique genetic materials, each of which has potential for development of medicines and foods and associated enterprises that create jobs. Conserving biodiversity is especially crucial in developing countries where people's livelihoods are directly dependent on natural resources such as forests, fisheries and wildlife. As natural systems breakdown, people around the world are forced to find alternative and often more costly ways to maintain adequate supplies of clean water or to deal with increasingly polluted air. The net economic benefits of biodiversity are estimated to be at least $3 trillion per year, or 11 per cent of the annual world economic output.

History of Convention

The Convention on Biological Diversity was opened for signature at the Earth Summit in Brazil on 5 June 1992 and entered into force on 29 December 1993. It contains three national level obligations: to conserve, to sustainably use and to share the benefits of biological diversity. The Convention reflects the policy and scientific recommendations of a number of groups, beginning with substantive inputs from the IUCN. Formal negotiations began in November 1988 when United Nations Environment Program (UNEP) convened a series of expert group meetings pursuant to Governing Council decisions 14/26 and 15/34 of 1987. The initial sessions were referred to as meetings of the Working Group of Experts on Biological Diversity. By the summer of 1990, a new Sub-Working Group on Biotechnology was established to prepare terms of reference on biotechnology transfer. Other aspects of biodiversity were included, such as *in situ* and *ex situ* conservation of wild and domesticated species; access to genetic resources and technology, including biotechnology; new and additional financial resources, and safety of release or experimentation on genetically-modified organisms (also known as biosafety).

In 1990, UNEP's Governing Council established a Working Group of Legal and Technical Experts to prepare a new international legal instrument for the conservation and sustainable use of biological diversity. In May 1993, UNEP's Governing Council established the Intergovernmental Committee on the Convention on Biological Diversity (ICCBD) to prepare for the first meeting of the Conference of the Parties (COP) and to ensure effective operation of the Convention upon its entry into force. The ICCBD made progress in addressing the long list of tasks mandated to it. The ICCBD established two Working Groups. Working Group I dealt with the conservation and sustainable use of biological diversity, the scientific and technical work between meetings and the issue of biosafety. Working Group II tackled issues related to the financial mechanisms, the process for estimating funding needs.

The treaty defines biodiversity as "the variability among living organisms from all sources including, interalia, terrestrial, marine and other aquatic ecosystems and the ecological complexes of which they are part; this includes diversity within species, between species and of ecosystems." Parties to the Biodiversity Treaty "affirm sovereign rights over the biological resources found within their countries, while accepting responsibility for conserving biological diversity and using biological resources in a sustainable manner," according to an International Union for the Conservation of Nature (IUCN) assessment of the treaty.

One hundred and eighty nine nations have signed the treaty since it was opened for signature at UNCED. A United Nations press release listing nations that have ratified the treaty as on December 1993 has been given in Table 2.1.

The convention has altogether 42 articles. The initial articles (1-6) clarify general conditions like the objectives, definition of terms, jurisdictional scope, cooperation, general measures for conservation and sustainable use etc. Articles 8-10 give guidelines for *in situ, ex situ* conservation and sustainable use of biodiversity components. Articles 11-17 deals with related aspects of biodiversity conservation like incentive measures, research and training, public education and awareness, impact assessment of biodiversity threatening projects, measures for minimizing impacts and measures to ascertain State's access to genetic resources and technology and its transfer, exchange of information. Articles 18- 22 give out guidelines for technical and scientific cooperation among the member nations, how they should make us of biotechnology and its benefits, how financial support for the convention should be extended, what shall be the financial mechanism of the Convention, what is its relationship with other international conventions etc. Articles 23–25 outline guidelines for conferencing of member nations, their secretariat, and the subsidiary body for scientific, technical and technological advice. Articles 26–31 narrate guidelines for reports to be submitted to the Conference of the Parties, settlement of disputes among member nations, adoption of protocols, adoption and amendment of annexes and norms for the right to vote. Articles 32- 42 explain the relationship between this convention and its Protocols, who shall have the right to be signatories for the Convention, ratification, acceptance and accession, time of entry into force for the Convention, norms for withdrawal, interim arrangements for financial matters and secretariat, who shall be the depository and what shall be the authentic texts of the convention.

Table 2.1

Countries that have Signed and Ratified COB
189–Afghanistan, Albania, Algeria, Andorra, Angola, Antigua and Barbuda, Argentina, Armenia, Australia, Austria, Azerbaijan, The Bahamas, Bahrain, Bangladesh, Barbados, Belarus, Belgium, Belize, Benin, Bhutan, Bolivia, Bosnia and Herzegovina, Botswana, Brazil, Brunei Darussalam, Bulgaria, Burkina Faso, Burma, Burundi, Cambodia, Cameroon, Canada, Cape Verde, Central African Republic, Chad, Chile, People's Republic of China, Colombia, Comoros, Democratic Republic of the Congo, Republic of the Congo, Cook Islands, Costa Rica, Côte d'Ivoire, Croatia, Cuba, Cyprus, Czech Republic, Denmark, Djibouti, Dominica, Dominican Republic, Ecuador, Egypt, El Salvador, Equatorial Guinea, Eritrea, Estonia, Ethiopia, EU, Fiji, Finland, France, Gabon, The Gambia, Georgia, Germany, Ghana, Greece, Grenada, Guatemala, Guinea, Guinea-Bissau, Guyana, Haiti, Honduras, Hungary, Iceland, India, Indonesia, Iran, Ireland, Israel, Italy, Jamaica, Japan, Jordan, Kazakhstan, Kenya, Kiribati, Kuwait, North Korea, South Korea, Kyrgyzstan, Laos, Latvia, Lebanon, Lesotho, Liberia, Libya, Liechtenstein, Lithuania, Luxembourg, The Former Yugoslav Republic of Macedonia, Madagascar, Malawi, Malaysia, Maldives, Mali, Malta, Marshall Islands, Mauritania, Mauritius, Mexico, Federated States of Micronesia, Moldova, Monaco, Mongolia, Morocco, Mozambique, Namibia, Nauru, Nepal, Netherlands, New Zealand, Nicaragua, Niger, Nigeria, Niue, Norway, Oman, Pakistan, Palau, Panama, Papua New Guinea, Paraguay, Peru, Philippines, Poland, Portugal, Qatar, Romania, Russia, Rwanda, Saint Kitts and Nevis, Saint Lucia, Saint Vincent and the Grenadines, Samoa, San Marino, São Tomé and Príncipe, Saudi Arabia, Senegal, Serbia and Montenegro, Seychelles, Sierra Leone, Singapore, Slovakia, Slovenia, Solomon Islands, South Africa, Spain, Sri Lanka, Sudan, Suriname, Swaziland, Sweden, Switzerland, Syria, Tajikistan, Tanzania, Thailand, Togo, Tonga, Trinidad and Tobago, Tunisia, Turkey, Turkmenistan, Tuvalu, Uganda, Ukraine, United Arab Emirates, United Kingdom, Uruguay, Uzbekistan, Vanuatu, Venezuela, Vietnam, Yemen, Zambia, Zimbabwe.
Countries that have signed, but not yet ratified–(1) United States

Institutional Structure

The institutional structure of the Biodiversity Convention includes a number of bodies. These include the COP, a Subsidiary Body on Scientific, Technical and Technological Advice (SBSTTA), and the Secretariat. The Conference of the Parties which brings together all member states is generally mandated with keeping the implementation of the Convention under review. More specifically, it reviews progress under the Convention, identifies new priorities to be pursued, sets work plans for members, amends the Convention, creates expert advisory bodies, reviews progress reports by member nations and collaborates with other international organizations and agreements. Periodic state reports to the Conference of the Parties constitute one of the main monitoring instruments instituted under the Convention. State parties must report on the means they have adopted to implement the objectives of the Convention and the level of success of such measures. The Conference of the Parties has launched a number of thematic programmes covering, for instance, the biodiversity of inland waters, forests, marine and coastal areas, dry lands and agricultural lands, agricultural biodiversity and cross-cutting issues like the control of alien invasive species, strengthening the capacity of member countries in taxonomy and the development of indicators of biodiversity loss.

The SBSTTA is established to provide other organs of the Convention Expert Advice. It is a multidisciplinary expert body which has the mandate to provide scientific and technical assessments of the status of biological diversity and of the effects of different measures taken in accordance with the Convention; identify innovative, efficient and state-of-the-art technologies and know-how relating to the conservation and sustainable use of biological diversity and advise on the ways and means of promoting development and/or transferring such technologies; providing advice on scientific programmes and international cooperation in research and development related to conservation and sustainable use of biological diversity. The Biodiversity Convention functions on a daily basis on the basis of a Secretariat based in Montreal which organizes meetings, drafts documents, assists member Governments in the implementation of programmes, and coordinates with other international organizations and collect and disseminate information. (Cullet, Philip 2003).

Biotechnology

One of the most controversial issues addressed by the Biodiversity Treaty is intellectual property rights related to biological and genetic resources. Groom Bridge (1992) reviews the state of relevant intellectual property law and its application to biodiversity (Groom bridge 1992) prior to the Biodiversity Treaty in an excerpt from *Global Biodiversity*. This issue is closely linked to equity issues: Although much of the Earth's biodiversity is found in developing countries, these countries have not necessarily benefited equally with developed countries from industrial, medical, agricultural and other uses of biological and genetic resources. The treaty, in its article on objectives, commits Parties to "fair and equitable sharing of the benefits arising out of the utilization of genetic resources."

The development of the genetic resources of biodiversity is known as biotechnology. Broadly defined, biotechnology includes any technique that uses living organisms or parts of organisms to make or modify products, to improve plants or animals, or to develop microorganisms for specific uses (Congress of the United States, Office of Technology Assessment, 1990). Intellectual property protection for biotechnology is currently in a state of flux. Whilst it used to be the case that living organisms were largely excluded from protection, attitudes are now changing and increasingly biotechnology is receiving some form of protection. There is at the moment no clear international consensus on how biotechnology should be treated. Although bodies like the World Intellectual Property Organization (WIPO, the United Nations permanent body primarily responsible for international cooperation in intellectual property), and the Organization for Economic Cooperation and Development (OECD) conducted separate studies and produced various reports, these have only sought to make governments more aware of the potential problems and to offer some suggested solutions.

There are currently two main systems of protection for biotechnology: rights in plant varieties and patents. Keeping biotechnology 'secret' can also be a valuable form of protection. National treatment of trade secrets is diverse, and attempts to harmonize trade secret laws are likely to fail. Most jurisdictions provide some form of protection against those who steal or use others' trade secrets unfairly. However, the problem with this form of protection is that the secret generally becomes public once the biotechnology is used commercially and thus the protection is lost. The law of copyright could offer some protection for biotechnology. Trademarks are unlikely to be of much use in protecting biotechnology, since an attempt to register the name of a plant or an animal as a trade mark is unlikely to be successful as public policy would prevent it.

CITES

CITES is the most widely accepted of international treaties on the conservation of natural resources. The number of Parties is steadily increasing from the initial signing of the convention in 1973.

(Groombridge, 1992). CITES (the Convention on International Trade in Endangered Species of Wild Fauna and Flora) is an international agreement between governments. Its aim is to ensure that international trade in specimens of wild animals and plants does not threaten their survival. Because the trade in wild animals and plants crosses borders between countries, the effort to regulate it requires international cooperation to safeguard certain species from over-exploitation. CITES was conceived in the spirit of such cooperation. Today, it accords varying degrees of protection to more than 30,000 species of animals and plants, whether they are traded as live specimens, fur coats or dried herbs.

CITES is an international agreement to which States (countries) adhere voluntarily. States that have agreed to be bound by the Convention ('joined' CITES) are known as Parties. Although CITES is legally binding on the Parties–in other words they have to implement the Convention–it does not take the place of national laws. Rather it provides a framework to be respected by each Party, which has to adopt its own domestic legislation to ensure that CITES is implemented at the national level.

For many years CITES is among the conservation agreements with the largest membership, with now 172 Parties. CITES works by subjecting international trade in specimens of selected species to certain controls. All import, export, re-export and introduction from the sea of species covered by the Convention has to be authorized through a licensing system. Each Party to the Convention must designate one or more Management Authorities in charge of administering that licensing system and one or more Scientific Authorities to advise them on the effects of trade on the status of the species.

The species covered by CITES are listed in three Appendices, according to the degree of protection they need (Table 2.2).

Table 2.2: The Approximate Numbers of Species that are Included in the CITES Appendices as of Present

	Appendix I	*Appendix II*	*Appendix III*
Mammals	228 spp. + 21 sspp. + 13 popns	369 spp. + 34 sspp. + 14 popns	57 spp. + 11 sspp.
Birds	146 spp. + 19 sspp. + 2 popns	1401 spp. + 8 sspp. + 1 popn	149 spp.
Reptiles	67 spp. + 3 sspp. + 4 popns	508 spp. + 3 sspp. + 4 popns	25 spp.
Amphibians	16 spp.	90 spp.	-
Fish	9 spp.	68 spp.	-
Invertebrates	63 spp. + 5 sspp.	2030 spp. + 1 ssp.	16 spp.
Plants	298 spp. + 4 sspp.	28074 spp. + 3 sspp. + 6 popns	45 spp. + 1 ssp. + 2 popns
Totals	827 spp. + 52 sspp. + 19 popns	32540 spp. + 49 sspp. + 25 popns	291 spp. + 12 sspp. + 2

Financial Mechanism

The Indian delegation proposed the inclusion of the following ideas in the financial resources and financing mechanism articles:

1. The additional burden on developing countries arising out of their protection of their biological diversity should be met by new and additional funding to be provided by the developed countries through a new multilateral fund to be set up under the convention;
2. Funding for developing countries should be "adequate, new and additional"; and

3. In the convention, there must be equal clarity about and emphasis on the rights and obligations of developing countries. The special situation of developing countries should be recognized in all relevant provisions of the convention to ensure a clear distinction between the obligations and responsibilities of developing countries and those of developed countries.

The Indian delegation's suggestions were incorporated in part in the final form of the funding articles. (Roberts, 1992).

The U.S. has signed the Convention on Biological Diversity but has not ratified it and at this point is not a party to the Bio-Safety Protocol

TRIPS and CBD

The extension of the application of the intellectual property rights (IPRs) system to living things– a process which has been globalised by the World Trade Organisation through the Trade-Related Aspects of Intellectual Property Rights (TRIPS) Agreement–has resulted in injustice to local and indigenous communities. By sanctioning the patenting of varieties of genetic material developed over generations by such communities and enabling large congolmorates to secure monopoly control over them, the TRIPS Agreement can undermine the concept of equitable benefit-sharing envisaged in the Convention on Biological Diversity.

The arguments for this are mainly around four premises. (Tewolde Berhan G Egziabher).

1. The royalties are insignificant compared to the expenses incurred in industrial espionage and in legal fees so that only big companies which wanted to control the market, rather than individual inventive minds, became the beneficiaries.
2. Particularly in seed production and marketing, the seed companies, dropped good seeds that were not IPR-protected and focused only on IPR-protected seeds, again to control the market. The companies promoted their own seeds, or seeds that were under royalty-sharing arrangements with IPR holders. Seed production, therefore, might move from the hands of the farmer to the corporations.
3. Breeders' Rights laws are becoming more and more restrictive.
4. Patenting of living things that has become acceptable, especially in the context of biotechnogy.

A TRIPS Council meeting was held at the WTO on 13 February, during which the issue of the TRIPS/CBD relationship and the issue of enforcement of IPR laws were the main topics.

On the TRIPS/CBD relationship, a group of developing countries and Norway reiterated their call for the TRIPS Agreement to be amended so that patent applicants are required to disclose the origin of genetic material and traditional knowledge. These countries include China, Brazil, Ecuador, China, India, Cuba, Thailand, Kenya, Pakistan, Norway, Venezuela, Colombia and Turkey. Several other countries were of the view that negotiations to amend the TRIPS Agreement were premature. The US, Korea, Canada, Australia and New Zealand were in this group. (Kanaga Raja 2007).

Biosafety Protocol

COP V saw the historical signing of the Cartagena Protocol on Biosafety which was concluded in January 2000. The Biosafety Protocol is the first internationally binding legal instrument that regulates the handling, use and transboundary movement of genetically modified organisms (GMOs) (known as living modified organisms in the Protocol). The Protocol is significant for establishing the basis of international law to regulate the trade in GMOs, which it recognises as inherently different and carrying special risks and hazards.

The Biosafety Protocol was officially opened for signature during the High-Level Segment of COP V. At the end of the special signing ceremony, a total of 68 countries signed the Protocol. Nearly two-thirds of the signatories were developing countries.

During the course of the Protocol negotiations, most of the developing countries began to negotiate as a block known as the Like-Minded Group. The LMG was resolute in its aim of achieving a strong Protocol that would include, inter alia, socio-economic considerations, and liability and redress, and include all GMOs under its scope and procedures for international trade. Foremost among their concerns were also worries that developing countries would become the dumping ground for GMOs rejected by consumers in the Developed countries and the serious threats to the their biological diversity.

However, a small group of six countries known as the Miami Group, led by the US, attempted to water down the Protocol to ensure that it did not disrupt their exports of GMOs. During the protracted negotiations, more and more scientific evidence of risks and hazards of GMOs on human health, the environment and biological diversity surfaced. Public pressure was instrumental in tipping the balance in favour of securing an international agreement that goes some way towards protecting health and the environment.

During the course of COP V, many government delegates and observers were filled with a sense of deja vu when the polarised interests of the same groupings of different countries led to disagreement over the scope of work of the Intergovernmental Committee of the Cartagena Protocol (the body that will undertake preparations for the first meeting of the Parties to the Protocol).

The countries of the Miami Group wanted a narrow workplan that would only address information-sharing and the biosafety clearing house mechanism, and capacity-building, while most countries supported the full proposed workplan, which included compliance, liability and redress and handling, transport, packaging and identification. In the end, the proposed workplan was adopted with minor changes. Also a hot issue of debate was the involvement of the private sector in capacity-building. The workplan, however, retains this element.

The Biosafety Protocol entered into force on 11 September 2003, three months after the fiftieth country, the Republic of Palau, acceded to the Protocol. As of 13 February 2004, 86 countries and the European Community have ratified the Protocol, with the latest countries to do so being: Belize, Tajikistan and Latvia.

NGOs and the CBD

The NGOs were not invited for the Convention meetings initially. But after widespread complaints internationally including a world wide signature campaign from the NGOs, the Secretariat of the Convention agreed to invite them to the Vienna meeting held during 15-19 Sep 1999.

Protection of Traditional Knowledge

The protection of the traditional knowledge, innovations and practices of indigenous people and local communities was taken a step further with the setting up of a Working Group to address the key issues in this important and controversial area. The focus was on implementation of Article 8(j) of the Convention which obligates Parties to: "respect, preserve and maintain knowledge, innovations and practices of indigenous and local communities embodying traditional lifestyles relevant for the conservation and sustainable use of biodiversity", to promote the wider application with the approval and involvement of the holders of such knowledge, innovations and practices, and to encourage the equitable sharing of the benefits arising from the use of such knowledge, innovations and practices. The working group came into action with the COP 5. (Chee Yoke Ling 1999). The functions of the working group are four fold:

1. The WG provides advice on the application and development of legal and other forms of protection for indigenous/local knowledge, innovations and practices. It is significant that legal protection is accepted for the first time at the international level since the encroachment of IPR regimes developed for industrial application of knowledge is increasing.
2. The WG advises the COP on the implementation of Article 8(j) and related provisions, in particular on a work programme at the national and international levels.
3. It identifies objectives and activities under the Convention and recommend priorities as well as identify which work should be directed to the Convention's scientific and technical advisory body and which to other international bodies and processes. This is to ensure collaboration and coordination rather than duplication of work in different flora.
4. The WG l also advices the COP on how to strengthen cooperation at the international level among indigenous peoples and local communities.

Individual Country Implementation

Several of the signatory countries have established Biodiversity Action Plans to implement the outcome of the convention. A Biodiversity Action Plan (BAP) is an internationally recognized program addressing threatened species and habitats, which is designed to protect and restore biological systems. As of 2006, 188 countries ratified the Convention on Biodiversity (CBD), but only a few of these have developed substantive BAP documents.

The principal elements of a BAP typically include: (*a*) preparing inventories of biological information for selected species or habitats; (*b*) assessing the conservation status of species within specified ecosystems; (*c*) creation of targets for conservation and restoration; and (*d*) establishing budgets, timelines and institutional partnerships for implementing the BAP. For example, the United Kingdom, New Zealand and Tanzania carried out elaborate responses to conserve individual species and specific habitats. The United States of America, a signatory who has not yet ratified the treaty, has produced one of the most thorough implementation programs through Species Recovery Programs and other mechanisms long in place in the USA for species conservation.

2010 Targets

In 2002, the Conference of the Parties of the Convention on Biological Biodiversity adopted the target to achieve a "significant reduction of the current rate of biodiversity loss at the global, regional, and national level as a contribution to poverty alleviation and to the benefit of all life on earth" by 2010. The Conference of the Parties identified seven focal areas and agreed on a limited number of trial indicators, to assess progress at the global level towards the 2010 target, and to effectively communicate trends in biodiversity related to the three objectives of the Convention.

1. Reducing the rate of loss of the components of biodiversity, including: (i) biomes, habitats and ecosystems; (ii) species and populations; and (iii) genetic diversity;
2. Promoting sustainable use of biodiversity;

3. Addressing the major threats to biodiversity, including those arising from invasive alien species, climate change, pollution, and habitat change;
4. Maintaining ecosystem integrity, and the provision of goods and services provided by biodiversity in ecosystems, in support of human well-being;
5. Protecting traditional knowledge, innovations and practices;
6. Ensuring the fair and equitable sharing of benefits arising out of the use of genetic resources; and
7. Mobilizing financial and technical resources, especially for developing countries, in particular the least developed countries and small island developing States among them, and countries with economies in transition, for implementing the Convention and the Strategic Plan.

It also established eleven more specific goals and a series of sub-targets focusing on certain aspects of biodiversity.

Meeting the targets would require unprecedented efforts. Given appropriate actions at global, regional, and especially national level, it is possible to achieve, by 2010, a reduction in the rate of biodiversity loss for certain aspects of biodiversity and in certain regions. Several sub-targets set under the Convention on Biological Biodiversity could thus be met.

The 2010 target implies that the rate of loss of biodiversity–as indicated by measures of a range of components or attributes–would need to be significantly less in 2010 than the current or recent trends. This is unlikely to be achieved globally for various reasons: current trends show a few indications of slowing the rate of loss; most of the direct drivers of biodiversity loss are projected to increase; and inertia in natural and human institutional systems implies lags of years, decades, or even centuries between actions taken and their impact on biodiversity and ecosystems (Chee Yokes Ling, 1999). With appropriate responses at global, regional, and especially national level, it is possible to achieve, by 2010, a reduction in the rate of biodiversity loss for certain components of biodiversity or for certain indicators, and in certain regions, and several of the 2010 sub-targets adopted by the CBD could be met.

Conclusion

Overall, the Biodiversity Convention is part of a web of treaties dealing with the management of species, habitats and ecosystems. The Convention provides the general framework for the conservation and use of all biological and genetic resources. It could be called as a global level novel attempt and the whole hearted response and collaboration from the many countries involved itself are indications that if humanity could do anything significant to lessen the destruction of environment, the convention would be the means for achieving that.

References

Bernstein *et al.*, 1993. A Brief History of the Convention on Biological Diversity, presented in the *Earth Negotiations Bulletin*.

Chee Yoke Ling, 1999. Delayed, but better, Biosafety Protocol TWN Third World Network Newsletter.

Cullet, Philippe C., 2003 IELRC Briefing Paper 2003-1

Groombridge, B. (Ed), 1992. Intellectual property rights for biotechnology. (In) *Global biodiversity: Status of the Earth's living resources,* 495-99. London: Chapman and Hall.

(http://www.cbd.int/convention/guide.shtml).

(http://www.usaid.gov/our_work/environment/biodiversity/)

(http://www.usaid.gov/our_work/environment/biodiversity/)

(http://www.iisd.ca/vol09/0918001e.html).

Kanaga Raja, 2007. South-North Development Monitor. TRIPS/CBD, enforcement dominate talks at TRIPS Council. 14 Feb 2007 TWN, Geneva.

Roberts, P. 1992. International funding for the conservation of biological diversity: Convention on Biological Diversity. *Boston University International Law Journal* 10 (2): 303-49.

Tewolde, Berhan G., Egziabher; The TRIPS Agreement of the WTO and the Convention on Biological Diversity: The need for coordinated action by the South. In third world net work (http://www.twnside.org.sg/title/berhan-cn.htm)

Biodiversity in Horticultural Crops Vol. 2 *Pages* **63–78**
Editor: **Professor K.V. Peter**
Published by: **DAYA PUBLISHING HOUSE**

Chapter 3

Sacred Groves for Biodiversity Conservation

Savita Bisht
Department of Botany, Government P.G. College, HNB Garhwal University, Kotdwara, Garhwal, Uttarakhand

There are many communities in the world who have shown close interaction with nature and nature conservation and there are many instances of sacred mountains, lakes, forests, and in fact of sacred landscapes. The communities associate these landscapes with their cultural values and do not harm them but contribute towards their conservation. One very old form of sacred landscape is sacred groves where a patch of forest often with water bodies is dedicated to local deities and none is allowed to cut plants or kill animals or any form of life in the grove. The harm to any living forms is forbidden because of cultural beliefs and customs etc., which together with sacred status given to many natural sites help greatly towards biodiversity conservation. These exist in many parts of the world and almost all parts of India unique examples of traditional ecological understanding and management.

Indigenous cultural and ritual practices like the sacred groves serve as tool for conserving biodiversity. These practices have made it possible for man today to co-exist with all the living forms. The understanding of nature and man's association with it were very prudent in the ancient time. This resulted in a countryside dotted with innumerable small and medium sacred groves. The sacred groves, one of major forms of informal protected areas, exhibit rich floral and faunal diversity with some rare and threatened species. The modern system of protected areas has something from the concept of sacred landscapes with legal framework replacing for the cultural and social beliefs. Sometimes, it is suggested that the existence of large tracts of protected areas could serve as indicators for biodiversity conservation status and survival of biodiversity in forests and agricultural land outside protected area network is relatively unimportant. But the fact remains that most of the world's biodiversity continues to exist outside protected areas. The biological resources outside the protected area network shall remain very important. These informal protected areas are equally important from the conservation point of view.

In 1992, the World Heritage Committee adopted cultural landscape categories, which paved the way for international recognition of sacred natural sites. The Operational Guidelines of the convention (an explanatory text to the convention providing detailed information on the specific criteria for selection of world heritage sites) stipulates in paragraph 39 (iii): "The final category is the associative cultural landscape. The inclusion of such landscapes on the World Heritage List is justifiable by virtue of the powerful religious, artistic or cultural associations of the natural element rather than material cultural evidence, which may be insignificant or even absent." (Schaaf, 2003)

Globally, there is increased focus on traditional natural resource management and conservation systems in recent times. There is a wide recognition that ecological prudence exhibits symbiotic relationship between ecosystems and social systems. Convention on Biological Diversity (CBD), signed at the United Nations Conference on Environment and Development (UNCED) in 1992, was the first international environmental convention to develop measures for use and protection of traditional knowledge, related to conservation and sustainable use of biodiversity. The International legal strategies and instruments have increasingly used terms like "traditional knowledge, innovations and practices" (Convention on Biological Diversity, Preamble, Article 8j, 10c and United Nation Convention to Combat Desertification, Articles 16g, 17c). Over 190 Nations at the meeting of the Convention on Biological Diversity in 2001 identified and adopted the indicator 'status of indigenous and traditional knowledge' under the focal area 'status of traditional knowledge, innovations and practices' under the Target 2010 which is a commitment to achieve by 2010, a significant reduction of the current rate of biodiversity loss at the global, regional and national level as a contribution to poverty alleviation and to the benefit of all life on earth, which was also endorsed by the World Summit on Sustainable Development in 2002 at Johannesburg.

The people initiated systems for management of natural resources like cultural protected sites and sacred groves are widely acknowledged as an indicator of their sustainable management. Existence and status of sacred groves are indicators of sustainability recognised by various international initiatives as an indicator of sustainable forest management (FAO, 1997, 1998, 2000 and ITTO, 1998). It is also adopted as an indicator for sustainable forest management in the Indian initiative of criteria and indicator approach assessment of sustainable forest management (Anon., 2001 and Anon., 2005).

India as a classical example showed significant wildlife population in spite of dense human populations. This perhaps can be equally attributed to widespread traditions of protection and conservation. The extent of beliefs, myths and cultural practices for relationship of life forms with society and individual well-being are perhaps the most developed in Indian culture. The cultural traditions in India are full of such instances, traditions and cultural practices, which help to maintain rich biodiversity that nature has bestowed us over thousands of years. One such tradition of sacred groves in India is one of the best traditional conservation practices. The sacred conservation practices followed by local people are in focus of late due to their importance for protecting several delicate ecosystems and threatened species. With the rising man-animal conflicts; present situation of scarce resources and modern wildlife management regimes, they present a unique opportunity of people oriented conservation. It is discussed here, the importance of sacred groves as people's contribution in maintaining these ecosystems and biodiversity conservation.

What are Sacred Groves?

The sacred groves are segments of landscape, containing vegetation and other forms of life and geographical features that are delimited and protected by human societies under the belief to keep them in a relatively undisturbed state as an expression of an important relationship of humans with the divine or with nature (Hughes and Chandran, 1998). The sacred groves are pockets of more or less climax vegetation preserved for religious grounds often found in remote areas. These sacred groves, dedicated to local deities or ancestral spirits, are protected by local communities through social traditions and taboos that incorporate spiritual and ecological values. In India these sacred patches of vegetation are scattered throughout, and are dedicated to the forest gods or other deities, revered by local inhabitants as the deity's sacred territory (Vartak and Suryanarayana, 1995). Sacred groves are forest patches conserved by the local people due to their socio-cultural and religious practices. In India, most of the sacred groves are associated with temples as existence of Hindu Gods and deities, though they are found in all religions including Islam. The cultural and social component of the sustainable use of natural resources has been most neglected though they are equally important (Martin, 1994). Figure 3.1 shows the three aspects associated with sacred groves leading to sustainable resource use.

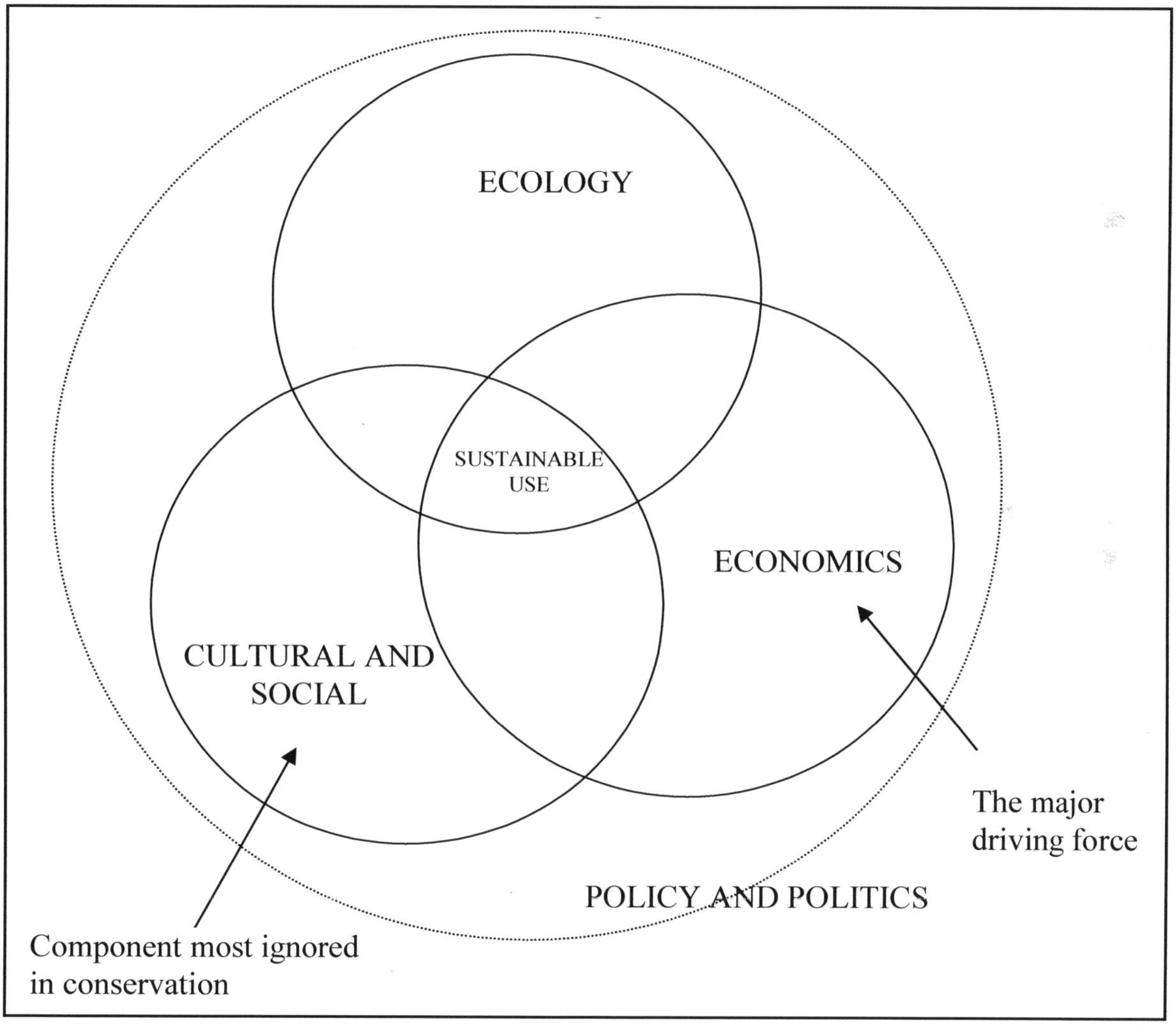

Figure 3.1: Sustainable Use of Resources and its Relationship with Social Sciences, Economics and Ecological Studies within the Policy and Political Framework (Martin, 1994)

Over a hundred of years ago, D. Brandis, the Inspector General of Forests in India who reported sacred groves for the first time in India, remarked that these sacred groves were never touched by the axe and its wood was used only for religious buildings.The sacred groves, however, are supposed to be pre-Vedic in origin (Dwivedi, 2000); to about 3000 B.C.E. (Apffel-Marglin and Parajuli, 2000). Sacred groves in the hills of Garhwal and Kumaon are mentioned in old Hindu scriptures like the Puranas. We find many references to sacred groves in the Vedic literature also.

The origin of sacred groves is generally based on traditional religious and cultural beliefs. The beliefs of the communities together with other dynamic forces give rise to unique myths and rituals associated with sacred groves. In many sacred groves, it is common belief that the deity and the spirits of their ancestors live in these groves from time immemorial and conservation of such landscapes bring them blessings of the deity.

The sacred groves find differences in distribution in spatio-temporal dimensions. It is estimated that about 15,000 sacred groves existed prior to 1800 AD in Kerala (Veluppillai, 1940) and they were innumerable in Malabar also (Logan, 1887). We find many references of sacred groves in written history. The Indian communities interacted with many other communities since prehistoric times and it led to dynamic interactions in culture, which might also have affected conservation practices and beliefs. By and large, the Indian traditions for conservation were much intact up to late Mughal period and imperialist British rule. After independence, there are other forces in this dynamic relationship as these are ever evolving. The monetary driven economy, its move from primary to secondary and to tertiary sector and globalisation may have lead, to a certain degree, loss of values towards these conservations traditions in recent years (Bisht *et al.*, 2007).

Sacred Groves and Their Role in Biodiversity Conservation

India is a vast country with rich biodiversity due to varied physiographic and climatic regimes. It is one of 12 mega-diversity countries in the world (Myers *et al.*, 2000) and encompasses a wide spectrum of habitats, *viz.* tropical rainforests to alpine vegetation and coastal wetlands to desert scrubs. It occupies 2.5 per cent of the world's land area and 1.8 per cent of the global forest area (Veach *et al.*, 2003). In India, a total of 572 protected areas, including national parks and wildlife sanctuaries (150,809.6 km square) cover 4.8 per cent of the total geographical area of the country (Ramanujam *et al.*, 2002). The limited resources and enormous demand for them in the recent past are the major forces threatening our biodiversity resources. There is little doubt that our cultural and social setup contributed significantly to help biodiversity conservation in these adverse conditions. There are many areas like the community forests, reserve forests, people recognised conservation forests like sacred groves, etc. which are also rich in biodiversity and wildlife. Studies on sacred groves show that these relatively small forest patches harbour rich biodiversity and play a significant role in the conservation of biodiversity.

The sacred groves also bring us many other benefits. The ameliorating effects of sacred groves to general aesthetics and ecosystem maintenance are well known. The rich biodiversity supported by sacred groves is helpful to us in many ways. The groves support many species of insects, helpful in acting as vector for pollen and causing fertilization of plants, including farm and horticultural crops. The groves also help lots of birds, which in turn feed on many pests, insects on the farmlands. Just like an ecotone supports lots of biodiversity, the same is true for these groves to a larger extent. There is movement towards less chemical intensive, organic production in agriculture and horticulture. Pushpagandan *et al.* (1998) showed that the floral biodiversity in small sacred groves represent the normal range of floral species richness of tropical rainforests. For example, 960 angiosperm species found in the 90 km^2 area of Silent Valley, of which 722 species were recorded from only 1.4 km^2 of sacred groves in Kerala.

As an ecosystem the sacred groves play an important role in soil and water conservation. This ecosystem can be compared to sponge soaking water during wet period and releasing it slowly during drought. Many sacred groves contain water resources like ponds and streams, and the vegetative mass that covers floor of a grove can absorb water during rainy seasons and release it during times of drought hence helping in its conservation. Trees also improve soil stability, prevent topsoil erosion and provide irrigation for agriculture in drier climates. Sacred groves also provide for water needs of nearby communities and for maintenance of ecosystem functions hence helping in biodiversity conservation. In Rajasthan, where water is scarce for drinking, animal husbandry and farming, the importance of some forest patches and very close association of people with these resources is understandable. Forests in hills reduce the runoff and help in ground water recharge, which thus become available in the wells during dry months.

The Importance of Sacred Groves

Man always learned to have these groves along with his farms for various benefits. The sacred groves are also storehouses of medicinal plants valuable to village communities as well as modern pharmacopoeia, and they contain wild relatives of crop species that can help to improve cultivated varieties. The sacred groves can also be helpful in providing and strengthening employment opportunities to the rural folk and conserving biodiversity at the same time. Collection of honey from honeybees nest in large trees in sacred groves is one such example. Present research on shelterbelts and windbreaks established their utility for protection of our farms from adverse effects. The sacred groves provide for this function to our farms. In this context, it is said that the sacred groves in our societal setup are beneficial to us in many ways. They improve environment, help to conserve

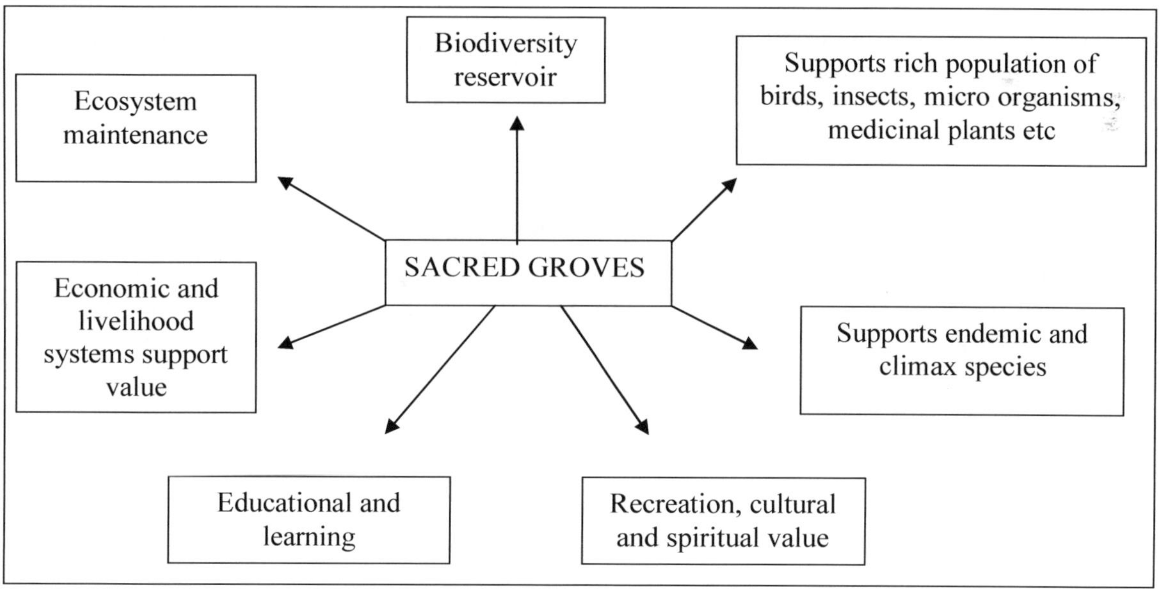

Figure 3.2: The Importance of Sacred Groves

biodiversity, maintain ecosystem and improve quality of life in urban as well as rural setup. In some cases, traditional respect for environment and access restrictions to sacred sites often led to well-conserved areas with high biological diversity within otherwise degraded environments. A protected forest in a plantation patch is a vital component of such a sustainable model. Where else other than sacred groves can we look for a widely accepted and successful conservation model. The maintenance of biodiversity through sacred groves can be of great help in this approach.

The conservation of biological diversity is not only important for purely ecological concerns but also because it sustains livelihoods of million of people all over the world. The biodiversity is important not only in protected areas but in every sphere of life like agriculture which helps millions of people to make a living and sustain all the human society on the earth. The ecosystem it supports ultimately helps stabilize climate, improve rainfall and enrich soil and water table.

Many workers enumerated the biodiversity value of sacred groves and analysed their conservation importance. Vartak (1983) enumerates 38 species of endemic and endangered plant species from 12 selected sacred groves along the Western Ghats. It is concluded that the climax type of vegetation in the sacred groves exhibits diversity in species of trees and other various life forms dependent for their very existence on trees, huge climbers, epiphytes and other shade loving plants. Studies showed that these sacred groves show optimum growth of vegetation relevant to local terrain and climate. Some workers also showed that the sacred groves show that these patches of culturally protected ecosystem show rich varied growth of plants of great dimensions. The sacred groves are home for the local flora and fauna and represent a mini biosphere reserve, making them an essential part of the conservation process. In the biodiversity hotspots like the Western Ghats, the sacred groves may be smaller in size but are ecologically and genetically very important to serve the vital function of preservation of plant species that are rare, endemic and endangered.

The present extent of forest cover and biodiversity are attributed to sacred grove status of the site, which deserves protection. Some work suggested similarity between the Raunkiaer's normal spectrum and sacred grove's spectrum which indicates sacred grove vegetation stands as a relic of the evergreen tropical rainforest. The biological spectrum of the sacred grove of different bio-climatically divergent regions indicates that this is a potent tool for understanding the general structure and functioning of these ecosystems, especially to analyse the human impact. It can be concluded that these sacred groves remain as treasure houses of a large number of endemic flora and fauna and the maintenance of the functional dynamics of these sacred groves is always related to a balance in the litter production and decomposition process of its floristic components.

The sacred groves represent a variety of formations ranging from typically evergreen to dry-deciduous forest types of Himalayan ranges and Western Ghats, swaps of coastal plains along the Western coast, oases of thorny scrubs in Aravalli ranges and scrub woodlands of Coromandel coastal belt, corresponding to the different climatic zones (Ramanujam *et al.*, 2002).

The Botanical Survey of India rediscovered some of rare specimens of orchids in the sacred groves of Khasi hills in Assam (Gadgil and Vartak, 1976). Chandrasekara and Sankar (1998) enumerated about 800 flowering species in *Kavus* of Karnataka, 150 of them endangered and 40 per cent having medicinal value. Chandran and Gadgil (1993) listed 134 rare and endemic taxa in Andhra Pradesh sacred groves. Induchoodan (1996) enumerated 722 species in 474 genera belonging to 129 families with 154 endemic species in 361 sacred groves in Kerala. Mohanan and Nair (1981) discovered a new taxon *Kunsteleria keralensis* in a sacred grove in Kerala. Nair (1985) discovered *Dalbergia benthami* in Kerala from a sacred grove. A 1.4 km^2 *Kavu* contained 722 species as against 960 in the 90-km^2 Silent Valley. They also harbour many wild relatives of cultivated species (Induchoodan, 1996).

The groves are also equally rich in fauna. There are some groves where a particular animal species may be treated as sacred and get absolute protection like *Nagadevaru* in Uttara Karnataka. People give protection to all the species in the sacred grove. The population of large species may not be significant in the small sized sacred groves as these groves may not support them. But the groves still support significant faunal biodiversity like birds, reptiles, amphibians, arboreal mammals etc. Nearly half of the bird species enumerated in Siddapur district of Karnataka are found in groves. The lion-tailed macaque, which is an endangered species, thrives in sizeable population in Katlekan sacred grove in Karnataka (Gadgil and Chandran, 1992).

In Yalong and Raliang sacred groves in Jainitia Hills, Meghalaya mere one-hectare plots contain 123 woody species, which act as a storehouse of tribal and folk medicines as well (Bhakat and Pandit, 2003). Sacred groves not only conserve useful species, but people tended to discover medicinal values more often among plants unique to sacred groves, than those found in other landscapes. Sacred groves had higher densities of the threatened flora than the state managed reserve forest and have more biodiversity (Boraiah *et al.,* 2003). Sacred groves are one of the finest instances of traditional conservation practices and it is very important to protect these sacred groves to maintain the ecological balance. These sacred groves are living museums of plants and other natural resources and supply fresh medicinal plants to villagers.

Social Dimensions and its Role in Conservation of Sacred Groves

Orans are islands of vegetation set aside by the Bishnois of Rajasthan for worship under community protection and management near each of their villages. Nobody can axe a tree or kill an animal here. If Orans still survive in the Thar Desert, it is because of the traditional commitment of the Bishnois, 363 of whom died in 1730 AD to protect Oran trees in Khejrali village near Jodhpur.

In many parts of Kerala, these sacred groves are found in the backyards of numerous traditional houses where a large part of the land is used for growing trees, vegetables and fruits. Traditional folklore is full of such references and these groves evolved as a part of daily life.

Villagers are using sacred groves for their meetings, community gatherings and fairs of the deity and for taking important decisions like time of sowing and planting rice in the monsoon. The sacred groves play important role in balancing the ecosystem including agro-ecosystems of the region. There are many studies on sacred groves on last a few decades. The sacred groves are reported from all parts of India. They are known by different names in different parts of the country. The names of sacred groves in different parts of the country are given below.

Origin of sacred groves is generally based on traditional religious and cultural beliefs. The beliefs of the communities together with other dynamic forces give rise to unique myths and rituals associated with sacred groves. In a number of indigenous societies,sacred groves are regarded as abode of deities that could assure welfare of the community and good harvest of crops but any harm to these groves will invite the wrath of concerning deity which could bring epidemic diseases, famine, natural calamities and invasion by enemies.

Depending upon the socio-cultural set up the rituals of the sheltered deity vary widely from sacred grove of one region to another. For example, in some areas of North-East India, the tradition of establishing small grove graveyard is still in trend and people believe that the souls of their departed relatives would rest in peace in these groves. The concept of sacred grove is even broader than this perception as the human societies recognised their ecosystems and realised that they cannot live in disjunction from plants and their natural environment. It is true that such beliefs, myths and rituals are acting as prime motivating factors for conservation and have kept sacred groves alive all over the world.

Table 3.1: Name of Sacred Groves in Different Parts of the Country

Sl.No.	*States/Areas*	*Name of Sacred Groves*
1.	Assam	*Madaico*
2.	Manipur	*Gamkhap, Mauhak* (bamboo grove)
3.	Meghalaya	*Lakyntang* (in Khasi Hills)
4.	Mizoram	*Mawmund* (bamboo reserves)
5.	Sikkim, Arunanchal	*Gumpas* (Buddhist)
6.	Himachal, Uttaranchal	*Dev Van*
7.	Jharkhand, Central India including Madhya Pradesh and Chhattisgarh	*Sarana* or *Jaherthan/Jeher/Jankor*
8.	Orissa	*Jahera, Thakuramma*
9.	Rajasthan	*Orans, Vanis* (Mewar), *Kenkris* (Ajmer), *Shamlat Dehs/Devbanis* (Alwar)
10.	West Bengal	*Garamthan, Shitalathan, Harithan, Sabitrithan, Santalburithan Deorali* (Darjeeling) and *Jahera*
11.	Karnataka	*Devarabana, Devarakadu, Hulidevarakadu, Nagabana, Bhutappanbana, Jataka-Ppanbana, Chowdibana*
12.	Kerala	*Kavu*
13.	Maharashtra	*Devrai* or *Devrahat, Devgudi*
14.	Tamil Nadu	*Kovilkadu, Swami Shola*

There is evidence of existence of sacred groves in many parts of the world among people with different religions and forms of social and economic organisation (Hughes and Chandran, 1998). The sacred groves are reported significantly from the regions wherever indigenous societies existed. Traditionally, nature worship is predominant in many societies of the world. In Africa, sacred groves exist widely in almost every sub-Saharan section of the continent. There people may not axe any tree, break no branch, gather no firewood, burn no grass or harm wild animals. In West Africa as well, the area of holiness is instrumental in conservation of these botanically rich groves from degradation (Lebbie and Guries, 1995). In some cases, like Ghana, there are efforts at the policy level towards recognition of the socio-cultural and religious importance of sacred groves (Spindel, 1989). There are many sacred groves associated with villages in the nations of Africa, such as Zimbabwe, Egypt and South Africa.

In Asia also, there are a few convincing evidences of wide spread of sacred groves. In Japan, Shinto shrines are as a rule surrounded by trees and often by extensive thick forests including sacred trees (Ono, 1962). Many Buddhist temples and monasteries in Japan, China, Korea and Southeast Asia have carefully tended gardens, including trees that have the aspect of sacred groves (Frazer, 1935;Yen,1994).

Europe had thousands of groves in ancient times, although most of them disappeared. There, as elsewhere, groves were originally associated with nature worship (Frazer, 1935; Vest, 1985). The sacred groves are also reported from Central Australia, where the Dieri tribe regard certain trees as sacred and they take care to see they are not harmed (Roy Burman, 1992).

The Extent of Sacred Groves in India

There are many attempts towards study of spatial distribution and number of sacred groves in India. Rao (1996) puts the number of sacred groves at 13,270 in India, of these 79 ranged from 0.01 to 900 ha covering a total area of 10,511 ha. The number of sacred groves reported in various regions of the country total at least 13,720 (Malhotra *et al.*, 2001), but experts estimate that the actual number is likely 100,000 to 150,000 (Vajpeyi, 2000). Gaikwad *et al.* (2004) report the development of sacred grove information system and a database for cursory information for about 3000 sacred groves in India with proposed collation along with cross-linking of select information parameters with other databases like regional/global species databases, genomic, molecular, biogeography, climate, meteorological, community, and agricultural or forestry databases, and information on its various dimensions like diversity. The number of recorded sacred groves in India is summarised below. There is urgent need to strengthen the documentation of sacred groves because not only less information is available but it is widely scattered (Ramakrishnan *et al.*, 1998).

Table 3.2: Recorded Number of Sacred Groves in India (adapted from Malhotra *et al.*, 2001)

States	*Reported no. of Sacred Groves*	*Source*
Andhra Pradesh	750	Anon., 1996
Arunachal Pradesh	65	–do–
Assam	40	Malhotra *et al.*, 2001
Chhattisgarh	600	–do–
Gujarat	29	Gupta *et al.*, 2000
Haryana	248	NAEB, 1995
Himachal Pradesh	5000	Sharma, 2000; Malhotra *et al.*, 2001
Jharkhand	21	–do–
Karnataka	1424	Chandran and Gadgil, 1993; Kalam, 1996; Chandran *et al.*, 1998; Gokhale, 2000
Kerala	2000	Rajendraprasad, 1995; Pushpangadan *et al.*, 1998
Maharashtra	1600	Gadgil and Vartak, 1981; Deshmukh *et al.*, 1998
Manipur	365	Devi, 2000; Khumbongmayum *et al.*, 2004;
Meghalaya	100	Tiwari *et al.*, 1999; Anon., 1999
Madhya Pradesh	275	Patnaik and Pandey, 1998
Orissa	322	Kulkarni *et al.*, 1993; Malhotra *et al.*, 1997
Pondicherry	149	Kadamban, 1998; Ramanujam *et al.*, 2002.
Rajasthan	1036	Pandey, 2007
Sikkim	56	Chhatterjee *et al.*, 2000, Dash, 2005.
Tamilnadu	448	Swamy *et al.*, 2003; Amrithalingam, 1998
Uttarakhand	32	Sinha and Maikhuri, 1998
West Bengal	670	Deb *et al.*, 1997; Bhakat and Pandit, 2003
Total	18907	

Like the Indian population itself, these forests and cultural aspects associated with them are tremendously diverse, but they share an important commonality: they are all held sacred. Like many countries throughout the world, India has a long history of nature worship, and that practice continues today, especially through the veneration of forest groves.

In many parts of India, the landscape is dotted by innumerable sacred groves of all sizes–big and small. Majority of the sacred groves are under the control of rural communities in India and they present an ideal case for people's participation in the conservation of natural resources. Involving people in actively participating in economic benefit sharing and ecological restoration shall be a good strategy in this context.

The Present Threats

A few authors reported increasing anthropogenic pressure on these storehouses of biodiversity, not particularly because of resource depletion by use *per se* but also by increasing urbanisation and subjugation of priority vis-à-vis economic gains. We may feel the immediate need to conserve these storehouses of biodiversity. The conservation of sacred groves is important not only for the biodiversity values and services they represent but also for the simple reason that they represent those important beliefs between man, nature and conservation on which our survival depends. It reaffirms our relationship with nature. The sacred groves in India exist with varying degree of people's participation, ownership patterns, and management. Cultural causes alone cannot rejuvenate conservation ethos in the present situation unless conservation is promoted as a means of sustainable economic development.

In recent times, some authors reported the depleting values in the countryside, which is causing many sacred groves to perish. The scenario has changed as modern era introduced a rapid decline in the traditional value system (Chandran and Gadgil, 1993; Chandran, 1997; Singh *et al.*, 1998). With improved accessibility and urbanization, sacred areas turned into tourist places to serve economic interest (Saxena *et al.*, 1998). The sacred groves survival is threatened by a gradual change in belief systems by forces of modernisation, deforestation, encroachments and political interference. Similarly economic development, population pressure and changing value systems and beliefs have undermined the status of sacred groves across India from Kerala (Chandrashekara and Sankar, 1998) to the Garhwal Himalayas (Sinha and Maikuri, 1998).

The number of groves is declining as social values and religious beliefs are changing due to urbanization and expansion of market economy. Habitat destruction is on rise in these once socially guarded ecosystems. No concrete steps are being initiated to protect and conserve these groves, mainly due to lack of accurate and adequate information. Therefore, in-depth assessment of these natural resource pockets can be an excellent tool for any regional biodiversity planning and it shall go a long way if we build a comprehensive database documenting various aspects and status of sacred groves.

Involving people in actively participating in economic benefit sharing and ecological restoration shall be a good strategy in this context. The local concerning communities always give a faithful protection to these sacred groves and never cause destruction. Their self-imposed rules are not less significant than any government policies regarding conservation of forest.

The Current Challenges and Future Conservation Strategies

A sacred natural site can constitute an effective means for environmental conservation as it is deeply embedded in local and traditional belief systems. These sites have great value for conservation ecology, as sanctuaries for rare or threatened species; as sites that protect freshwater sources; as indicator sites showing potential natural vegetation in areas subject to environmental degradation, which is important for restoration and rehabilitation of degraded ecosystems.

The two predominant land uses in many inhabited areas are forestry and agriculture. The agriculture landscapes are man-made and are often poor in biodiversity compared to natural systems like forests. These two landscapes have a close relationship since the evolution of agriculture based civilizations. The practice of setting aside patches of forests as sacred groves would have strengthened with the spread of agriculture, when slashing and burning of forests began on a massive scale. The reasons for this practice could be religious and cultural compulsions as well as subsistence and ecological needs. The sacred groves existed as a refuge for biodiversity that helped maintenance of agricultural systems through ages. With the passage of time, people might have been encouraged to increasingly associate with these groves socially and culturally and considered them as sacred. In the Western Ghats for example, sacred groves are very characteristic of the agricultural landscape. (Chandran *et al.*, 1998).

The sacred groves are helpful in achieving some degree of harmonisation in agriculture and forestry. With extensive monoculture these groves help in maintenance of ecosystem functioning. This function may also be achieved to some extent with the modern concepts like agro-forestry but with the difference that the groves mainly support natural vegetation though we may still find some planted groves. The sacred groves have cultural aspects associated with them by local population and hence have extensive acceptance. Although an estimated 7000 plant species are collected and cultivated for food, only 30 contribute over 90 per cent of the entire global population energy needs. The case of the banana is illustrative. Bananas are the fourth most important food source in the tropics after rice,wheat and corn. They are cultivated in nearly 120 countries and farmers use only about 25 edible sterile banana varieties. The number of varieties is diminishing due to the spread of pests and diseases and the deterioration of the resource.

Believing trees to be the abode of gods and ancestral spirits, many communities set aside sanctified areas of forest and established rules and customs to ensure their protection. We still have to answer that, whether environmental conservation can be based effectively on cultural values and traditional belief systems. We would have to look into the question of whether such an approach is sustainable in today's world, especially in the light of changing societies and cultures, and globalization processes. Can cultural values and traditional belief systems, which respect the environment, be a more powerful or at least an equally powerful means to conserve nature than protected areas with legal framework? Such an approach appears promising in many traditional societies of the world where the concept of a "sacred nature" is well embedded in societal norms (Schaaf, 2003).

The Way Ahead

The sacred groves are important in conservation of all forms of life outside protected area network. The Indian culture is particularly rich in this age-old tradition of protection of forest patch on cultural grounds. Present estimates put the number of sacred groves in India at more than 18,900 but their number may be much more as many sacred groves are not documented and reflects further need of their complete enumeration. We need to study the status of sacred groves in India including dynamic relationship of people oriented conservation traditions, cultures and practice of sacred groves in various Indian communities. It becomes important to have a comprehensive documentation about the status of sacred groves in India. In this context, traditional values which help in conservation should be properly recognised and steps taken for its protection. There have been some efforts for development of a factual database to support the development of strategies for conservation and protection of these unique heritage ecosystems (Gaikwad, 2004). There is an urgent need to strengthen these efforts.

Some authors discussed about importance of natural habitats including sacred groves for sustainable management of agriculture habitats. Dudley *et al.* (2005) discussed about indicators for sustainable management of agricultural habitats, which include Indicator 3: Area of semi-natural habitats within farmland. The sacred groves are also known to enhance the suitability of habitats for biodiversity conservation. They are directly beneficial by helping our crops and providing economic benefits but also more importantly, indirectly as for biodiversity conservation, ecosystem functioning and maintenance. There is an urgent need for recognising these traditionally valued natural systems at various levels and planning for their better management ultimately aiming for conservation of biodiversity outside protected areas.

Regarding importance of sacred groves in biodiversity conservation and community development, M.S. Swaminathan of M. S. Swaminathan Research Foundation,Chennai observes "the sacred groves are a biological heritage and a system that has helped to preserve the representative genetic resources existing in the surrounding regions for generations. By conserving the flora and fauna these communities have conserved valuable genetic resources, which can be used in further afforestation, programmes and also certain endemic and rare plants are being conserved in the sacred groves. A study to identify such existing groves and the threats faced by these groves has to be undertaken on a priority basis".

The very fact that the sacred groves existed in almost all the parts of India today despite diverse socio-cultural and ecological conditions only gives some indications of their potential to evolve them as ideal sustainable ecosystem management model. Since the existence of sacred groves, majority of them are under control of the rural communities in India and they present an ideal case for the people's participation in the conservation of natural resources. The only way to preserve biodiversity is to give communities the right to look after it. They need to be made stakeholders where they stand to benefit in terms of livelihood and recognition. Their knowledge needs to be updated and they should be sensitised into protecting it from being stolen by outsiders for commercial purposes.

The concept of sacred groves has not realised its true potential as a tool for biodiversity conservation and ecosystem management. There may be good focus on protected area network presently, but the concept of sacred gorves is more close to the people with very deep association unlike the protected areas. This may help in strengthening the whole conservation movement for the years to come.

References

Amrithalingam, M., 1998. Sthala Vrikshas of Tamil Nadu. (In) Krishna, N. and J. Prabhakaran (Eds.), *The Ecological Traditions of Tamil Nadu*. CPR Environment Education Centre, Chennai.

Anonymous, 1996. Sacred and Protected Groves of Andhra Pradesh, World Wide Fund for Nature–AP State office.

Anonymous, 1999. The Sacred Groves. *Down To Earth,* 8 (3), June 30.

Anonymous, 2001. Bhopal–India process: sustainable forest management in India, Indian Institute of Forest Management, Bhopal, August.

Anonymous, 2005. Proceedings–National workshop on refining indicators of Bhopal–India process and implementation strategy of criteria and indicators for sustainable forest management in India, Indian Institute of Forest Management, Bhopal.

Apffel-Marglin, F. and Parajuli, P. 2000. "Sacred Grove" and ecology: ritual and science. In: Chapple, C. K. and Tucher, M. E. (eds.), *Hinduism and Ecology–The Intersection of Earth, Sky and Water*, Oxford University Press, New Delhi.

Bisht, S., Ghildiyal, J. C. and Rawat, T. S. 2007. The sacred traditions of conservation: Status of sacred groves in India. Abstract. International seminar on forest products and services, HNB Garhwal University, Srinagar, Garhwal.

Bhakat, R. K. and Pandit, P. K. 2003. Role of sacred grove in conservation of medicinal plants. *Indian Forester*, 129(2): 224-232

Boraiah, K. T., Vasudeva, R., Bhagwat, S. A. and Kushalappa, C. G. 2003. Do informally managed sacred groves have higher richness and regeneration of medicinal plants than state-managed reserve forests? *Current Science*, 84 (6).

Chandran, M. D. S. 1997. On the ecological history of the Western Ghats. *Current Science*, 73: 146-155

Chandran, M. D. S. and Gadgil, M. 1993. Kans–safety forests of Uttara Kanada. In: Brandil, M. (ed.), Proc. IUFRO Forest History Group Meeting on Peasant Forestry, Freiburg, September 2-5, 1991. pp 49-57.

Chandran, M. D. S. and Gadgil, M. 1993. *Sacred groves and sacred trees of Uttara Kannada (a pilot study)*. Mimeograph. Centre for Ecological Sciences, IISc., Bangalore.

Chandran, M.D.S., Gadgil, M. and Hughes, J.D. 1998. Sacred groves of the Western Ghats of India. (In) P.S. Ramakrishnan, K.G. Saxena and U.M. Chandrashekhara (Ed.), *Conserving the sacred for Biodiversity Management*, Oxford and IBH, New Delhi, pp. 211-231.

Chandrashekara U. M. and Sankar S. 1998. Ecology and management of sacred groves in Kerala, India. *Forest Ecology and Management*, Volume 112, Issues 1-2, Pages 165-177.

Chhatterjee, S., Sastry, A. R. K., Roy, B. N. and Lohan, R. 2000. Sacred groves of Sikkim and Arunachal Pradesh. Abstract. National workshop on community strategies and management of natural resources. Bhopal.

Dash, S. S. 2005. Kabi sacred grove of North Sikkim. *Current Science*, 89(3): 427-428.

Deb, D., Deuti, K., and Malhotra, K. C. 1997. Sacred grove relics as bird refugia. *Current Science*. 73(10): 815-817

Deshmukh, S., Gogate, M. G. and Gupta A. K. 1998. Sacred groves and biological diversity: Providing new dimensions to conservation issues. (In) P.S. Ramakrishnan, K.G. Saxena and U.M. Chandrashekhara (Ed.) *Conserving the sacred for Biodiversity Management*, Oxford and IBH, New Delhi. pp. 415-421.

Devi, S. 2000. Sacred groves of Manipur. Abstract. National Workshop on Community Strategies on the Management of the Natural Resources. Bhopal.

Dudley, N., Baldock, D., Nasi, R. and Stolton, Sue. 2005. Measuring biodiversity and sustainable management in forests and agricultural landscapes. *Phil. Trans. R. Soc. B*, 360, 457–470.

Dwivedi, O. P. 2000. Dharmic Ecology. (In) Chaple, C. K. and Tucker, M. E. (Ed.) *Hinduism and Ecology–The intersection of Earth, Sky and Water*. Oxford University Press, New Delhi.

FAO, 1997. *State of World Forests 1997*. Food and Agricultural Organization of the United Nations. Rome, Italy.

FAO, 1998. *Asia-Pacific Forestry Commission Development of national level criteria and indicators for sustainable forest management of dry forests in Asia–Background papers*. Regional Office for Asia and the Pacific, Bangkok, Thailand.

FAO, 2000. *Report of the FAO/UNEP/ITTO/IIFM/USFS Workshop on Regional initiative for the development and implementation of national level criteria and indicators for the sustainable management of dry forests in Asia*. Bhopal, India, 30 November–3 December 1999. FAO Regional Office for Asia and the Pacific, Bangkok, Thailand. FAO–RAP Publication, 2000.

Gadgil, M. and Chandran, M. D. S. 1992. Sacred Groves. (In) Sen, G. (Ed), *Indigenous Vision: People of India Attitudes to the Environment.* Sage Publications and India International Centre, New Delhi.

Gadgil, M. and Vartak, V. D. 1976. The sacred groves of Western Ghats in India. *Econ Bot*. 30: 152-160.

Gadgil, M. and Vartak, V.D., 1981. Sacred groves in Maharashtra–An inventory, In: S.K. Jam (Ed.), *Glimpses of Indian Ethnobotany*, Oxford and IBH, New Delhi, pp. 279-294.

Gaikwad, S. S., Paralikar, S. N., Chavan V., and Krishnan S., 2004. Digitizing Indian sacred groves–An information model for web interfaced multimedia database. (In) Ghate, V. Sane, H., and Ranade, S. S. (Ed), *Focus on Sacred Groves and Ethnobotany*, Prisam Publications, Mumbai, India, pp. 123-128.

Gupta, A., Shukla S., Koradiya D., Bhavsar P., Anil R., Patel P., and Taviyad R. 2000. A cultural and ecological study of sacred groves in Balaram Ambaji and Jessore sanctuary in Banaskantha district of Gujarat. Abstract. National Workshop on Community Strategies on the Management of Natural Resources. Bhopal.

Hughes, J. D. and Chandran, M.D.S. 1998. Sacred groves around the earth: An overview. (In) P.S. Ramakrishnan, K.G. Saxena and U.M. Chandrashekhara (Ed.), *Conserving the Sacred for Biodiversity Management*, Oxford and IBH, New Delhi.

Induchoodan, N. C. 1996. *Ecology of the Sacred Groves Kerala*. Ph.D. Thesis, Pondicherry University, Pondicherry.

ITTO, 1998. *Criteria and indicators of sustainable forest management of natural tropical forests.* ITTO Policy Development Series No 7, International Tropical Timber Organization, Yokohama, Japan.

Kadamban, D. 1988. *Biocultural perspective and plant diversity of sacred groves and traditional medical knowledge in Pondicherry Environs*. Ph.D. Thesis. Pondicherry University. Pondicherry.

Kalam, M. A. 1996. Sacred groves in Kogadu district of Karnataka (South India)–A Socio-histological study, Institute Francais de Pondicherry, Pondicherry.

Khumbongmayum, Devi A., Khan M. L., and Tripathi, R.S. 2004. Sacred groves of Manipur–ideal centres for biodiversity conservation. *Current Science*, 87 (4), 25 pp. 430-433.

Kulkarni D.K., Barve J.P., Jagdale R.P. and Inamdar A.C., 1993. Floristics of a sacred forest patch from Sundargad district, Orissa. *J. Econ. Tax. Bot.*, 17(2): 415-419.

Lebbie, A. R. and Guries R. P. 1995. Ethnobotanical value and the conservation of sacred groves of the Kpda Mende in the Moyamba District, Sierra Leone. *Economic Botany* 49 (3): 294-308.

Logan, W., 1887. *Manual of Malabar District*. Madras Govt. Press (reprinted 1906), pp. 185-186.

Malhotra, K. C., Gokhale, Y., Chatterji, S. and Srivastava, S., 2001. *Cultural and ecological dimensions of sacred groves in India*. INSA, New Delhi.

Martin, 1994. Alternative approaches to sustainable use: What does and doesn't work. Paper presented at the conference on conservation, sustainable use of wildlife. University of Queensland, Brisbane.

Malhotra, K. C., Stanley S., Hemam N. S., and Das K. 1997. Biodiversity conservation and ethics: sacred groves and pools, (In) Fujiki N. and Mecer, R. J. (Ed), *Bioethics in Asia,* Eubois Ethics Institute, Japan, pp 338-345

Mohanan, C.N and Nair C.N. 1981. Kunstleria Prain–A new genus record for India and a new species in the genus. *Proc. Indian Acad. Sci,* (Plant Sci.) 90: 207-209.

Myers, N., R. Muttermeier A., C. A. Muttermeier, G. A. Da Fonseca B., and Kent J. 2000. Biodiversity hotspots for conservation priorities. *Nature,* 403, 853-858.

NAEB, 1995. *Sacred groves of Kurukshetra, Haryana.* Agricultural Finance Corporation Ltd., Bombay Regional Centre, National Afforestation and Eco- development Board. MoEF, Govt. of India, New Delhi.

Nair, K. K. N. 1985. Unrecorded species of Dalbergia L. F. Fabaceae in Kerala. *J. Enon. Tax. Bot.* 730-734.

Pandey, D. N. 2007. Sacred Forestry: the case of Rajasthan, India, Sustainable development international, available online at URL http://www.infinityfoundation.com/ECITsacredforestryframeset.htm

Patnaik, S., and Pandey A. 1998. A study of indigenous community based forest management system: Sarna (sacred grove), (In) Ramakrishnan, P.S., Saxena, K.G., and Chandrashekhara, U.M. (Ed.), *Conserving the Sacred for Biodiversity Management,* Oxford and IBH, New Delhi, pp. 315-321.

Pushpangadan, P., Rajendraprasad M. and Krishnan P. N. 1998. Sacred groves of Kerala–A synthesis of the state-of-art knowledge. (In) *Conserving the Sacred for Biodiversity Management* P.S. Ramakrishnan, K.G. Saxena and U.M. Chandrashekhara (Ed.), Oxford and IBH, New Delhi. pp. 193-209.

Rajendraprasad, M. 1995. Sacred Groves of Kerala–A Synthesis on the state of-art-of knowledge, (In) P.S. Ramakrishnan, Saxena, K.G. and Chandrashekhara, U.M. (Ed.), *Conserving the sacred for Biodiversity Management,* Oxford and IBH, New Delhi.

Ramakrishnan, P.S., Saxena K.G. and Chandrashekhara U.M. (Ed.) 1998. *Conserving the Sacred for Biodiversity Management,* Oxford and IBH, New Delhi.

Ramanujam, M. P., Kadamban D., Kumaravelu, G., and Praveenkumar, K. 2002. Sacred groves–An overview, In: Trivedi, P. C. (Ed), *Ethnobotany: An Overview,* Avishkaar Publishers, Jaipur. pp. 13-53.

Rao, P. 1996. Sacred groves and conservation. *WWF–India Quarterly,* 7: 4-8.

Roy Burman, J. J. 1992. The institution of sacred groves, *Journal of Indian Anthropological Society* 27: 219-238.

Saxena, K. G., Rao, K. S., and Maikhuri, R.K.1998. Religious and cultural perspective of biodiversity conservation in India: A Review. (In) P.S. Ramakrishnan, K.G. Saxena and U.M. Chandrashekhara (Ed.), *Conserving the Sacred for Biodiversity Management,* Oxford and IBH, New Delhi.

Schaaf, T., 2003. UNESCO's Experience with the Protection of Sacred Natural Sites for Biodiversity Conservation, (In) Cathy Lee and Thomas Schaaf (Ed.), *The Importance of Sacred Natural Sites for Biodiversity Conservation*–Proceedings of the International Workshop on the Importance of sacred natural sites for biodiversity conservation, China, 17–20 February, UNESCO.

Sharma, B.R. 2000. Sacred groves and their role in social life in Himachal Himalayas. Abstract. National workshops on community strategies on the management of natural resources. Bhopal.

Singh, G. S., Rao, K. S. and Saxena, K. G. 1998. Eco-cultural analysis of sacred species and ecosystems in Chhakinal Watershed, Himachal Pradesh. (In) *Conserving the Sacred for Biodiversity Management* (Ed.) P.S. Ramakrishnan, K.G. Saxena and U.M. Chandrashekhara), Oxford and IBH, New Delhi.

Sinha, B. and Maikhuri, R. K. 1998. Conservation through "Socio-cultural-religious practice" in Garhwal Himalaya: A case study of Hariyali sacred site. (In) Ramakrishnan P.S., Saxena K.G. and Chandrashekhara U.M. (Ed.), *Conserving the Sacred for Biodiversity Management*, Oxford and IBH, New Delhi. Pp. 289-299.

Swamy, P. S., Kumar M. and Sundarapandian, S. M. 2003. Spirituality and ecology of sacred groves in Tamil Nadu, India. *Unasylva*, 233, Vol. 54.

Tiwari, B. K., Barik S. K. and Tripathi R. S. 1999. Sacred Forests of Meghalaya: Biological and Cultural Diversity. Regional Centre, National Afforestation and Ecodevelopment Board, North-Eastern Hill University, Shillong.

Vajpeyi, Y. 2000. Tree of Life. *Indian Express* (Sunday Magazine), September 3.

Vartak, V. D. 1983. Observations on rare imperfectly known and endemic plants in the sacred groves of Western Maharashtra, (In) Jain, S.K., and Rao, R.R. (Ed.), *Assessment of Threatened Plants in India*, Botanical Survey of India. Calcutta, pp. 169-178.

Vartak, V. D. and Suryanarayana M. C. 1995. *Sacred groves–an information kit* 9-6; pp. 1-2, Sponsored by WWF; IIR; SIDA and Brit. Hist. Comm., Goa.

Veach, R., Lee, D. and Philippi, T. 2003. Human disturbance and forest diversity in the Tansa Valley, India. *Biodivers. Conserv.*, 12: 1051-1072.

Veluppillai, T. K. 1940. The Travancore State Manual. Govt. of Travancore, Vol. I: 623-628.

Part II

Biodiversity of Vegetables

Biodiversity in Horticultural Crops Vol. 2 *Pages* **81–98**
Editor: **Professor K.V. Peter**
Published by: **DAYA PUBLISHING HOUSE**

Chapter 4

Amaranth

V.A. Celine
Department of Olericulture, College of Agriculture,
Kerala Agricultural University, Vellayani, Trivandrum

Introduction

Amaranths (*Amaranthus spp.*) are a group of under utilized or rather under exploited crops consisting of grain, leaf, forage and ornamental types. Over the past decade amaranth has become a focus of intensive nutritional and agronomic research and it is on the verge of commercial exploitation (Tucker, 1986). The current status of amaranth is as a crop which has great potential, a variety of possible uses, and more than a decade of research behind it. Great genetic diversity, wide adaptability, tolerance to biotic and abiotic stress conditions and high nutritional value of both grain and vegetable types makes it a potential candidate for poverty alleviation among the rural poor.

Classification of Amaranthus Species

It is estimated that there are 87 species of *Amaranthus*: 17 in Europe, 14 in Australia and 56 in America (Mujica and Jacobsen 1999; Jacobsen *et al.*, 2000; Mujica *et al.*, 2001; Jacobsen and Mujica 2001), however, due to the few studies on Amaranthus, systematics, the number of species is still tentative. Brenner *et al.* (1999) mention 60, and USDA, ARS (1999) 86 species. Of the American species 10 are dioecious, only found in North America, and 46 monoecious (Hunziker, 1992). *A. hybridus* L. whose synonym may be *A. quitensis* S. is a common ancestor to all the three cultivated grain species (Chan and Sun, 1997).The different species can be grouped into grain type, vegetable type, and weedy types. The important species under each category are as follows:

Grain Amaranth

1. *A. caudatus* L., whose synonyms are: *A. edulis Spegazzini, A. mantegacianus Passerini*
2. *A. hypochondriacus L.,* whose synonyms are: *A. leucocarpus* S.Wats, *A. flavus* L.
3. *A. cruentus* L., whose synonym is *A. paniculatus* L.

Vegetable Amaranth

1. *A. tricolor* whose synonyms are: *A. gangeticus* L., *A. tristis* L., *A. mangostanus* L.,
2. *A. blitum* L. synonymous of *A. lividus* L.
3. *A. dubius* Mart.ex Thell.

Weedy Type

1. *A. viridis* L. synonymous of *A. gracilis* Desf.
2. *A. spinosus*
3. *A. retroflexus*
4. *A. hybridus*
5. *A. powellii*

Wild and cultivated species are both useful for genetic improvement.

Grain Amaranth Type

The grain amaranths are native to the New World. Pre-Columbian civilizations grew thousands of hectares of this pseudo-cereal. It is cultivated in USA, China, India, Kenya, Mexico, Nepal Peru, Russia and several other European countries. There are three species of the genus *Amaranthus* which produce large seedheads of edible, light-colored seeds.

History

The most significant historical use of amaranth as a grain crop was in central Mexico during the Aztec civilization (the Aztec name for amaranth was *huahtli*). Amaranth was both an important food grain for the Aztecs, and a part of their religious practices. By the 1700s, amaranth had spread throughout Europe for use as a herb and ornamental. In the late 1800s, amaranth was reportedly being grown in mountain valleys of Nepal and parts of east Africa. During the 20th century it has been grown in China, India, Africa, and Europe, as well as North and South America.

Sauer (1993) evaluated the evidence on amaranth's domestication, and concluded that amaranth came into use as a grain at least 6000 years ago in Central America. *A. cruentus* L. and *A. hypochondriacus* L. are native to Mexico and Guatemala whreas, *A. caudatus* L. is native to the Andean regions of Ecuador, Peru, and Bolivia (Sauer 1967). Cultivated land races of the three species have a wide degree of genetic diversity which can be utilized to develop selections of improved lines with suitable characteristics for modem grain production.

An overview on grain amaranth with an emphasis on currently available information about the utilization of germplasm to promote more efficient production of the crop is given by many workers (Feine *et al.*, 1979; National Academy of Sciences, 1984; Kauffman and Weber, 1990; and Myers, 1996).

Food and Industrial Uses

The seeds are popped or ground and used in many preparations by the Aztecs. Nowadays, it is used in various processed foods such as breads, crackers and cookies. There has been steady use of the crop for breakfast cereals, snack foods, and more recently, in mass produced multi-grain bread products. Wu and Corke (1999) suggested that *Amaranthus* starches can be developed for a wide range of food uses.

In India, *A. hypochondriacus* L. is known as "rajgeera" (the King's grain) and is often popped to be used in confections called "laddoos," which are very similar to Mexican "alegria" In Nepal, amaranth seeds are eaten as a gruel called "sattoo" or milled into a flour to make chappatis (Singhal and Kulkami 1988).

Extensive reviews on the nutritional components, digestibility and potential problems that will be encountered by those who intend to use grain amaranth as a food product were published (Becker *et al.*, 1981, Teutonics and Knorr 1985, Bressani *et al.*, 1987a, Saunders and Becker 1984, Pedersen *et al.*, 1987). The most studied nutritional aspect concerning the food value of grain amaranth is the identification of the limiting amino acids of the protein component. The crude protein content of selected light-seeded grain amaranths has been reported to range from 12.5 to 17.6 (Teutonico and Knorr 1985, Becker *et al.*, 1981, Lorenz and Gross 1984, Sanchez Marroquin *et al.*, 1986, Pedersen *et al.*, 1987, Correa *et al.*, 1986). Amaranth grain is reported to have high levels of lysine, a nutritionally critical amino acid, ranging from 0.73 to 0.84 per cent of the total protein content (Bressani 1987a). The limiting amino acid is usually reported to be leucine (Singhal and Kulkarni 1988), although some reports indicate that threonine actually may be the amino acid which is more biologically limiting than leucine (Bressani 1987b). The presence of saponins, phytic acid, and trypsin inhibitors in the concentrate made from A cruentus flour favor the metabolism of lipids, suggests that consumption of the concentrate might reduce the risk of heart disease (Escudero *et al.*, 2004).

Amaranth, like most grains, has potential for use in industrial products. The oil fraction of the grain is unusually high in squalene, a chemical that sells for thousands of dollars per ton. However, the per cent of squalene in the grain is still small, and may not be economical to extract. Greater promise lies with the starch fraction of the grain. The physical characteristics of the starch grains have been cited as being of potential value for both industrial and food product uses, though none has been commercialized to date.

When the leaves, stem, and head are used for forage, the product will range from 15 per cent to 24 per cent protein. Palatability is not well known. Within the available amaranth germplasm are genotypes that show potential for significant forage production, but more research in this area is needed.

Genetic Diversity and Evolution of Grain Amaranth Species

Amaranth species have different centres of domestication and origin, being widely distributed in North America (Canada, United States), Central America (Mexico, Guatemala), and the South American Andes (Peru, Bolivia, Ecuador), where also the greatest genetic diversity is found. The amaranth has a broad genetic variability, with diversity in plant type (erect to prostrate), number of inflorescences (one to several), seed colour, earliness, protein content, plant height, seed and green matter yield, resistance to pests and diseases, and adaptation to soil type, pH, climate, rainfall and day-length (Kulakow 1990; Mujica and Berti 1997; Jacobsen *et al.*, 2001).

Studies on isoenzymes have demonstrated a high degree of polymorphism between populations, in which it is possible to find intermediate stages of domestication, such as wild types of amaranth that live without human intervention, amaranths that survive despite the efforts to eradicate them, and cultivated species that require special care for survival (Cox and Atkins 1979). A great deal of variation has also been found within populations. Evidence of introgression of wild genes into cultivated populations has been found from isoenzymes typical of *A. hybridus* (wild) in *A. caudatus* (cultivated), in a frequency less than 2 per cent (Hauptli 1986). Moreover, the presence of plants homozygous for an isoenzyme, despite having the normal morphology of the cultivated species, indicates introgression of wild genes in the cultivated population. The high level of allogamy explains the variation (Hauptli 1986).

Vegetable Type

Vegetable amaranths, known as 'poorman's spinach' is perhaps the most widely eaten leaf vegetables in the humid tropics due to due to its many advantages like high yields, ability to grow in hot weather, and very high nutritive value. In many tropical regions of India, China, Southeast Asia, and the South Pacific islands, amaranths, such as *A. tricolor* and *A.dubius*, are grown. They have been cultivated for more than 2,000 years, and many different cultivars, some with brightly colored leaves, have been developed. In humid tropical Africa, *A. cruentus* is extensively grown as a leaf vegetable. In much of the Caribbean, *A. dubius* and other amaranths are considered among the best of leafy greens. Vegetable amaranth is a promising C_4 crop for semi-arid regions due to its high nutritive value and an ability to adapt to drought stress (Liu and Stützel, 2004).

Leaf amaranth is considered as a crop to alleviate rural poverty and malnutrition in the rural areas of the developing countries. Nutritionally, it is rated superior to spinach as it is high in minerals (Makus, 1984; Makus, 1990; Rangarajan and Kelly, 1998; Rangarajan *et al.*, 1998; Shukla, 2006a), vitamins (Watt and Merrill, 1975; Shukla, 2006b) and protein Grubben, 1976). Vegetable amaranths provide a high concentration of vitamin A, important in preventing eye diseases (Martin and Telek, 1979).

Morphological Characterization

Biometric Characters

Characterization of vegetable amaranth germplasm based on morphological characters like leaf weight, stem weight, leaf length, leaf breadth, stem diameter and plant height were carried out by various workers.

Performance of Vegetable Amaranth Species, Varieties and Accessions

Vegetable amaranths belonging to *A. dubius, A. tricolor, A. blitum, A. hypochondriacus, A. cruentus, A. tristis* and *A. edulis* vary in growth and yield characters (Mohideen and Rajagopal, 1974; Vijayakumar,1980; Olufolaji and Tayo, 1980; Campbell and Abbott, 1982; Igbokwe *et al.*, 1988; Norman and Sichone, 1993; Allemann *et al.*,1996; Singh and Whitehead,1996). *A. dubius* out yielded the other species, probably due to its polyploidy nature.

The different accessions and varieties within *A. tricolor* were also studied for yield and yield attributes by various workers (Subbiah, 1979; Mohideen and Muthukrishnan, 1981; Kauffman and Gilbert, 1981; Devadas, 1982; Varalakshmi and Reddy, 1994; Sirohi and Sivakami, 1997; Mohanalekshmi *et al.*,1998; Priya,1998; Hossain and Rahman, 1999; Kanthaswamy *et al.*, 2000; Shukla and Singh, 2000; Shukla and Singh, 2002a;)

Wu *et al.* (2000) recorded wide diversity for stem and leaf colour while evaluating the amaranthus genetic resources from China. Xiao *et al.* (2000) classified 31 vegetable amaranth varieties based on 17 biological characters, of which leaf shape and colour were most significant.

Variability among *A. dubius* accessions for biometric characters were studied by Sindhu (2002) and reported that the yield varied from 1555.94 g to 464.80 g and that that the leaf/stem ratio ranged from 0.93 to 2.48. Days to 50 per cent bolting showed a variation from 47.80 days to 75.13 days.

Quality Characters

Nutrient Factors

Amaranth is one of the few vegetables in which several essential dietary components such as protein, vitamin C, vitamin A, Fe and Ca occur together (Mohideen and Subramanian, 1974; Martin and Telek, 1979). Sudhir Shukla *et al.* (2006) reported that vegetable amaranth is a rich source of minerals like calcium (1.7±0.04 g/100 g), iron (1233.8±50.02 mg/kg), and zinc (791.7±28.98 mg/kg). The studies conducted by various workers are given in Table 4.1.

Table 4.1: Nutritive Value of Vegetable Amaranth (*A. tricolor*)

Nutrient	*Content*	*Reference*
Protein	3.5 g/100 g (fresh wt)	Rajagopal *et al.* (1977)
	37.19 g/100g (dry wt)	Mathai *et al.* (1980)
	12.5 g/100g (dry wt)	Mohideen *et al.* (1985)
	12.5–14.5/100g (dry wt)	Vijayakumar and Shanmughavelu (1985)
	29.3 g/100g (dry wt)	George *et al.* (1989)
	12.74g–14.65 g/100g (dry wt)	Calderon *et al.* (1991)
	1.03–5.23 g/100 g (fresh wt)	Raja *et al.* (1997)
	4.0 g100 g (fresh wt)	Hemalatha *et al.* (1999)
	12.97 g/100g (dry wt)	Hossain *et al.* (1999)
	1.4–3 g/100 g (fresh wt)	Pal (1999)
	1.5–3.2 g/100 g (fresh wt)	Tewari *et al.* (2002)
	0.7–3.6 g/100 g (fresh wt)	Celine *et al.* (2007)
Carotene	11.04 mg/100 g	Mohideen *et al.*,1985
	9–10.9 mg/100g (dry wt)	Vijayakumar and Shanmughavelu (1985)
	36.1 mg/100g (dry wt)	George *et al.* (1989)
	9- 20 mg/100g (dry wt)	Prakash and Pal (1991)
	4.95 mg/g	Kononkov *et al.* (1995)
	9.0 to 20.0 mg/100 g (fresh wt)	Pal (1999)
	1.14–14.52 mg/100g (dry wt)	Yadav and Sehgal (1999)
	40 mg/100g (dry wt)	Gins *et al.* (2000)
	11.5-20.2 mg/100 g (fresh wt)	Tewari *et al.* (2002)
	1. 27-4.65 mg/100 g (fresh wt)	Celine, *et al.* (2007)
Vitamin C	325–1250 mg/100 g	Grubben (1976)
	32.9–44.2 mg/100g (dry wt)	Vijayakumar and Shanmughavelu (1985)
	120 mg/100 g (fresh wt)	Hemalatha *et al.* (1999)
	12.46-61.86 mg/100g (dry wt)	Yadav and Sehgal (1999)
	105–180 mg/100 g (fresh wt)	Tewari *et al.* (2002)
	54.9–151.2 mg/100 g (fresh wt)	Celine, *et al.* (2007)

Contd..

Table 4.1–Contd..

Nutrient	Content	Reference
Iron	19 mg/100 g	Rajagopal *et al.* (1977)
Calcium	310 mg/100 g	Rajagopal *et al.* (1977)
	2.30–2.52 g/100g (dry wt)	Vijayakumar and Shanmughavelu (1985)
	340.0 mg/100 g (fresh wt)	Hemalatha *et al.* (1999)
	1.54 g/100g (dry wt)	Hossain *et al.* (1999)
Phophorus	39.38 mg/100 g	Rajagopal *et al.* (1977)
Fibre	1.3 g/100 g	Rajagopal *et al.* (1977)

Leaves of *A. hybridus* contained 28.2-31.6 per cent protein, 25.2-37.3 mg per 100 g beta-carotene and 455-535 mg per 100 g vitamin C (Mziray *et al.*, 2001)

A. dubius cv. Ibondwe proved to be superior among different species studied for carotene content (15.4 mg g^{-1} fresh weight) and low oxalate content (Sealy *et al.*, 1988). Sindhu (2002) noted that the minimum amd maximum protein contents were 9.03 per cent and 23.00 per cent respectively in *A. dubius* accessions whereas the vitamin A content ranged from 4331.50 I.U. to 8915.96 I.U.

Antinutrient Factors

Presence of antinutrients like nitrates and oxalates is a main problem limiting the large scale consumption of amaranth (> 100g/day) according to health experts. However, they are not harmful under normal conditions of dietary intake (Kauffman and Gilbert, 1981). Excess of oxalates can combine with the dietary calcium forming calcium oxalate crystals which leads to kidney stone problems. Nitrates are also highly hazardous as they get converted into nitrites and join with the blood haemoglobin thereby blocking its function as an oxygen transporter. Nitrate content above 1- 3 per cent (dry wt) can cause acute toxicity in animals (Adams, *et al.*, 1992). Der Marderosian *et al.* (1979) reported nitrate levels of 0.43 per cent for leaves and 1.72 per cent for stems on dry weight basis. Oxalate content of leaves and stems were 5.00 and 0.63 per cent respectively. In a study using eight accessions of *A. tricolor*, Makus (1984) recorded that the contents of nitrates and soluble oxalates were 1.1 and 2.3 per cent respectively. According to Teutonico and Knorr (1985), oxlates can comprise 0.2 to 11.4 per cent of the dry matter in vegetable amaranths.

After conducting studies in 61 landraces involving vegetable and grain amaranthus, Prakash and Pal (1991) reported that the nitrate and oxalate contents varied from 1.8 to 9.2 g kg^{-1} and 3.0 to 19.2 g kg^{-1} respectively on fresh weight basis. Oxalate content increased with advance in the growth period while nitrate content remained constant. According to Tewari *et al.*,(2002), nitrate and oxalate contents were found to range from 0.29 to 0.89 per cent and 0.8 to 1.9 per cent respectively in vegetable amaranths.

The species other than *A. tricolor* is not an exception for the presence of antinutrient factors.George *et al.* (1989) compared the nutritive aspects of three species of vegetable amaranthus namely, *A. tricolor, A. dubius* and *A. cruentus* and noted that all green entries had low oxalate contents, while red and green-red entries had high oxalate contents ranging from 3.04 to 6.8 per cent. The oxalate content was found to range from 0.94 to 1.29 per cent and the least content of 0.94 per cent was seen in variety Co-3 (Devadas and Mallika, 1991). In the genus *Amaranthus*, section Blitopsis had higher content of antinutrients than section Amaranthus. Red pigmented lines contained higher oxalate content when compared to green pigmented amaranthus lines (Devadas *et al.*, 1993; Priya and Celine, 2001). After comparing 41 amaranthus lines, Thamburaj *et al.* (1994) reported that the oxalate content ranged from 0.820 to 0.921 per cent and that red types had higher oxalate content when compared to the green ones. The oxalate and nitrate levels vary from 0.62–3.85 per cent and 0.25–1.09 per cent respectively in *A. dubius* accessions as reported by Sindhu (2002). Celine, *et al.* Guill *et al.* (1997) reported that in *A. viridis*, the nitrate content was 597 mg 100 g^{-1}. Bianco *et al.* (1998) reported extremely high levels of soluble oxalates in *A. retroflexus*. Mziray *et al.* (2001) analysed freshly harvested leaves of *A. hybridus* for antinutrients and reported a nitrate content of 501 to 560 mg per 100 g and oxalate content of 3383 to 4333 mg 100 g^{-1}.

Biotic Stress Factors

Amaranth is susceptible to a number of insects(Grubben, 1976; Wilson, 1989 and El-Aydam and Burki, 1997) and identified sources of resistance. Resistance to root knot nematodeis available in some accessions of *A. cruentus* and *A. dubius* (Babatola and Awoderu, 1986). The infection of leaf webber will lower the yield and market value of the crop and it also makes the leaves unfit for culinary purposes.

The leaf webbers, *Psara basalis* Fab. and *Hymenia recurvalis* F. badly damage the leaves of vegetable amaranthus. (Bhattacharjee and Menon, 1964; Nair, 1980). *Hymenia recurvalis* was reported as an important defoliator of *A. viridis* (Pande, 1973). Leaf webber infestation on *A. hypochondriacus* cultivars was investigated and cultivars OGA-2, SKN-6, OGA-3 and Rasna-2 recorded lowest leaf infestation rates Jena *et al.* (2001).

Sealy *et al.* (1988a) screened amaranth germplasm for damping of resistance. Although amaranths are hosts for many viruses, (Sooby *et al.*, 1999). They have not been an economic problem. Fungal diseases are a problem in *A. tricolor* and will colonise aggressively on the leaves and stem and kill them (Walters *et al.*, 1988 and Bansal, 1996).

Leaf blight caused by *Rhizactonia solani* Kuhn is a major problem that adversely affects amaranthus cultivation in Kerala (Nayar *et al.*, 1996). It greatly reduces the yield and lowers the market value of the crop. The same pathogen causes collar rot also (Roy, 1975).Resistant sources were identified under field conditions and artificial epiphytotic conditions (Gokulapalan *et al.*,1997; Priya, 1998; Krishnakumary *et al.*, 2001; Celine *et al.*, 2002; Sindhu, 2002). *A. dubius* accessions were highly disease resistant whreas the *A. tricolor* accessions showed disease susceptibility of varying degrees (Celine, *et al.*, 2007).

Stem canker and wirestem caused by *Rhizoctonia solani* were observed on the stems of green amaranth (*Amaranthus viridis*) and Chinese amaranth (*Amaranthus tricolor*) in greenhouses at Ximao district in Yunnnan Province, China (Yang *et al.*, 2003).

Genetic Variability, Heritability and Correlation

Genetic variability, heritability and correlation studies were carried out in vegetable amaranth by various workers (Hauptli and Jain, 1980; Vijayakumar, 1980; Mohideen *et al.*,1982; Pan *et al.*,1991; Varalakshmi and Reddy, 1994; Revanappa and Madalgeri, 1997; Reddy and Varalakshmi, 1998; Priya and Celine, 2001; Shukla and Singh, 2002c; Vaidya and Jain, 2002; Shukla and Singh, 2003a; Rani, A. R. B., Veeraragavathatham, D. 2003)

Genetic divergence studies helped in grouping the accessions into different clusters (Devadas *et al*, 1992; Pan *et al.*, 1992; Verma *et al.*, 2002).

Vegetable Amaranth Species

Vegetable amaranth species include *A. tricolor, A. dubius, A. blitum, A. hypochondriacus, A. cruentus, A. tristis* and *A. edulis*

Amaranthus cruentus

This species has been described earlier as a grain type. A very deep red, dark-seeded form of the species, sometimes known as blood amaranth. It is an important crop in tropical Africa. Amaranthus cruentus was introduced to Africa by Europeans.

Amaranthus dubius

This species is used as a green vegetable in West Africa and the Caribbean and is found in Java and other parts of Indonesia as a home garden crop. It has distinctive dark-green, broad, ridged leaves. It is fast growing, high yielding, and has considerable morphological variation, resulting partly from its repeated hybridization with Amaranthus spinosus. The "greens" of this species are considered very palatable. This is the only tetraploid (2n = 64) species in the genus known so far.

Amaranthus hybridus

This weedy species is one of the most common leafy vegetables. It is an herb that grows up to 1.5 m tall. Originating in tropical America, it is now spread throughout tropical areas. In Indonesia it is often planted in kitchen gardens. It also grows wild on moist ground, in waste places, or along roadsides, and thrives in altitudes up to 1,300 m. The size and color vary greatly. Red-stemmed varieties are usually planted as ornamentals; green varieties are commonly used as vegetables.

Amaranthus lividus

This widely distributed species (also known as *A. blitum*) is well adapted to temperate climates and has a number of weedy forms that come with either red or green leaves. It promises to allow the development of highly palatable crossbred vegetable amaranths. In Madhya Pradesh, the edible forms, known as norpa, are especially liked for their tender stems. This is the species widely eaten in Greece under the name vleeta. It is also grown in Taiwan, where it is known as horsetooth amaranth.

Amaranthus tricolor

Varieties of this species are native to a large area from India to the islands of the Pacific and as far north as China. It is probably the best-developed of the vegetable amaranth species. The plants are succulent, low growing, and compact, with growth habits much like spinach. They can be produced as a hot-season leafy vegetable in arid regions when few other leafy greens are available. In India a number of domesticated forms are available, especially in Andhra Pradesh, Karnataka, Tamil Nadu, and Kerala. Some ornamentals with very beautiful foliage also belong to this species.

Non Traditional Uses of Amaranths

A future promise of vegetable amaranths is the development of leaf protein concentrates. Compared with most other species, amaranth leaf protein is highly extractable. In one trial, amaranth had the highest level of extractable protein among 24 plant species studied. The protein quality of the amaranth leaf-nutrient concentrate is excellent.

Field and laboratory investigations revealed that *Amaranthus* pigments are red-violet betacyanins, similar to that produced in red beet (Beta vulgaris). The chemical compositions were being investigated in 42 accessions from 7 species (*A. caudatus, A. cruentus, A. hypochondriacus, A. hybridus, A. lividus,*

A. paniculatus, and *A. tricolor*), and the results showed that amaranthine, isoamaranthine, and proteins are the major components of the crude pigment extract. The high biomass production and extraction rate of *Amaranthus* pigments suggest that there is a great potential for their commercial development as natural food colorants (Cai *et al.*, 1998a, b). Amaranth pigment can also be produced in tissue culture (Trail, 1993).

Amaranths are used for phyoremediation for removing toxic cesium–137(Lasat *et al.*, 1998) and copper (Chen *et al.*, 1996).

Amaranths which produce a large amount of biomass in a short period of time, can be used as a forage crop for domesticated animals. In China, amaranth has been cultivated expressly for use as forage for cattle. There are several cuttings made per growing season. In Peru, the stover is grazed or milled for use as a feed supplement after the seed heads have been hand-harvested. Some types of amaranth accumulate toxic levels of oxalate and nitrate when grown under conditions of stress (Saunders and Becker 1984). Additional research is needed to better define the feeding value, potential toxicity problems and the most appropriate type of amaranth for use as forage (Weber *et al.*, 1988).

Botany

Taxonomy

Amaranthus is often difficult to characterize taxonomically, due to the similarity between the large number of species, small, difficult-to-see diagnostic parts, intermediate (hybrid) forms and the broad geographical distribution, which is the reason for many synonyms.

Most of the taxonomic keys are controversial. Some of the taxonomic names are synonyms which can be investigated with computer data bases like GRIN, USDA, ARS (1999). Descriptive taxonomic keys for all the cultivated species of the genus *Amaranthus* was developed by many workers (Sauer, 1967; Feine-Dudley, 1981; Covas, 1992; Williams and Brenner, 1995 and Mosyakin and Robertson,1996).

Phylogenetic Relationships Among the Species and Gene Pools of Amaranthus

Phylogenetic studies were carried out to unveil the species relationships (Hauptli and Jain, 1984; Jain *et al.*, 1984; Kulakow and Hauptli, 1994). Molecular studies have been conducted on genetic diversity among wild amaranth species and grain amaranths using Random Amplified Polymorphic DNA (RAPD) and isozyme markers (Chan and Sun 1997; Mandal and Das, 2002). The evolutionary origins of the three grain amaranths were further studied using repetitive DNA as fingerprinting probes. Pairwise comparisons of genetic similarity indices identified *A. hybridus* as the most likely common progenitor of grain amaranths, among the three putative progenitor species, *A. hybridus*, *A. powellii* and *A. quitensis* (Sun *et al.*, 1999). *A. hypochondriacus* and *A. caudatus* are genetically closer than *A. cruentus*, despite having originated in different areas (Transue *et al.*, 1994). Kirkpatrick(1995) used morphological traits, isozymes, and nuclear ribosomal DNA sequences to study the species relationships. Though Lanoue *et al.* (1996) identified potential mutants, the overall variations at restriction sites were very low resulting in poorly resolved phylogenetic trees. Ranade*et al.* (1997) analyzed interspecific relationships in Amaranthus and their RAPD data ruled out a putative hybrid origin of *A. dubius* from *A. hybridus* and *A. spinosus*.

Dixit, (1998) described a simple, efficient and reliable method is for isolation of total DNA from young leaves of *Amaranthus* species. This procedure yields a high amount (600–800 g DNA/g fresh leaf tissue) of good quality DNA. The DNA is suitable for digestion with several restriction endonucleases, preparation of Southern blots, and PCR amplification. The DNA has been successfully used for generating DNA fingerprint profiles and RAPD banding patterns in two species of *Amaranthus*. The procedure is suitable for processing of a large number of samples simultaneously.

Electrophoretic analysis of seed proteins proved to be useful for distinguishing species and cultivars of Amaranth, for describing similarity between species and for estimation its outcrossing rate. (Zheleznov, 1997; Drzewiecki, 2001). Low-Cot DNA sequences are suitable probes for a fingerprinting analysis of plant germplasm diversity and for determining phylogenetic relationships (Sun *et al.*, 1999).

The AFLP-based phylogenetic relationship of 'Morelos' and delimitation of *A. cruentus* and *A. caudatus* are further supported by micromorphology, showing that the combination of these techniques can provide more reliable data for germplasm identification than each method used alone (Costea *et al.*, 2006).

Jung-Ro Lee *et al.* (2004) reported isolation and characterization of 12 polymorphic microsatellite markers for 19 amaranth species representing cultivated grain and vegetable species, their putative progenitors and wild species. These results demonstrate wide potential applicability of these markers for the study of intra- and inter-specific genetic diversity as well as evolutionary relationships among cultivated and wild amaranths.

Steinau *et al.* (2003) based on their molecular studies reported that transposon-like elements may play an important role in the formation of new fragments in *Amaranthus* hybrids derived from interspecific hybridization, suggesting that considerable instability of the hybrid genome may occur.

RAPD analysis is especially valuable for classifying germplasm collections that lack sufficient isozyme variation. Isozyme analysis appears to be more suitable for phylogenetic investigations of interspecific relationships and, in particular, for studying the evolutionary origin of crop species (Chan and Sun 1997). Studies were initiated by Wetzel *et al.* (1999) to develop a molecular marker identification system utilizing restriction enzyme analysis of amplified ribosomal DNA (rDNA) for the identification of weedy Amaranthus species. Although useful, molecular techniques should be used with carefully defined population samples together with other sources of data, including morphological studies, as RAPD and isozyme data have not always the same dendrograms (Brenner *et al.*, 2000).

Xu and Sun (2001) used internal transcribed spacer (ITS), amplified fragment length polymorphism (AFLP) and double-primer fluorescent inter simple sequence repeat (ISSR) to reexamine the taxonomic status and phylogenetic relationships of grain amaranthus and their wild relatives. Low ITS divergence in these taxa resulted in poorly resolved phylogeny. But, extensive polymorphisms exist at AFLP and ISSR loci both within and among species.

A Principal Coordinate Analysis was made in Amaranth germplasm based on the presence or absence of 23 polymorphic bands provided a general overview of genetic diversity within the germplasm. These multivariate approaches grouped most cultivars and strains according to taxonomical classifications and illustrated interrelationships among accessions. This study suggests that isoenzymatic seed characteristics can be a profitable complement of other morphological and biochemical markers for practical applications in amaranth breeding programs allowing the characterisation of cultivars and new breeding lines. Renzo *et al.* (2001)

Polymerase chain reaction (PCR) amplification of specific alleles(PASA) was adapted as a molecular marker based method for the rapid detection of point mutations in *A. retroflexus* and *A. rudis* leading to acetolactate synthase (ALS) inhibitor resistance. PASA is useful for identification of resistant weed biotypes (Wagner *et al.*, 2002).

High degree of polymorphism exists both within and between populations of amaranth, even neighbouring ones, characterized by different alleles or isozymes {Jacobsen and Mujica,2003). Manilay (2003) has assessed the genetic diversity in amaranthus using RAPD markers.

Wassom and Tranel (2005) performed AFLP based UPGMA analysis on eight such amaranth species represented by 98 accessions.

Yudina*et al.* (2005) studied the electrophoretic patterns of alcohol dehydrogenase (ADH), glutamate dehydrogenase (GDH), malate dehydrogenase (MDH), isocitrate dehydrogenase (IDH) and malic enzyme (Me) in various populations of the cultivated and weedy amaranth species. In total, 52 populations and two varieties (Cherginskii and Valentina) have been examined. Allozyme variation of this material was low. Irrespective of species affiliation, 26 populations and two varieties were monomorphic for five enzymes; a slight polymorphism of three, two, and one enzymes was revealed in three, nine, and fourteen populations, respectively.AFLP data were useful for grouping several weedy *Amaranthus* species in UPGMA analysis

Floral Biology and Pollination

The flowers appear as terminal or axial panicle. The smallest unit of inflorescence is the glomerule. The first flower in the glomerule is generally staminate and the later pistillate. (Pal, 1972; Joshi and Rana, 1991). Most species are monoecious (Table 4.2), whereas few are dioecious (Sauer, 1957) and few with separate staminate and pistillate glomerules (Pal, 1972).

Table 4.2: Monoecious Species of Amaranth

Geographical Distribution	*Species*
America	1. *A. albus*; 2. *A. blitoides*; 3. *A. californicus*; 4. *A. caracasanus*; 5. *A. crassipes*; 6. *A. cruentus*; 7. *A. deflexus*; 8. *A. hybridus*; 9. *A. lividus*; 10. *A. polygonoides*; 11. *A. powelli*; 12. *A. spinosus*; 13. *A. squamulatus*; 14. *A. retroflexus*; 15. *A. urceolatus*; 16. *A. viridis*
Only North America and Mexico	1. *A. acutilobus*; 2. *A. brandegei*; 3. *A. chihuahuensis*; 4. *A. fimbriatus*; 5. *A. lepturus*; 6. *A. mexicanus*; 7. *A. obcordatus*; 8. *A. pringlei*; 9. *A. pumilis*; 10. *A. scariousus*; 11. *A. torreyi*; 12. *A. viscidulus*; 13. *A. wrighii*
Only Antilles, Central America and South America	1. *A. asplundii*; 2. *A. cardenasianus*; 3. *A. crispus*; 4. *A. furcatus*; 5. *A. kloosianus*; 6. *A. lombardoi*; 7. *A. looseri*; 8. *A. minimus*; 9. *A. muricatus*; 10. *A. persimilis*; 11. *A. peruvianus*; 12. *A. quitensis*; 13. *A. rosengurttii*; 14. *A. scariosus*; 15. *A. sclerantoides*; 16. *A. standleyanus*; 17. *A. vulgatissimus*

Both species are primarily wind pollinated (Murray, 1940). Being dioecious (separate male and female plants), *A. tuberculatus* is an obligate outcrosser. *A. hybridus* has separate male and female flowers (monoecism); however, due to the proximal arrangement of these flowers, it is thought to be mostly self-pollinated. Intercrossing within amranth genepool was tried by many workers with low success rate. Use of bridge species and colchiploidisation was tried. New techniques like protoplast fusion were tried (Mastuiti, *et al.*, 1997).

Pratt and Clark (2001) hypothised that there is a single, polymorphic species of waterhemp, *A. tuberculatus*. Principal components analysis of pistillate morphological, staminate morphological, and isozyme data sets all show a single waterhemp continuum exhibiting geographic variation.

The breeding systems of Amaranthus species are influenced by genetic and environmental variations. The mean outcrossing rates under field conditions has been estimated at 3.5 to 34 per cent (Jain *et al.*, 1982; Haupltli and Jain, 1985 and Espitia–Rangel, 1994). Agong and Ayiecho (1991) studied intra specific and inter specific out crossing rates using populations of *A. hypochondriacus* and *A. cruentus*. The mean intra specific out crossing rate estimates for *A. hypochondriacus* and *A. cruentus* were 10.4 per cent and 10.9 per cent respectively. Lower rates of outcrossing was reported in vegetable species like *A. tricolor* and *A. lividus* (Pal and Khoshoo, 1973).

Of the four interspecific hybrids, three (*A. graecizans* var. *graecizans* X *A. tricolor* cv. 'Purple leaf', *A. lividus* var. *lividus* X, *A. tricolor* var. *viridis* and *A. gracilis* X *A. tricolor* cv. 'Purple leaf') were studied cytologically. In all the three, differentiation between the parents is chiefly a result of interchanges and paracentric inversions. The interchange complexes may involve from four (*A. gracilis* X *A. tricolor* cv. 'Purple leaf') to fourteen (*A. lividus* var. *lividus* X *A. tricolor* var. *viridis*) chromosomes, indicating that the parents differ from each other in 1 to 6 interchanges. Because of the small size of the chromosomes, it is possible that crossing-over in interchange for small segments is restricted. The particular parental species representing the ancestral condition from which others were derived or compounded is difficult to pin-point. With preferential pairing and the restoration of fertility in the amphididiploid *A. lividus-tricolor*, it became clear that it is very likely that the interchanged segments are small and sterility in the hybrid is entirely chromosomal.

This situation is in strong contrast to that in Sect. *Amaranthus*, where the single male flower per glomerule and huge showy inflorescences lead to more cross-pollination; coupled with this are strong morphological divergences between taxa and less genetic differentiation. In the Sect. *Blitopsis*, there are a number of male flowers per glomerule and small non-showy axillary inflorescences, leading to self-pollination, and less morphological but strong genetic differentiation by interchanges and inversions.

Cytology and Cytogenetics

Most Amaranthus species are n = 16 or n = 17 with the only polyploidy species *A. dubius* with n = 32 (Grant, 1959).The relationship between the two basic numbers (n = 16 and n = 17) in the grain group of species of the genus *Amaranthus* was determined through an interspecific dibasic cross involving white seeded cultivated *A. hypochondriacus* (n = 16) and black seeded wild *A. hybridus* (n =17, African). In the hybrid at metaphase I 98 per cent of pollen mother cells had 15 bivalents + 1 trivalent; the remaining cells had 16 bivalents + 1 univalent. Chromosome counts in the 55 hybrid plants revealed the distribution of plants with $2n$ = 32, 33, and 34 in the ratio of 1:2:1, respectively. The origin of n = 17 from n = 16 through primary trisomy is suggested. (Pal *et al.*, 1982).

Detailed cytogenetic studies, have revealed a very interesting series of reproductive barriers like unidirectional incompatability, male sterility, hybrid sterility and 'virus' like syndromes, the last being unique and unravelled for the first time in this group of plants (Pal, 1999). The cytogenetic investigation of F_1 and F_2 progenies of the dibasic interspecific crosses involving *A. hybridus* (2n=16), *A. hypochondriacus* (2n=16), *A. cruentus* (2n=17) and *A. retroflexus* (n=17) have shown that x=16 is ancestral and x=17 derived through primary trisomy. A close genetic homology has been found between the progenitors like *A. hybridus* and *A. quitensis* on the one hand and between them and their respective domesticates-namely *A. hypochondriacus*–cruentus and *A. caudatus edulis.* The domesticates and the progenitor species thus constitute essentially a single 'gene pool' in which despite be achieved through the wild progenitor species. The discovery for the first time of the existence of a wild species *viz. Amaranthus retroflexus L.* in the Ladakh and adjacent regions of India and evaluation of its cytogenetic relationship with the cultivated grain amaranth species *viz. A. cruentus* L. (n=17) is of special significance in as much as the fertility shown by the hybrid could be used for evolving a grain amaranth crop for cold dry desert regions of India and other adjacent countries. Earlier *Amaranthus retroflexus* was not considered a close relative of grain species and thus the study has also helped in broadening the 'gene pool' of grain amaranths.

Tatum *et al.* (2005) reported the *in vitro* method for cytological studies. *In vitro* root propagation allowed large numbers of roots to be obtained from a single plant, thereby resulting in increased probability of obtaining cells with metaphase chromosomes that reflected the original plants' chromosome numbers and therefore may be used for molecular cytogenetic analysis.

Genes Identified

Genes coding for different characters were identified (Brenner *et al.*, 2000) and the important ones are given in Table 4.3.

Table 4.3: Important Genes Coding for Different Characters

Trait	*Gene*	*Source*
Male fertile/sterile	Ms 1/ms 1	*A. hypochondriacus*
Leaf blade red/green	R/r	*A. tricolor*
Stem red/green	A/a	*A. tricolor*
Stem green/white	B/b	*A. tricolor*
Seed coat black/pale	P/p	*A. hypochondriacus*
Tall/dwarf	Dw/dw	*A. caudatus*

Germplasm Collections

Amaranth germplasm is catalogued and stored in germplasm banks in 71 institutions spread over 43 countries totaling to 14,165 accessions belonging to different species (Bioversity International, 2007). Most collections have less than hundred accessions. The important ones are given in Table 4.4.

Gaps in collection exist for wild species and for regions where amaranthus is a minor crop Amaranths are well suited to gene bank conservation, because the seeds are small enough to need little storage space, and long-lived in cool, dry storage. Fresh seed stocks can be regenerated efficiently. Brenner and Widrlechner (1998) described an efficient protocol for regenerating seeds of Amaranthus germplasm and maintaining genetic integrity *ex situ*.

The National Academy of Science of the United States has recently been funding studies of amaranth in Mexico, Guatemala, Peru, Thailand and Kenya, in order to develop improved varieties for human consumption. Genetic material of amaranth has been collected widely in the Andean area and exchanged between Peru, Bolivia and Ecuador, where important trials have been conducted, such as the American Amaranth Test by the Food and Agriculture Organization (FAO) (Mujica *et al.*, 1999), and through the Programa Cooperativo de Investigación en Cultivos Andinos (PROCIANDINO) (Mujica and Berti 1997).

Table 4.4: Main Gene Banks of Amaranth Germplasm

Sl.No.	Name of the Institute	No. of Accessions	References
1.	Institute of Crop Germplasm Resources (CAAS), Beijing, China	438	IPGRI, 1999; Bioversity International, 2007
2.	Universidad Nacional del Altiplno Puno, Peru	440	Bioversity International, 2007
3.	Instituto Nacional de Investig Forestales y Agropecuarias (INIFAP), Mexico	700	Bioversity International, 2007
4.	Univ. Nacional San Antonio Abad del Cuzco (UNSAAC), Peru	1600	Bioversity International, 2007
5.	National Bureau of Plant Genetic Resources (NBPGR), Regional Station, Shimla, India	3,000	Joshi, (1995)
6.	North Central Plant Introduction Station (NCRPIS), Ames, Iowa, United States	3,380	USDA, ARS (1999)
7.	Catedra de Cerealicultura, Facultad de Agricultura, Univ Nacional de la Pampa, Argentina	700	Bioversity International, 2007
8.	Estacion Experimental Santa Catalina, DENAREF, INIAP, Ecuador	435	Bioversity International, 2007
9.	Organic Gardening and Farming Research Center, Kutztown, United States of America	1200	Bioversity International, 2007
10.	National Botanical Research Institute of India (NBRI), Lucknow, India	400	Pal, 1999.

Role of Rodale Research Centre (RRC)

The RRC under the leadership of Robert Rodale started the germplasm collection. Since 1977, RRC has maintained a working germplasm collection. The germplasm collection includes representatives from 45 amaranth species (Table 4.5). Descriptions of all species and the grain types can be found in the RRC germplasm catalogue (Kauffman and Reider 1986). Germplasm enhancement has been conducted at RRC using recurrent single plant selection and mass selection. Selections continue to be made for the development of ideotypes which meet the needs of modern agriculture.

Table 4.5: Taxa Held by the North Central Regional PI Station, Iowa

1.	*Amaranthus acanthochiton* (2 Accessions)
2.	*Amaranthus acutilobus* (2 Accessions)
3.	*Amaranthus albus* (7 Accessions)
4.	*Amaranthus arenicola* (6 Accessions)
5.	*Amaranthus asplundii* (1 Accessions)
6.	*Amaranthus australis* (2 Accessions)
7.	*Amaranthus blitoides* (7 Accessions)
8.	*Amaranthus blitum* (7 Accessions)
9.	*Amaranthus blitum* subsp. *blitum* (1 Accessions)
10.	*Amaranthus blitum* subsp. *oleraceus* (3 Accessions)
11.	*Amaranthus blitum* var. *pseudogracilis* (1 Accessions)
12.	*Amaranthus californicus* (1 Accessions)

Contd..

Table 4.5–Contd..

13.	*Amaranthus cannabinus* (3 Accessions)
14.	*Amaranthus caudatus* (552 Accessions)
15.	*Amaranthus crassipes* (2 Accessions)
16.	*Amaranthus crispus* (1 Accessions)
17.	*Amaranthus cruentus* (362 Accessions)
18.	*Amaranthus deflexus* (5 Accessions)
19.	*Amaranthus dubius* (44 Accessions)
20.	*Amaranthus fimbriatus* (2 Accessions)
21.	*Amaranthus floridanus* (1 Accessions)
22.	*Amaranthus graecizans* subsp. *aschersonianus* (8 Accessions)
23.	*Amaranthus graecizans* subsp. *silvestris* (3 Accessions)
24.	*Amaranthus graecizans* subsp. *thellungianus* (1 Accessions)
25.	*Amaranthus greggii* (1 Accessions)
26.	*Amaranthus hybrid* (147 Accessions)
27.	*Amaranthus hybridus* (198 Accessions)
28.	*Amaranthus hypochondriacus* (1,493 Accessions)
29.	*Amaranthus muricatus* (1 Accessions)
30.	*Amaranthus palmeri* (16 Accessions)
31.	*Amaranthus powellii* (3 Accessions)
32.	*Amaranthus powellii* subsp. *bouchonii* (7 Accessions)
33.	*Amaranthus powellii* subsp. *powellii* (8 Accessions)
34.	*Amaranthus pumilus* (6 Accessions)
35.	*Amaranthus quitensis* (49 Accessions)
36.	*Amaranthus retroflexus* (26 Accessions)
37.	*Amaranthus* sp. (72 Accessions)
38.	*Amaranthus spinosus* (24 Accessions)
39.	*Amaranthus standleyanus* (2 Accessions)
40.	*Amaranthus tamaulipensis* (1 Accessions)
41.	*Amaranthus tenuifolius* (1 Accessions)
42.	*Amaranthus tricolor* (187 Accessions)
43.	*Amaranthus tuberculatus* (50 Accessions)
44.	*Amaranthus viridis* (18 Accessions)
45.	*Amaranthus wrightii* (2 Accessions)

A multilingual Multiscipt Amaranth name Database is given by Porcher Michel H. *et al.* (1995).

Germplasm characterization is being conducted at a number of locations around the world. Extensive, well-documented amaranth germplasm characterization has been conducted by organizations in India, Peru and Mexico (Joshi 1981, Sumar *et al.*, 1983, Espitia 1986).

Grain Amaranth Characters for Improvement

Dr. Subodh Jain has led an active program to identify genetic traits for the improvement of grain amaranth at the University of California-Davis (Jain *et al.*, 1986). Efforts have been made to determine the efficacy of different breeding strategies (Kulakow and Jain 1987, Ayiecho 1986). Additional techniques have been studied to determine how to manipulate plant height and increase grain yields (Vaidya and Jain 1987).

The ideal plant type of grain amaranth is as follows:

1. Reduced plant stature (1.0 to 1.5 m)
2. Flowering above the leaf canopy
3. Lack of branching
4. White or gold seed
5. 100 to 120 day maturity
6. Higher yield as a result of traits such as a large flower head with a high ratio of pistillate flowers that set seed.

However, a number of other important traits still need to be improved are seed size, tolerance to insects and diseases, seedling vigor, reduced shattering, standability (resistance to lodging), synchronous drydown of the plant and seedhead, ability to be easily threshed, high protein content and functional attributes of the grain.

Genetic investigations have been conducted in India to learn more about the traits which control yield, including harvest index, yield per plant and weight per 1000 seeds (Pandey 1984).

The first released line of amaranth named 'MT-3' was announced by the Montana Agricultural Experiment Station (Cramer 1988). It was selected from a segregating accession of *A. cruentus* L. originally collected in Mexico. Similar programs are presently in progress in Mexico and Kenya. The University of Cuzco in Peru has released three cultivars of *A. caudatus* L. which have been named 'Oscar Blanco, 'Noel Vietmeyer' and 'Alan Garcia'. These three cultivars are being grown commercially on hundreds of hectares by farmers in Peru (Sumar K., L. personal communication, 1988). A cultivar of *A. hypochondriacus* L., named `Annapurna' has been released in India (Joshi 1985).

The first released cultivars have helped to promote the commercialization of this crop. However, the gene pools of all of the cultivated amaranth landraces lack one or more desirable traits which are perceived as requirements for commercial production. For example, with adequate moisture and fertility, all of the improved varieties which mature and produce good seed yields of light-colored grain at northern latitudes reach a height of 2.5 m at maturity, too tall to harvest with modern grain combines.

Plant Breeding

The cultivated species of grain amaranth are monoecious. The basic techniques for making emasculations and pollinations have been described (Murray 1938) and refined (Kauffman 1981). Crossing requires agitating synchronously flowering plants (Webber and Kauffman, 1990; Lehman *et al.*, 1991). Genetic markers can be used to distinguish hybrids and male sterile genes helps to avoid emasculation.

The potential for the development of hybrids has been advanced by the identification of gene markers in the program at Davis, and other locations. Plant and seed pigment patterns, leaf characteristics, time of flowering, and determinate vs. indeterminate growth, have been identified (Kulakow *et al.*, 1985, Kulakow and Jain 1985 Kulakow 1987). An additional marker that has been observed in breeding trials at RRC is bract length. The perisperm and starch of the grain can be used as potential gene markers (Okuno and Sakaguchi 1982).

Markers can be used to identify and discard self-pollinated plants in the F1 generation. In addition, markers allow breeders of grain amaranth to avoid the process of emasculation when making crosses. The identification of three sources of cytoplasmic male sterility paves the way for additional hybridization techniques (Peters and Jain 1985, Peters and Jain 1987, Gudu and Gupta 1988).

There have also been some unique investigations to identify the mode of inheritance of traits that affect the nutritional characteristics of the grain. The identification of factors related to the unique starch characteristics of the grain. (Okuno 1985, Konishi *et al.*, 1985). In India, the mode of inheritance of grain protein percentages has been studied (Pandey and Pal 1985).

A series of advanced breeding lines (F6-generation, and higher) were developed at RRC are being distributed "K-series." One of the lines, `K-343', is the progeny of a cross of a white seeded *A. hypochondriacus* L. from Mexico and a black seeded *A. hybridus* L. from Pakistan. It has an unbranched short plant habit at high plant densities, and slightly earlier maturity, as compared to `MT-3' and other selected Mexican landraces which have been grown by farmers.

The future of amaranth is dependent on the careful recombinant selection of genotypes from the germplasm collections. We have had to use unusual sources of germplasm to find traits such as early maturity and shorter plant stature. The ideotype required for mechanized production in the USA has a number of characteristics in common with the landraces from Mexico. Dark-seeded *A. cruentus* accessions ("African" grain type) from West Africa is a source for early maturity while short plant stature was found in *A. hybridus* L. ("Prima" grain type) accessions from Asia.

Very few successful crosses have been made between Amaranthus species and the three cultivated grain amaranths (Khoshoo and Pal 1972). Interspecific hybridization barriers exists between crosses of *A. cruentus L., A. hypochondriacus,* L. and *A. hybridus* L. However, the cultivated grain species of amaranth have crossing compatibility with some species (Brenner *et al.*, 2000). Both inter- and intraspecific hybridization within the grain amaranths are potentially useful improvement methods (Lehmann and Clark 1990). Crossing between *A. cruentus* and the other two grain species is usually unsuccessful.

To assist in the development of selected progeny from breeding programs, RRC has distributed segregating material from F_2 and F_3 generations of our breeding progeny. The segregating material is grown at the location where the advanced lines will be grown to allow for selection pressure at that location. This technique is currently in progress in Peru, Thailand, China and Kenya, where new progeny are being selected.

To date, the most inclusive evaluations of segregating progeny of interspecific crosses have been conducted in Peru. Most of the first progeny of interspecific crosses observed in Peru did not perform well. However, a few accessions of *A. caudatus* L. crossed with two accessions of the Aztec grain type of *A. hypochondriacus* L. produced progeny that included segregates which were shorter and earlier maturing than the commonly available varieties of *A. caudatus* L. that are presently being grown in Peru. The performance of the segregating progeny and the selections that have been made will provide important clues to help select the most appropriate sources of genetic material to use in future breeding programs in Peru. Similar approaches will be useful at other locations.

Amaranth stands to benefit from some of the new techniques which are being developed in molecular genetics. Callus formation has been documented (Flores *et al.*, 1982) which will allow in-vitro cultivation for advancement of breeding lines. The gene expression of certain carboxylases is also undergoing continuing studies (Berry *et al.*, 1986). In the future hybrids might be created easier by new techniques such as protoplast fusion (Mastuti *et al.*, 1997).

Conclusion

Amaranth with its great diversity in germplasm resources and uses stand out as the future crop for poverty alleviation and nutritional security of humankind. Only modest research has been carried out so far which needs to be strengthened with the leadership of RRC and American Amaranth Institute (AAI) with a networking of the institutions and scientists working in this area.

Biodiversity in Horticultural Crops Vol. 2 *Pages* ***99–109***
Editor: **Professor K.V. Peter**
Published by: **DAYA PUBLISHING HOUSE**

Chapter 5

Drumstick

B. Varalakshmi
Division of Vegetable Crops, Indian Institute of Horticultural Research, Hessaraghatta Lake Post, Bangalore – 560 089

Drumstick or Horseradish tree is a vitamin-rich, mineral-packed and nutritious vegetable indigenous to North West India. It grows wild in the sub-Himalayan tract from river 'Chinab' eastwards to 'Sarda' and in the *tarai* tract of Uttar Pradesh in India. Stretches of wild forms of the tree can be seen in the *tarai* belts of Uttarakhand and Bihar. Drumstick is cultivated in rural areas throughout the Indian Subcontinent as well as much of Southeast Asia. Although widespread in Africa, drumstick is seldom cultivated intensively in Africa. There are about 13 species of *Moringa* within the monogeneric family, Moringaceace. *Moringa oleifera* (the drumstick tree) is the only species, which has been accorded much research and development.

The highly valued fruits and seed products of *M. oleifera* facilitated its rapid dispersal from its native sub-Himalayan tracts of northern India to many parts of the tropical world over the last century. The tree is widely distributed in Egypt, Sri Lanka, Thailand, Malaysia, Burma, Pakistan, Singapore, Cuba, Jamaica, Nigeria, Kenya and the Philippines. In addition it was introduced to the West Indies and tropical America (from Mexico to Peru, Paraguay, and Brazil).

Nutritional Composition

Drumstick is a nutritious tree; leaves contain high amounts of vitamin A (four times more than carrots), vitamin C (seven times more than oranges), protein (5-10 per cent), calcium (four times more than milk) and potassium (three times more than bananas). The young pods, seeds and flowers (which taste like mushrooms) are edible. Per 100 g, the pod contains 86.9 g moisture, 2.5 g protein, 0.1 g fat, 8.5 g total carbohydrate, 4.8 g fiber, 2.0 g ash, 30 mg Ca, 110 mg P, 5.3 mg Fe, 184 IU vitamin A, 0.2 mg niacin, and 120 mg ascorbic acid, 310 mg Cu, 1.8 mg I. Leaves contain 7.5 g H_2O, 6.7 g protein, 1.7 g fat, 14.3 g total carbohydrate, 0.9 g fiber, 2.3 g ash, 440 mg Ca, 70 mg P, 7 mg Fe, 110 mg Cu, 5.1 mg

I, 11,300 IU vitamin A, 120 mg vitamin. B, 0.8 mg nicotinic acid, 220 mg ascorbic acid, and 7.4 mg tocopherol/100 g.

Uses

Moringa leaf powder is being used in sauces. The seed is often used to purify dirty or cloudy drinking water. Drumstick seed oil is often used for cooking and in rare cases used for even lubrication purposes. The trunk is gaining importance as a raw material for paper making. Young seedlings are pulled up, boiled, and eaten whole. The flowers (which occur year round in some places but are more often seasonal) are cooked as a vegetable and are sometimes steeped in boiling water to yield a fragrant tea. In Oaxaca, Mexico, poor people have adopted the tree solely as a source of white flowers for decorating churches and houses on religious festival days. Livestock relishes the foliage so much that in some regions moringa is an important fodder. In India, for example, water buffaloes are fed the chopped up leaves and branches, which boost milk production. The root bark, for instance, contains two alkaloids–moringine and spirochine–that act on the nervous system. The roots contain pterygospermin, which inhibits growth of gram-positive and gram-negative bacteria.

Climate

It is well adapted to hot, humid conditions, at least in certain areas, it has thrived under annual rainfall exceeding 3,000 mm. Moringa is especially adapted to dryness and resists several months of drought, probably because its swollen roots store water. It has survived, for example, blistering heat, desiccating drought, and depleted dunes in Sudan and the Sahel. Originally, moringa was recommended for planting only at altitudes below 600 m. However, the discovery of healthy stands at elevations of 1,200 m in Mexico and over 2,000 m in Zimbabwe demonstrated that it is much more adaptable than supposed. It withstands light frost but is seriously harmed if temperatures dip 5°C below freezing for even one evening.

Systematics

Drumstick is a medium-sized tree that attains about 10 m height. It has a straight trunk (10-30cm thick) with bark that is whitish or gray, corky, with longitudinal cracks. It also has a tuberous taproot, whose presence helps explain the species' tolerance to drought conditions. Normally umbrella shaped trees have tri-pinnate leaves. Those leaves are densely crowded at the tops of the branchlets. Depending on climate, the foliage is evergreen or deciduous. In season the tree is enshrouded in creamy white, honey-scented flowers arranged in drooping panicles of 10-30 cm long. Flowers are insect pollinated and require a large number of insect visitations; carpenter bees are the most common visitors. Flowers and fruits (pods) occur twice a year; in many places flowering and fruiting occur year-round. The fruits are initially light green, slim, and tender, eventually turning dark green and firm. Depending on genotype, they are up to 120 cm long. While most are straight a few are wavy and some curly. In cross-section most are rectangular but a number are triangular and some are round. Fully mature, the dried seeds are surrounded by a lightly wooded shell with three papery wings.

Biodiversity in Moringa

Drumstick belongs to family Moringaceae, a monogeneric family. Botanically it is called *Moringa oleifera* Lam. The genus *Moringa* has 13 species mostly spread over South East Asia and Africa. They are; *Moringa drouhardii* (Madagascar), *Moringa concanensis* (mostly India), *Moringa arborea* (northeastern Kenya), *Moringa hildebrandtii* (Madagascar), *Moringa oleifera* (India), *Moringa borziana* (Kenya and Somalia), *Moringa ovalifolia* (Namibia and extreme southwestern Angola), *Moringa peregrina* (Horn of

Africa, Red Sea, Arabia), *Moringa longituba* (Kenya, Ethiopia, Somalia), *Moringa stenopetala* (Kenya and Ethiopia), *Moringa pygmaea* (Northern Somalia), *Moringa rivae* (Kenya and Ethiopia), *Moringa ruspoliana* (Kenya).

The above 13 species of *Moringa* can be fitted into three broad categories that reflect their life form and geography.

Bottle Trees

Massive trees with bloated water- storing trunks and small radially symmetrical flowers.

1. *M. drouhardii*; Madagascar
2. *M. hildebrandtii*; Madagascar
3. *M. ovalifolia*; Namibia and extreme SW Angola
4. *M. stenopetala*; Kenya and Ethiopia

Slender Trees

Trees with a tuberous juvenile stage and cream to pink slightly bilaterally symmetrical flowers

1. *M. concanensis*; mostly India
2. *M. oleifera*; India
3. *M. peregrina*; Red Sea, Arabia, Horn of Africa

Trees, Shrubs and Herbs of NE Africa

The eight *Moringa* species found in northeast Africa span the whole range of life form variation found in *Moringa*. All but *M. peregrina* are endemic to northeast Africa, that is, found nowhere else on earth. These species are tuberous adults or tuberous juveniles maturing to fleshy-rooted adults; having colorful, bilaterally symmetrical flowers.

1. *M. arborea*; NE Kenya
2. *M. borziana*; Kenya and Somalia
3. *M. longituba*; Kenya, Ethiopia, Somalia
4. *M. pygmaea*; N Somalia
5. *M. rivae*; Kenya and Ethiopia
6. *M. ruspoliana*; Kenya, Ethiopia, Somalia

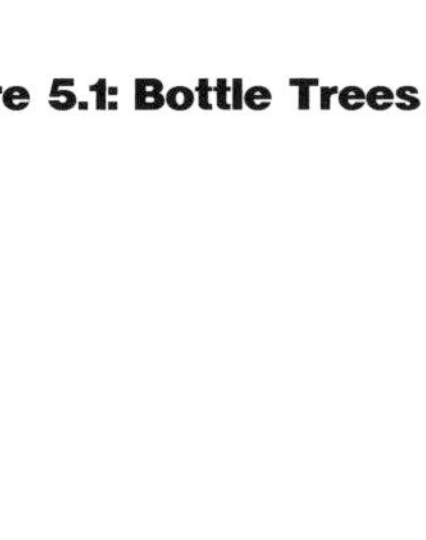

Figure 5.1: Bottle Trees

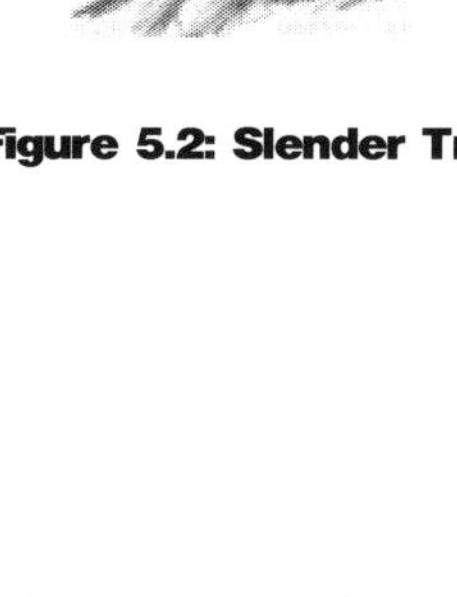

Figure 5.2: Slender Trees

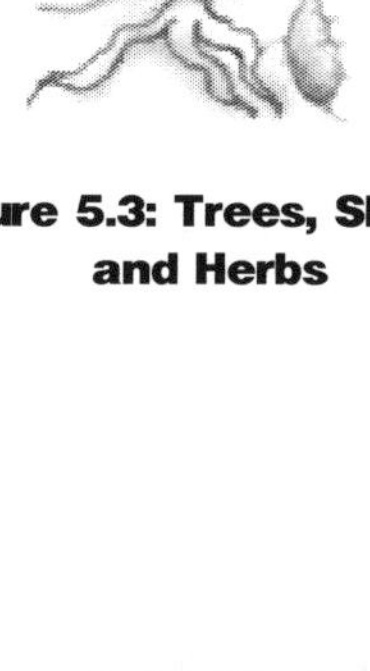

Figure 5.3: Trees, Shrubs, and Herbs

Geography of the Moringa Species

Arabia and India: 3 Slender Trees

The three species that occur in the Red Sea area, Arabia, and the Indian subcontinent are all slender trees. This group includes the family's best-known and most economically valuable species, *M. oleifera*. Now cultivated in all the countries of the tropics, *M. oleifera* is native to Sub-Himalayan India. In India, over 85 types of moringa are developed to reflect local tastes in pod size and shape. Many of these types are grown for one to three years, and then replaced with new trees on a regular rotation, almost as if they were biennials.

M. concanensis is mainly from India and Pakistan, barely reaching into Bangladesh. *M. peregrina* has the widest range of all, growing from the Dead Sea area sporadically along the Red Sea Coasts to Northern Somalia and around the Arabian Peninsula to the mouth of Persian Gulf.

Southern Hemisphere: 3 Bottle Trees

The three species restricted to the southern hemisphere are all bottle trees. Namibia and Angola in southwestern Africa are home to *M. ovalifolia*, while *M. drouhardii* and *M. hildebrandtii* are endemic to Madagascar.

Moringa Diversity is the Highest in the Horn of Africa

With nine species, the Horn of Africa is the center of *Moringa* diversity and includes a variety of life forms. *M. peregrina* is found in northern Somalia, Arabia, and the Red Sea coasts north to the Dead Sea. *M. stenopetala* is widely cultivated but only known from five localities in the wild. The densiest concentration of *Moringa* species is in Mandera District, in the extreme northeast of Kenya, where *M. arborea, M. longituba, M. rivae*, and *M. ruspoliana* can be found, though the species do not grow intermingled with one another. The other two species, *M. borziana* and *M. pygmaea* are distributed in Kenya and North Somalia.

Detailed description of the 13 Moringa species is given below:

Moringa oleifera Lam. (Figures 5.4–5.6)

Native to India, Arabia, and possibly Africa and the East Indies; widely cultivated and naturalized in tropical Africa, tropical America, Sri Lanka, India, Mexico, Malabar, Malaysia and the Philippine Islands. Moringa oleifera thrives in subtropical and tropical climates, flowering and fruiting freely and continuously. It grows the best on a dry sandy soil and is drought resistant.

This species is one of the world's most useful plants. Though apparently native only to restricted areas in the southern foothills of the Himalayas, *M. oleifera* is cultivated in all the countries of the tropics. *M. oleifera* is cultivated for its leaves, fruits, and roots for a variety of food and medicinal purposes. The young fruits called "drumsticks" can be cooked in a number of different ways. An excellent oil is derived from the seeds, which is used for cooking and lubrication of delicate mechanisms. The leaves are extensively used as a vegetable in many parts of the world, and the root can be made into a condiment similar to horseradish. *M. oleifera* is also of interest because of its production of compounds with antibiotic activity such as the glucosinolate 4 alpha-L-rhamnosyloxy benzyl isothiocyanate. *M. oleifera* seeds and fruits are used in water purification. It is commonly and incorrectly known under the names *M. aptera* and *M. pterygosperma.*

Figure 5.4: Figure Leaf

Figure 5.5: *Moringa oleifera*

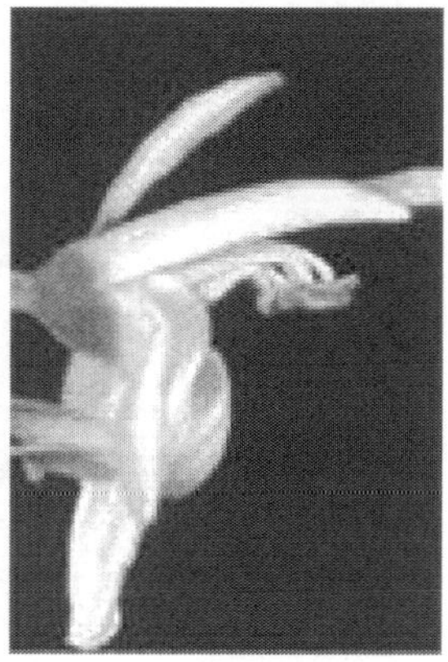

Figure 5.6: Flower

This crop has not been systematically improved, but several named cultivars with individual qualities are known in India. One called 'Bombay' has curly fruits and is considered among the best. Another, 'Jaffna,' is noted for having fruits of 60-90 cm long. A third, 'Chavakacheri murunga,' is known for its exceptionally elongated pods (90-120 cm long). Finally, there are the so- called baramassi varieties, which tend to flower continuously because they have been selected through the ages to provide buds, flowers, and fruits for food throughout the year. There is also a dwarf variety known as 'PKM1,' bred for short stems that make it easy to harvest the pods, which are exceptionally long. Some Indian varieties show exceptional promise at the International Institute of Tropical Agriculture in Nigeria.

Moringa concanensis Nimmo (Figures 5.7–5.9)

Moringa concanensis has a very strong central trunk that is covered with an extremely distinctive layer of very furrowed bark that can be more than 15cm thick. The flowers also have distinctive green

Figure 5.7: *Moringa concanensis*

Figure 5.8: Flower

Figure 5.9: Leaf

patches at the tips of the petals and sepals. *M. concanensis* occurs in tropical dry forest from southeastern Pakistan almost to the southern tip of India. It has recently been found in a western Bangladesh. Despite this wide range, *M. concanensis* habitat is under heavy pressure from agriculture and urbanization.

The medicinal properties of *M. concanensis* and the commonly cultivated *M. oleifera* (drumstick) are similar. Root and root bark of *M. olifera* are useful for the treatment of paralysis, intermittent fever, abscess, epilepsy, rheumatism, fainting and giddiness. Stem bark is abortifacient and fruits are used for curing liver and spleen diseases, joint pains and paralysis. Flowers are used as aphrodisiac. Seed oil is used for treatment of rheumatism. Gum is used for treating dental and ear problems.

Moringa peregrina Forssk. ex Fiori (Figures 5.10–5.12)

This native of Red Sea region is the strangest member of a strange group. When *Moringa peregrina* seedlings start out, they have broad leaflets and a large tuber. Through many dry seasons, the shoot dies back below ground to the tuber. As the plant gets older, the leaves get longer and longer, but the leaflets get smaller and smaller and more widely spaced. Adult trees produce leaves with a full complement of tiny leaflets, only to drop them as the leaf matures. However, the naked leaf axis remains, giving the tree a wispy look similar to *Tamarix* or *Cercidium microphyllum*. The pink zygomorphic flowers are sweetly scented and contrast with the blue leaves.

It is thought that *M. peregrina* oil may have been one of the important oils of ancient times. In the southern Arabian Peninsula, the tubers of saplings are roasted and eaten.

Figure 5.10: *Moringa peregrina*

Figure 5.11: Seed

Figure 5.12: Leaf

Moringa drouhardii Jumelle (Figures 5.13–5.15)

Moringa drouhardii with its bloated bright white trunks is a conspicuous tree species in the southern Malagasy dry forest. It grows in scattered stands that can number hundreds of individuals, usually on limestone. *M. drouhardii* grows extremely fast, surpassing three meters in its first year. Like other moringas in other parts of the world, *Moringa drouhardii* is used medicinally. Locals regularly remove the very strongly scented bark and wood for treatment of colds and coughs.

Moringa hildebrandtii Engler (Figures 5.16–5.17)

Moringa hildebrandtii is a beautiful tree with a massive, water-storing trunk that grows to 20 meters tall. This bloated trunk makes *Moringa hildebrandtii* strongly resemble the more well-known but unrelated boabab trees (*Adansonia*) that also occur in Madagascar. The pinnately compound leaves

Figure 5.13: ***Moringa drouhardii***

Figure 5.14: Flower Cluster

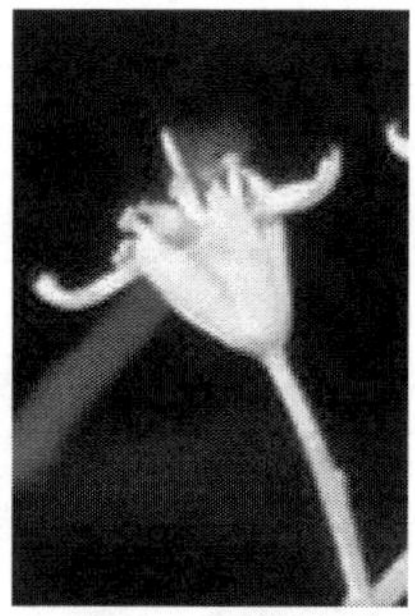

Figure 5.15: Flower

can be up to a meter long, and the leaf rachis and stem tip of young plants are often a distinctive deep red. The small whitish flowers are borne in large sprays.

The first collection of *Moringa hildebrandtii* by western botanists was made in 1880, by Hildebrandt which grew in the town of Trabonjy in northwestern Madagascar. It is frequently planted in villages all along the west coast of the island, and seeds abundantly. The trees are planted as ornamentals, for medicine, and to mark special occasions.

Moringa ovalifolia Dinter ex Berger (Figures 5.18–5.19)

After *M. oleifera*, this is perhaps the most familiar species of *Moringa*, growing in such well-visited Namibian parks as Etosha and Namib-Naukluft. Its bloated white trunks standing out on otherwise bare hillsides earned it the name "ghost tree". The species is from central Namibia to southwestern Angola, usually on very rocky ground. At the Sproukieswoud ("Fairy Forest") area of Etosha park, however, there is a rare stand of *M. ovalifolia* on soil. Along with other bottle trees in the genus and several others, the young leaves of *M. ovalifolia* are palmately compound.

Moringa stenopetala (Baker f.) Cufodontis (Figures 5.20–5.21)

Most research on applied uses of *Moringa* other than *M. oleifera* focuses on this species. It is an important food plant in southwestern Ethiopia, where it is cultivated as a crop plant. There are readily accessible populations of this species in the island of Lake Baringo, in the Rift Valley in Kenya. However, the distribution of the species through its range is very poorly known. After Lake Baringo, it

Figure 5.16: ***Moringa hildebrandtii***

Figure 5.17: Leaf

Figure 5.18: ***Moringa ovalifolia***

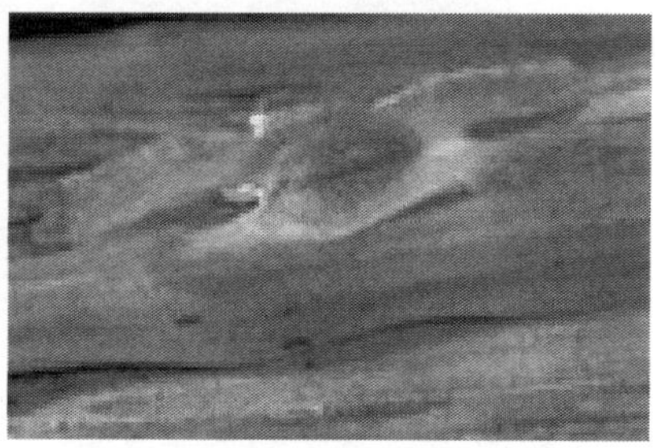

Figure 5.19:Fruits and Seed

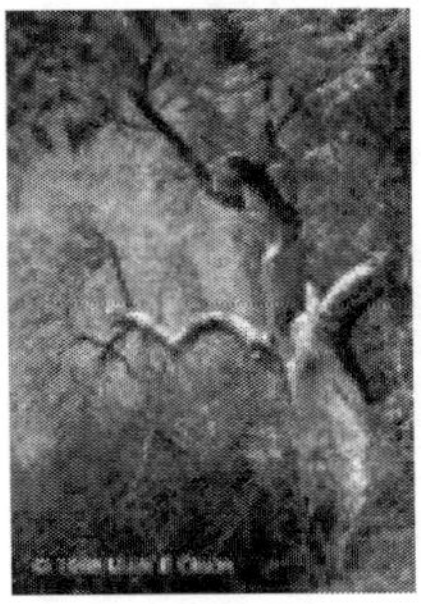

Figure 5.20: ***Moringa stenopetala***

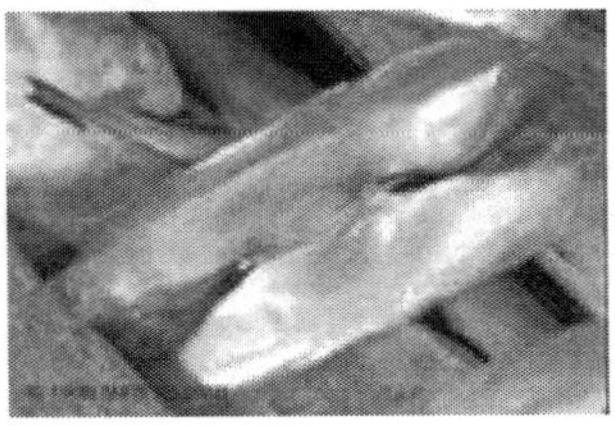

Figure 5.21: Seed

Figure 5.22: ***Moringa arborea***

Figure 5.23: Leaf

is known from only four other localities, all around Lake Turkana, a huge lake that reaching hundreds of miles into Kenya from the Ethiopian border. All moringas have nectaries at the bases of their leaves and leaflets. M. stenopetala quickly produces a large gray trunk and leaves covered with glistening nectaries.

Moringa arborea Verdcourt (Figures 5.22–5.23)

This species was discovered by Allan Radcliffe-Smith and Peter Bally in 1972. They found a single tree growing in a rocky canyon in northeastern Kenya near the Ethiopian border. *Moringa arborea* is a very beautiful tree, especially when covered with its large sprays of pale pink and wine red flowers. The young fruits resemble a yard-long string bean. Locals use the tree, like other species of *Moringa*, for medicine, especially the roots, which are thick and fleshy and pungent smelling.

Moringa borziana Mattei (Figures 5.24–5.26)

This small *Moringa* occurs from southern Kenya to the region of Kisimayu in southern Somalia. It usually only bears one or two stems, which typically reach about waist level. These shoots seem to die

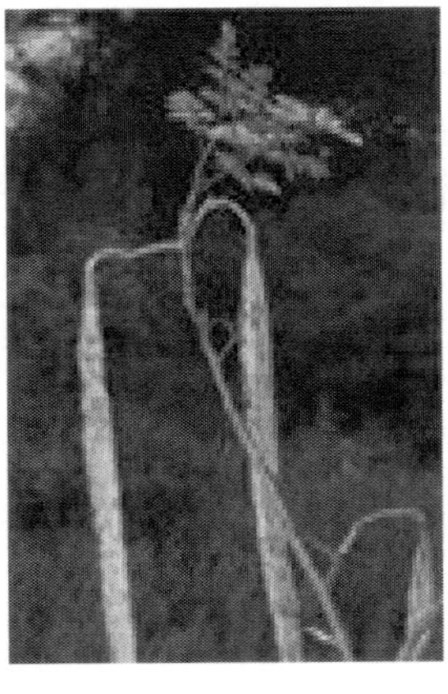

Figure 5.24: ***Moringa borziana***

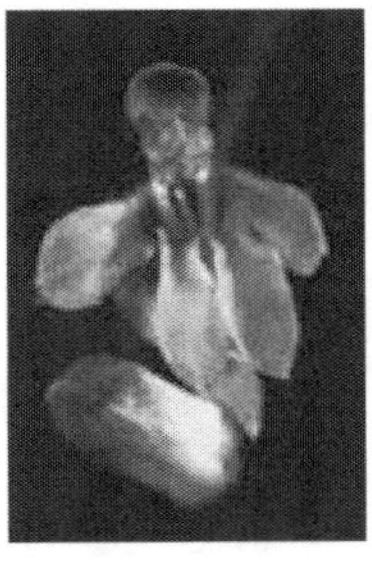

Figure 5.25: Flower

Figure 5.26: Leaf

back to the tuber every a few years, but occasionally the plant grows into a small tree. *M. borziana* will make a shoot 3 meters high in the course of 2 months. The tuber is often very large and can reach more than one and a half meters below the surface. The flowers of *M. borziana* are greenish cream to yellow with brown smudges on the petal tips, and *Moringa borziana* have an almost sickeningly sweet odor.

Moringa longituba Engler (Figures 5.27–5.30)

It is impossible to mistake this species for any other *Moringa*. No other species has bright red flowers or the petal and sepal bases fused to form a long, tubular hypanthium. It usually has a large tuber deep underground with one small shoot reaching knee-high above the soil. If an individual is well-sheltered under a tree, it can scramble through the branches for over three meters. At the beginning of this century, the botanist Engler placed this species in its own subgenus *Dysmoringa*.

Moringa longituba grows in northeastern Kenya, southeastern Ethiopia, and much of Somalia. Like other *Moringa* species in the Horn of Africa, *M. longituba* is used medicinally, particularly for treating intestinal disorders of camels and goats, for which the root is given internally.

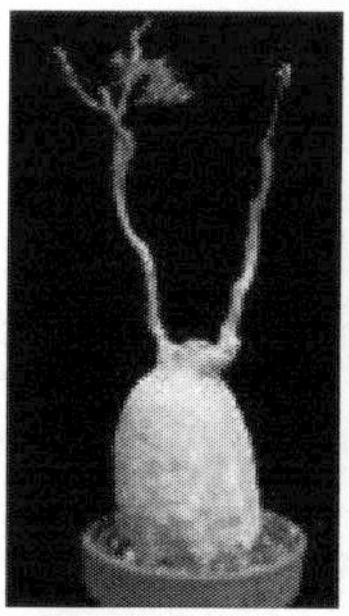

Figure 5.27: *Moringa longituba*

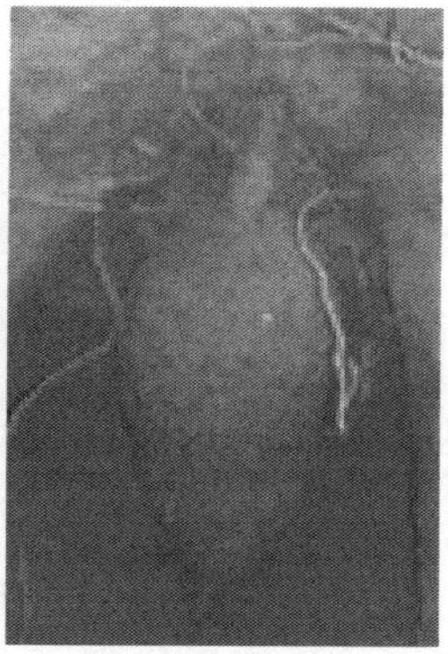

Figure 5.28: Tuber

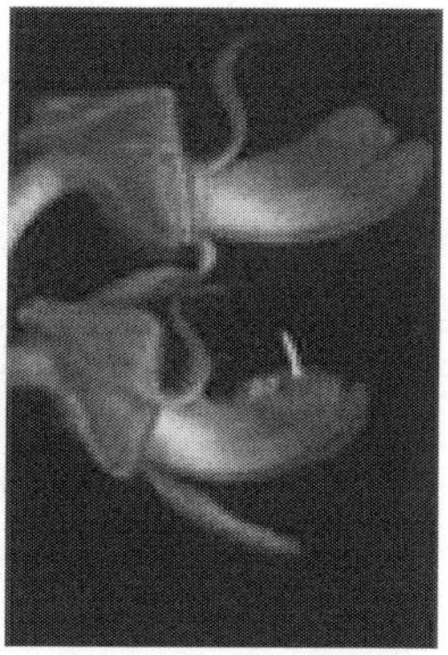

Figure 5.29: Flower

Figure 5.30: Inflorescence

Moringa pygmaea Verdcourt (Figures 5.31–5.32)

Moringa pygmaea is from northern Somalia. It is a delicate tuberous shrub or herb with very tiny leaflets on its tripinnate leaves. Its flowers are yellow in color.

Moringa rivae Chiovenda (Figures 5.33–5.35)

This is the most taxonomically frustrating taxon in the family. It is native from southern Lake Turkana to Mandera District in Kenya and throughout southeastern Ethiopia. Subspecies *rivae* is recognized as having creamy petals with brownish tips and short fruits. Subspecies *longisiliqua* has yellow flowers and very long fruits. These are large shrubs to 3 meters tall with

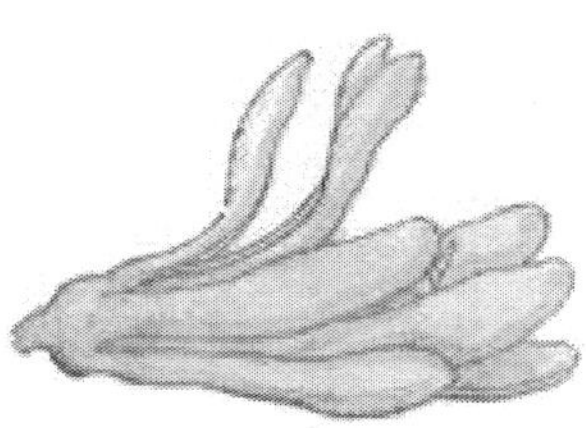

Figure 5.31: Flower

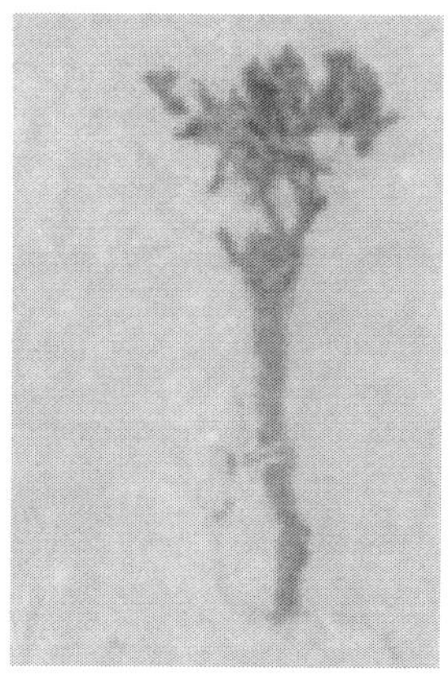

Figure 5.32: *Moringa pygmaea*

Figure 5.33: ***Moringa rivae***

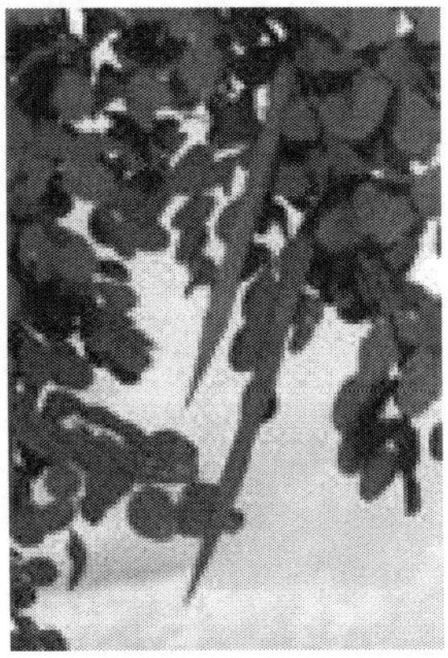

Figure 5.34: Fruits and Leaves

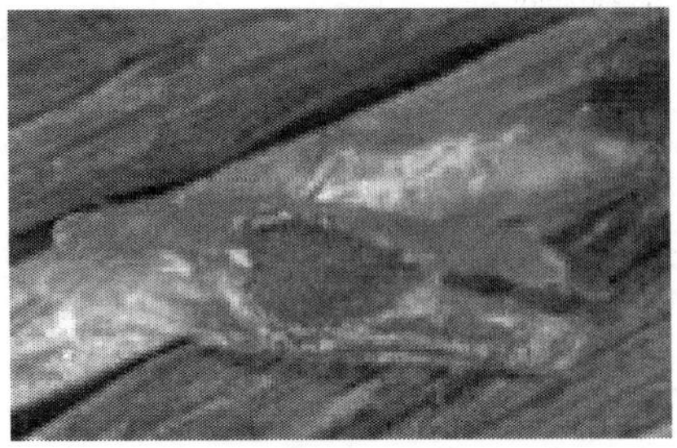

Figure 5.35: Seed

greatly swollen roots. Subspecies *longisiliqua* has an arborescent habit to over 6 meters tall. *Moringa rivae* is very similar to *M. arborea* and *M. borziana*.

Moringa ruspoliana Engler (Figures 5.36–5.37)

One of the most morphologically divergent species in the family, *M. ruspoliana* occurs from northern Somalia to southeastern Ethiopia. It is easily distinguished from all the other species in the family by having simply pinnate leaves. The leaflets are the largest in the family, reaching 15 cm in diameter, and are the thickest and toughest leaflets in the family. It also stands out for having the largest flowers in the family, reaching 3 cm long. The flowers are pink with green bases and when in flower. When young, it forms a thick taproot. As the plant ages, the root swells and becomes more globose. It begins to send out thick side roots, usually four. The adult plant is a small tree to 6 m tall with an octopus-like system of long, fleshy roots.

Figures 5.36: ***Moringa ruspoliana***

Figures 5.37: Heavily Pollarded Tree with Exposed Roots

Future Thrusts for the Research and Development of Moringa Species

1. Assessment of global biodiversity and conservation of the genetic resources
2. Breeding biology, genetic improvement and hybridisation studies
3. Building capacity at village or local level on processing of products to add value
4. Market surveys and intelligence to broaden market base as a way of increasing adoption of moringa technologies

Acknowledgements

The author wishes to thank Dr. M. Olson, Missouri Botanical Garden, USA for valuable photographs and for designing the MORINGA website, *Moringa* Home Page.

References

Odee, D.W., Muluvi, G.M., Machua, J., Olson, M.E. and Changwony, M. Domestication of M*oringa* species in Kenya. Department of Biochemistry, Kenyatta University, P.O. Box 43844, Nairobi, Kenya

Peter, K.V. 1998. *Drumstick*.(In) *Genetics and Breeding of Vegetables*. ICAR, Pusa, New Delhi-110 012. pp 193-198.

Ramachandran, C., Peter, K.V., Gopalakrishnana, P.K. 1980. Drumstick (*Moringa oleifera*): A multipurpose Indian vegetable. Econ. Bot. 34: 276-283.

Verma, S.C., Banerji, R., Misra, G., Nigam, S.K. 1976. Nutritional value of moringa. Current Sci. 45(21):769-770

Biodiversity in Horticultural Crops Vol. 2 *Pages* **110–138**
Editor: **Professor K.V. Peter**
Published by: **DAYA PUBLISHING HOUSE**

Chapter 6

Radishes (*Raphanus* spp.)

Amish K. Sureja[1], E. Sreenivasa Rao[2] and Anilabh D. Munshi[3]
[1]*Department of Vegetable Crops, Central Agricultural University, Pasighat, Arunachal Pradesh – 791 102*
[2]*Division of Vegetable Crops, IIHR, Hessarghatta, Bangalore – 560 089*
[3]*Division of Vegetable Science, IARI, New Delhi – 110 012*

Radish is a popular vegetable in both tropical and temperate regions of the world and is the best recognized for the crisp texture and pungent flavour of its enlarged storage roots and hypocotyls, cooked or salted as a vegetable, and also eaten fresh as grated radish, garnish and salad. Traditionally in the East Asian countries, and increasingly in the West, the sprouted seeds, young leaves and immature seed pods of radishes are eaten. The 'Japanese Bisai' type is particularly suited for sprouts and leaves (Crisp, 1995). Sprouted seedlings at the cotyledon stage are also used as garnish in many countries. In Japan, the young leaves and seedlings known as 'kaiware' are also eaten. The Indian 'rat-tail' or 'mougri' radish and German 'München beir' types, which are very similar to common radish, are grown for their long (about 30 cm) slender pods, eaten fresh, cooked or pickled, and as snack foods. Some types of mini-radish like 'Hatsuka-Daikon' which requires 20 days for harvest in Japan are also grown for salad in Europe and other countries (Kaneko and Matsuzawa, 1993).

The landraces of daikon, Japanese radish, *Raphanus sativus* L. Var. *hortensis* Becker, are used for several Japanese dishes like *Kayu* (rice gruel with vegetables or red beans), *Nitsuke* (boiled foods with soy source), *Sunomono* (vinegared foods), *Oroshi* (grated radish), *Takuan* (pickled radish), and *Sengiri* or *Kiriboshi* (radish strips dried for winter storage) (Yamaguchi and Okamoto, 1997). In Korea, Chinese cabbage and radish are the most widely used vegetables in making *kimchi*, a lactic acid fermentation product (Lee, 2003).

In India and Egypt, *R. sativus* var. *oleifer* is cultivated for vegetable oil and green manure (Iwasa, 1980). Radish is also grown successfully as a fodder crop in U.K. and South Africa. In South Africa, the giant radish of Japan is grown as a fodder crop. Fodder-radish bears little or no fleshy root and is grown for its foliage, used as fodder or green manure. It is gaining popularity in Western Europe due to its rapid growth.

Radish has refreshing, diuretic properties and is an appetizer. In Indian system of medicine, radish preparations are used in liver and gall-bladder troubles. In homeopathy, they are used for neuralgic headaches, sleeplessness and chronic diarrhoea. Roots, leaves, flowers and pods are effective against gram-positive bacteria. The roots are useful in urinary complaints, piles and in gastrodynia. Juice of fresh leaves is used as diuretic and laxative. Seeds are peptic, expectorant, diuretic and carminative (Kirtikar and Basu, 1935). Radish contains two important medicinal compounds– peroxidase and isothiocyanates. Isothiocyanates, a dietary glucosinolate (GL) have anticarcinogenic properties (Barillari *et al.*, 2002). 4-(Methylthio)-3-butenyl isothiocyanate is the principal antimutagen in daikon (Japanese white radish) (Nakamura *et al.*, 2001). Hyperlipidaemia, a condition associated with excess fat in the blood, is the main cause of coronary heart disease. Radish roots are a rich source of peroxidase, an oxidoreductase, which scavanges harmful free radicals (Curtis, 2003). Black radish (*Raphanus sativus* var. *nigra*) has antioxidant properties and the juice has beneficial health effect in the hyperlipidaemia.

Radish is an important vegetable crop in China, Japan, India, Korea, Egypt and many other countries of the world. In India, radish is cultivated in an area of 67345 ha. with annual production of 802529 t. having average productivity of 12.77 t/ha (Sidhu, 2002).

Origin, History, Evolution and Spread

Radish is one of the most ancient vegetables. Pliny recorded that radish was extensively cultivated in Egypt at the time of pharaohs and was also known to and highly prized by Greeks. The Greeks served them on dishes of gold in sacrificial offers to Apollo (Gill, 1993). Radish (probably *niger* type) was depicted on the inner walls of the pyramids in Egypt about 2000 B.C. (Banga, 1976). Herodotus (*c.* 484-424 BC) suggested that it was already an important food crop in Egypt nearly 5000 years ago *i.e.* about 2700 B.C. (Becker, 1962). It spread to China in about 500 B.C. and to Japan about 700 A.D. (Sirks, 1957). Radish was known in the eastern Mediterranean at least two millennia before it was known in China (Banga, 1976). In China and Korea there exist the oldest records on domestication, dated 400 and 100 BC, respectively (Kaneko and Matsuzawa, 1993). Radish was a well-established crop in eastern China over 2000 years ago, prior to the establishment of the 'Silk Road' to central Asia (Li, 1989) down which several occidental crops were introduced to China (Crisp, 1995). This crop was cultured in Europe in fifteenth or sixteenth century and was introduced to England, France and America in the beginning of 16th century. In 1806, eleven sorts were known in America (Gill, 1993).

Early work suggesting that variability of the crop is the greatest in Europe and the least in Japan (Sinskaia, 1928 and Sirks, 1957) now seems incorrect. There may be a much greater diversity of physiological and culinary types in Asia, particularly in China, than in Europe (Zhao, 1989), thus indicating east of Mediterranean, most probably China as the centre of origin of radishes. Some taxonomists consider that probably radishes originated as crop plants in both Asia and Europe, with subsequent gene flow between two major gene centres commencing with establishment of Silk Road (Crisp, 1995). A few taxonomists do not consider China to be the centre of origin of radish.

The area from eastern Mediterranean to Caspian sea is likely origin of European types, but Werth (1937) argued for a most easternly source. Indian group of radishes including rat-tail type originated in the Near Eastern and Indo-Burma centre of origin though their ancestors have been lost (Bailey and Bailey, 1956 and Mansfield, 1959). The rat-tail radish(*R. caudatus* L). is closely related to the Indian group of radishes but more tropical in distribution (Mansfeld, 1959).

The wild taxa of *Raphanus* occur through much of Europe and Asia, and are introduced weeds in America (Crisp, 1995). The greatest diversity of the important wild species of radish *R. raphanistrum* Linn., *R. maritimus* Smith, *R. landra* Morett. and *R. rostratus* DC., are found in between the Mediterranean and Caspian Seas, but 'wild radish' also occurs in China (Zhao, 1989) and Japan (Makino, in Pistrick, 1987). European wild radish, *R. raphanistrum* is widely distributed in Eurasia, Africa and North America, but is absent from China and India (De Candolle, 1883, Trouard-Riolle, 1914 and Hegi 1935). The *R. raphanistrum* L. is recently reported as a weed in wheat fields in Himachal Pradesh, India and is probably a recent introduction to the Western Himalayas from Europe (Sharma and Dhaliwal, 1997). *Raphanus raphanistrum* L. naturally occurs from Europe eastwards to the Volga and Mediterranean eastwards to the Caspian; *R. maritimus* Smith (sea radish) and *R. landra* Morett. (an inland form of *maritimus*) are distributed from Mediterranean eastwards to the Caspian and, in Europe, along the coasts of France, Belgium, Holland and Great Britain; and *R. rostratus* DC from Greece eastwards to the Caspian (Banga, 1976).

East Asian wild radish, *Raphanus sativus* L. var. *hortensis* f. *raphanistroides* Makino (=*R. raphanistroides* (Makino) Sinsk.) is a self-incompatible outcrossing species (Hinata and Nishio, 1980) and is called 'Hamadaikon' in Japanese. It grows primarily on dunes, sandy beaches, sandy cliffs or barren fields near the sea in East Asia, mainly China, Korea and Japan (Kitamura, 1958, Aoba, 1967, 1989, Lee *et al.*, 1992, Huh, 1995 and Yamane *et al.*, 2005). Natural populations of East Asian wild radish played important roles in establishment of Japanese landraces of Chinese radish (Kumazawa, 1961, Aoba, 1993), hence, are very important study materials in investigations on the origin of cultivated radish varieties and the genetic relationship between them (Kaneko and Matsuzawa, 1993) and from the viewpoint of crop evolution.

Owing to the long domestication history and diversity of cultivated radishes, origin of cultivated radish (*R. sativus*, including European radish, var. *sativus* and Chinese radish, var. *hortensis*) has not been clarified (Panetsos and Baker, 1967, Banga, 1976, Lewis-Jones *et al.*, 1982, Kaneko and Matsuzawa, 1993). The cultivated radish originated from a single wild species [*R. raphanistrum* (De Candolle, 1883 and Kobabe, 1959) or *R. maritimus* (Henslow, 1898, Schulz, 1919 and Thellung, 1927)], or through interspecific hybridization of two wild species [*R. landra* x *R. maritimus* (Schulz, 1919, Kitamura, 1958 and Clapham *et al.*, 1962)]. The Japanese types derived from East Asian wild radish *R. sativus* L. var. *hortensis* f. *raphanistroides* Makino (syn. 'Hama-Daikon', *R. raphanistroides* Sinsk.) (Kumazawa, 1956 and Yamagishi and Terachi, 1996). Another thearem is that *R. sativus* descended from a common ancestral species with *R. raphanistrum*, but that the ancestral species is now extinct. None of these thearems provide any decisive evidence, because they are mostly based on morphological observations and have no convincing genetic, cytological, or molecular data. Furthermore, the fact that all of the taxa of wild and cultivated species might be included in *R. sativus* (Pistrick, 1987) makes the study more complicated. Recently, Yamagishi and Terachi (2003) suggested that cultivated radish was of multiple origin as all of the mtDNA types present in wild species (*R. landra, R. maritimus,* and *R. raphanistrum*) were also present in modern cultivars of radish. AFLP markers variation showed a very close genetic relationship between *R. raphanistrum* and *R. sativus*, confirming the assumption that *R. raphanistrum* might be involved in the origin of *R. sativus* (Huh and Ohnishi, 2002). Yamane *et al.* (2005) reported that *R. raphanistrum* has a unique chloroplast haplotype not present in cultivated radish, hence was not the maternal ancestor of cultivated radish. Cultivated Chinese radish (var. *hortensis*) originated from East Asian wild radish (Sinskaya, 1931 and Yamane *et al.*, 2005) through a long history of gene flow between the two (Yamane *et al.*, 2005). Japanese cultivated radish varieties have a long history of more than a thousand years, and a large number of local varieties with various morphological characteristics are developed, thus making the phylogenetic relationships among them so complicated that the evolutionary process is difficult to analyze (Yamagishi and Sasaki, 2004).

From the available evidence, it is concluded that both Chinese and European centers of origin contributed immensely to evolution of present day cultivated radish.

The small European or table type radishes originated much later than the large types. A long white form of it appeared first in Europe at the end of 16th century, most probably from large types (Helm, 1957). White globular forms were developed in 18th century and later on red ones were also developed. In the earlier phases of the evolution of radish, there was great variation in root shape (long, half-long, globular, pear-shaped, even flattened) and colour (white, red, yellow and black) (Banga, 1976). Garden radish and three groups of north Chinese small radish, occidental small radish and black radish were native to the western region of the Central Asia or southwest China. The wild form grown in coastal areas of Japan and South China played a role in the origination of this group (Kaneko and Matsuzawa, 1993). North Chinese radish was differentiated in the secondary centre in North China, where the wild and cultivated forms from North China, Central Asia and the plains of Southwest China contributed in its origin (Kumazawa, 1965).

Origin of Japanese wild radish 'Hamadaikon' (*Raphanus sativus* forma *raphanistroides*) remains controversial. It might be a form that escaped from the cultivated common radishes (Makino, 1909), or it might be a descendant of the wild radish that came from Asian continent in ancient times (Aoba, 1989), or both (Kumazawa, 1956).

Taxonomy

The genus *Raphanus* is classified into two sections: *Raphanis* DC. and *Hesperidopsis* Boiss. The section *Raphanis* DC. consists of six species, *R. sativus, R. raphanistrum, R. rostratus, R. landra, R. maritimus* and *R. microcarpus* (Kitamura, 1958), which can cross reciprocally with each other (George, 1986). The section *Hesperidopsis* Boiss contains only one species, *R. aucheri* (Kitamura, 1958). Of the seven *Raphanus* species, only *R. sativus* is cultivated.

Banga (1976) grouped the commonly cultivated radishes (*R. sativus*) in the world into four types: small, cool season radish (*R. sativus* var. *radicula*); long types, adapted to wide range of temperature (*R. sativus* var. *niger*); rat-tail or mougri radish (*R. sativus* var. *mougri*; Syn. *R. sativus* var. *caudatus*) and fodder types, commonly grown in Africa (*R. sativus* var. *oleifera*). All these types intercross freely with each other and other related wild species. Both rat-tail radish and fodder radish do not produce fleshy root.

Radish cultivars can be ecologically classified into 5 main varieties: *R. sativus* var. *niger* (Mill.) Pers.; var. *radicula* DC.; var. *raphanistroides* Makino; var. *caudatus* L. and var. *oleifer* Netz (Hida, 1990). According to Rubatzky and Yamaguchi (1997) there are seven botanical varieties of *R. sativus*: var. *radicula* (Garden or European radish); var. *longipinnatus* (Daikon, Chinese winter radish, varied types); var. *raphanistroides* (Japanese winter radish); var. *niger* (Black or Spanish radish); var. *caudatus* (Rat-tail radish, grown for its long edible tender seed pods); var. *moughi* (fodder radish, no enlarged storage root) and var. *oleifera* (oilseed radish). A division of *sativus* into small-rooted (sometimes referred to as var. *radicula*) versus large-rooted types (including names like var. *nigra, niger, sinensis, acanthiformis* or *longipinnatus*) is too artificial to be useful as taxonomic character, although such a distinction may have phylogenetic relevance, and reflects commercial development (Crisp, 1995).

Pistrick (1987) divided cultivated forms into three groups: convar. *Oleifera*–the oilseed and fodder radishes; convar. *Caudatus*–the rat-tail radish (Syn var. *mougri*); and convar. *Sativus*–all forms with edible roots, which he divided into four geographic types–European, Chinese, Indian and Japanese. Recent study showed that all of the basic variation extant in the Japanese types is derived from Chinese gene base (Crisp, 1995).

A few authors group the Indian types of radish under *R. indicus* Sinsk, heat tolerant and pungent and are intermediate in position, with the European types and Japanese types forming the two extremes. The rat-tail radish *Raphanus sativus* var. *caudatus* is closely related to the Indian group of radishes but more tropical in distribution (Mansfield, 1959). The European types *R. sativus* L., are frost tolerant, rich in carotene, long day plants and strictly annuals. The Japanese types and the turnip-shaped giant radishes cultivated in the Sakurajima Island of Japan are derived from *R. sativus* var. *raphanistroides* which grows widely in coastal regions of Japan.

Cytology

Raphanus sativus and its wild relatives *Raphanus raphanistrum, R. landra, R. maritimus* and *R. rostratus* have 2n=2x=18, and are cross-fertile (Lewis-Jones *et al.,* 1982).

Cytological studies of interspecies hybrids between *R. sativus* and *B. rapa* and between *R. sativus* and *B. oleracea* demonstrated close chromosome homology as indicated by presence of up to six bivalents in the amphihaploid produced from each cross (Richharia, 1973). Recent work showed close affinities between *R. sativus* and *B. oleracea* genomes.

Botany

Radish is an annual or biennial herb depending on type and belongs to family Cruciferae. Leaves are alternate, lyrate-pinnatifid, of variable shape but most often oblong linear, and varies from 10 to 15 cm in some small-rooted cultivars to as long as 45 cm in large-rooted cultivars. The leaves are usually covered with stiff bristles and the growth habit is mostly rosette.

The edible portion of radish root develops from both primary root and the hypocotyls and varies greatly in size, shape, length and other external characters depending on cultivars. The shape varies from oblate to long tapering and the exterior colour may be white, black or various shades of scarlet.The flesh colour of most cultivars is white.

The inflorescence of radish is a terminal raceme. The flowers are small, white, rose or lilac with purple veins in bractless racemes and are self-incompatible. Sepals are erect and petals clawed. Fruit is indehiscent corky siliqua, rounded at base, with a long conical beak at apex, and smooth or rarely slightly constricted between seeds. Silique size differs among cultivar types, but is usually 3-7 cm long and up to 1.5 cm in diameter with 6-12 seeds. Pods of rat-tail radish are usually 20-30 cm in length. Seeds are globose or ovoid, about 3 mm in diameter and are larger than those of *Brassica* species. When matured, the seed is at first yellowish but turning reddish brown with age. About 100-120 seeds weigh 1g.

The principle difference used by taxonomists to separate the two species *R. sativus* and *R. raphanistrum* is pod morphology (de Candolle, 1886). In *R. sativus* the silique is continuous and forms a single cavity. *Raphanus raphanistrum,* in contrast, has siliques which are articulated with a single seed in each division. The seeds are dispersed when the divisions break apart. *Raphanus sativus* var. *hortensis* f. *raphanistroides* is herbaceous, diploid biennial, with 2n=18 (Kitamura and Murata, 1987). Three leaf types are found, of which the pinnatifid (PF) form is predominant, the lyrate (LY) form is less common and the pinnatisect form rare (Yamaguchi, 1987). The plant height is 30-60 cm with purple or whitish-pink flowers. Flowers are visited by insects and are cross-pollinated. Each silique contains 1-10 seeds and the matured siliques do not burst (Huh and Ohnishi, 2001).

Genetic Diversity

In radish, diversity exists with regard to fleshy root color, shape, and size, plant height, degree of division and size of leaves, flower colour, and fruit shape and size. The varieties also vary in other external characteristics like length, time of root formation and length of time they remain edible (Gill, 1993). Storage root length and width range from short to very long and slender to thick. Shapes may be spherical, cylindrical, tapering, or combination of these. The size may also vary from oblate to long tapering. Skin colours may be red, white, yellow, purple, black, green or various shades of scarlet. Flesh colour of almost all the cultivars is usually white. A red fleshed form is popular in China (Rubatzky and Yamaguchi, 1997). Several hundreds of land races or local cultivars in the daikon (Japanese radish, *R. sativus* L. var. *hortensis* Becker) are known in Japan (Furusato and Miyazawa, 1958) which show polymorphism for various traits like root colour (Yamaguchi and Okamoto, 1997). Although Japan is at the eastern end of distribution of cultivated radishes, the Japanese radish varieties display a large amount of morphological variations, not observed in radishes in other areas of the world (Kitamura, 1958). For the development of such variations, the genetic resources of radishes which respond to selection of Japanese local people should have been present originally (Yamagishi and Sasaki, 2004).

The large genetic variation in East Asian wild radish contributed to diversification of landraces of Chinese radish in Japan, not only through male sterile cytoplasm (Yamagishi and Terachi, 1996, Nakagawa *et al.*, 1998 and Yamagishi, 2001), but also by providing agronomically important characters like fragrance of roots and bolting resistance to landraces (Kumazawa, 1961, Aoba, 1993). These variations of the East Asian wild radish contribute to the provision of valuable genetic resources for future breeding of cultivated radish (Huh and Ohnishi, 2002).

In China, *Raphanus sativus* var. *longipinuatus* genotypes are cultivated in 7 regions and two climatic ecotypes, northern (cold-dry land) and southern (warm-moist climate) are recognized. The distribution of root colour indicates that the centre of genetic diversity is in the third cultivated region comprising Shangdong, Jiangsu, Anhui and Henan provinces. Vavilov inferred that the centre of origin of cultivated radish and Chinese radish were the Huanghuai plain and Shandong hilly areas (Zhou and Chen, 1991).

Recently, several studies on the genetic diversities in *Raphanus* were performed using molecular markers (Lewis-Jones *et al.*, 1982), allozyme variations (Huh and Ohnishi, 2001), RAPD analysis (Yamagishi *et al.*, 1998, Rabbani *et al.*, 1998), AFLP markers (Huh and Ohnishi, 2002, Kong *et al.*, 2005, Muminovic *et al.*, 2005) and ISSR markers (Muminovic *et al.*, 2005). Japanese and Korean populations of *R. sativus* var. *hortensis* f. *raphanistroides* exhibit a high level of genetic diversity because of insect pollination, gene flow from cultivated radish population, and a propensity for high fecundity (Huh and Ohnishi, 2001). The level of polymorphism exhibited by the various germplasm in the above studies could be exploited in genetic mapping populations to tag economically important traits.

Cultivar Groups

There are four broad cultivar groups of culinary radishes, small-rooted European radish, and large-rooted Asiatic radish or Chinese radish (daikon), leaf radish and rat-tailed radish.

Small-rooted or European Radish

The small, short season European types are exclusively grown in the field as garden salad crop throughout temperate regions, and can be produced under plastic or glass in Arctic regions. They are used as relishes, appetizers and to add variety and colour to green leaf salads. Intensive and early production under glass are done by specialist growers. The storage root sizes range 2 to 5 cm in diameter and length less than 15 cm. They differ in root shape–from highly elongated to flattened spheres, the most common being globe shape. Skin colours vary from red to white, or a combination of both. They are harvested within a short period that may range between 20 and 40 days. Most radish cultivars have a relatively short post-harvest life of 7-10 days.

Large-rooted or Asiatic Radish

The large-rooted radishes are mostly Asian and is an important vegetable in Japan, Korea, China, India and other eastern countries where it is eaten raw or cooked or preserved by storage, pickling, canning or drying (Banga, 1976). They are far more important and diverse than small European types and include types suited to temperate or tropical conditions. Root shapes vary from highly elongated to flattened spheres. Skin colours include white, red, green, yellow, purple, and black. Flesh can be white, red, purple or green. In Tibet and north-east China they are stored as intact roots as winter food and in Korea they are pickled with other vegetables in large underground vats. A few distinct winter storage types occur in Europe, but most have now virtually disappeared (Crisp, 1995). The large-rooted radish has a wider range of temperature adaptation, storage ability and reaction to various pests and diseases, which has not been fully exploited yet.

Chinese winter radish (daikon) is a major vegetable in Japan, China, Korea and many other Asian countries and is highly appreciated in Japanese diets. Daikon is eaten fresh, dried, pickled and can be sliced, diced and cooked,. They require a longer growth period to produce the large storage root and foliage biomass. Leaf blades are often strongly notched and may be erect, spreading, or prostate. Daikon root size ranges from 10 to more than 50 cm in length, 4 to 10 cm in diameter, and 1 and 4 kg in weight. Roots, as heavy as 20 kg (like *'Jaunpuri mooli'* or *'Newari'* of U.P, India), are also obtained. Edible quality diminishes with excessive size. Faster growth rate results in early pithiness. Under refrigerated storage conditions, many daikon cultivars can be stored for 3-4 weeks.

Cultivars or varieties of Chinese radish are classified into 4 ecological types, *i.e.* south Chinese, middle Chinese, north Chinese and western plateau Chinese. The south Chinese radish has low starch content and watery roots, hence not suitable for storage. They are suitable for salted vegetables and cooking. The north Chinese radish has a relatively smaller root with high starch content and a greenish or reddish skin colour, and can be consumed after storage (Kaneko and Matsuzawa, 1993).

The Japanese radishes are divided into 15 groups (Kumazawa, 1965), the major ones being 'Minowase', 'Nerima', 'Miyashige', 'Shogoin', 'Hatsuka' and 'Ninengo' (Kaneko and Matsuzawa, 1993). Japanese radish is derived from south Chinese radish and introduced to Japan from China. The north Chinese radish and Hatsuka radish were grown for salad at the beginning of The Meiji period (1868) (Aoba, 1979). The derivatives of north Chinese radish were domesticated predominantly in the restricted mountainous regions of central Japan such as Shinshu, Sanin, Hokuriku and Tohoku provinces, and now form some local varieties called 'Jidaikon' groups (Kaneko and Matsuzawa, 1993).

Leaf Radish or Fodder Radish

Fodder radish which bears little or no fleshy root and may be a derivative of an ancient oilseed form (*R. sativus* var. *oleiferous*), is grown in Southeast Asia, Africa and Europe for leafy fodder, and as green manure (Banga, 1976 and Crisp, 1995). Stewart and Moorhead (2004) reported a unique smooth-leaved, low-crowned, late-flowering fodder radish, 'Ceres Graza', originated from a complex series of crosses and selections, carried out over 17 years, involving vegetable garden radish (*Raphanus sativus*), cabbage (*Brassica oleracea*) and perennial seaside radish (*Raphanus maritimus*).

Rat-tail Radish or Mougri Radish

Rat-tail radish or Mougri-radish is grown in Southeast Asia specifically for its leaves and its young long immature seed pods, usually 30 cm but grows to 80-100 cm long.

Besides above, other cultivar groups include Spanish radish which, an annual winter radish type and produces round, usually black-skinned roots of 10 to 15 cm in diameter, maturing in 50-90 days after planting. Spanish radishes can be stored for 2-3 months under properly refrigerated condition.

Germplasm Exploration and Collection

In China, Yuan (1995) collected 156 radish cultivars in Anhui Province and divided into red, white and green types according to skin colour of the fleshy roots. The early cultivars were mostly white skinned whereas medium and late types were mostly red and green skinned. Du (1995) surveyed Hanzhong district, Shaanxi province and classified the varieties of radish into 4 groups autumn-winter types, winter-spring types, spring-summer types and summer-autumn types according to their growing season.

IPGRI (now Bioversity International) has recognized China, the Indian subcontinent and the Mediterranean as the first priority areas for collecting missions of radish germplasm (Toxopeus and van Sloten, 1981).

Germplasm Introduction

The Japanese radish variety 'Shiroaguri-Kyo' was introduced to India and released in the name 'Japanese White' by IARI Regional Station, Katrain. Chinese Pink is also an introduced variety, released by Dr. Y.S. Parmar University of Horticulture and Forestry, Nauni, Solan. Indian radish breeders utilized 'Japanese White' as one of the parents in hybridization programme and developed two varieties 'Pusa Himani' and ' Punjab Safed'. The varieties 'Pusa Chetki' and 'Arka Nishant' were developed by utilizing the introduced material from Denmark and Singapore, respectively. The cultivars Mino Early, Cherry Belle, Crimson French Breakfast and Chinese Pink were introduced to India.

Germplasm Evaluation

Sharma *et al.* (2002) evaluated 28 cultivars of radish for root yield and component traits in Palampur, Himachal Pradesh, India. Based on root weight, length, root diameter, girth and yield/plot, the superior Asian cultivars Mino Early White, Sutton's Long White and Nadauni, and European cultivar Palam Hirday were identified. Langthasa and Barah (2000) analysed performance of nine radish cultivars (Long White Icicle, Japanese White, Pusa Chetki, Pusa Rahsmi, Pusa Himani, Bombay Long White, Crimson French Breakfast, Punjab Selection and Bombay Red) and recommended cultivars Pusa Chetki and Punjab Selection as the most suitable for hill zone of Assam. Among radish cultivars Pusa Himani, Pusa Chetki, Japanese White, Pusa Reshmi, Bombay Red, Jaunpuri and Punjab Safed, the cultivar Bombay Red has almost consistent performance in both dry and rainy seasons in South Andaman (Dhar, 1998). The F_1 hybrid KSSC-03 has crimson-white highly pungent roots (Badiger *et al.*, 2001).

Germplasm Conservation

The accessions of cultivated and wild species of *Raphanus* are maintained both as base collection in the Genebanks (in long-term storage at–20⁰C) and as active collection (medium-term storage at 4⁰C) in various institutions in many countries worldwide. In India, the germplasm of Asiatic radishes are being maintained at NBPGR and various SAUs and research organizations (*viz.* Indian Institute of Vegetable Research, Varanasi) working for improvement in this crop. The genetic resources of temperate radishes are being maintained at IARI Regional Station, Katrain and Dr. Y. S. Parmar University of Horticulture and Forestry, Solan (Gill, 1993, Rana, 1993). NBPGR, New Delhi has 57 working collections of radish (Personal communication). The major institutions, holding the accessions, are listed below.

Table 6.1: Radish Germplasm Maintained at Major Centres in the World

Institutes	*Accessions of R. sativus and Rild Relatives*
Institute of Plant Genetics and Crop Plant Research (IPK), Gatersleben, Germany	836
Horticulture Research International (HRI), Wellesbourne, UK	714
N.I. Vavilov Institute of Plant Industry (VIR), St.Petersburg, Russia	626
USDA-ARS Germplasm Research Unit, NS Agricultural Experimental Station, Geneva, New York, USA	596
National Institute of Agrobiological Sciences, Tsukuba, Japan	418
Scottish Agricultural Science Agency (SASA), East Craigs, UK	250
Centre for Genetic Resources (CGN), The Netherlands	203
Centre de Recursos Fitogeneticos INIA Apdo., Alcalá de Henares, Spain	200
Institute of Plant Genetic Resources "K. Malkov" (IPGR), Sadovo, Plovdiv, Bulgaria	180
Nordic Gene Bank (NGB), Alnarp, Sweden	152
Research Institute of Vegetable Crops (RIVC), Skierniewice, Poland	87
Australian Temperate Field Crops Collection, Department of Primary Industries, Horsham, Vic 3401 Australia	56

Other major institutes where radish germplasm is held are:

1. Institute of Vegetable Crops and Flowers, CAAS, Beijing
2. National Duplication Gene Bank, Qinghai Academy of Agricultural Sciences, Xining
3. Pyongyang Crop Genetic Resources (PCGR), Institute of the Academy of Agricultural Sciences, Pyongyang, DPR Korea
4. National Institute of Agricultural Science and Technology (NIAST), RDA, Suwon, Republic of Korea
5. Aegean Agricultural Research Institute (AARI), Izmir, Turkey
6. National Research Council (CNR), Bari, Italy
7. Research Institute of Crop Production, Prague, Czech Republic
8. Universidad Politecnica, Madrid, Spain
9. Plant Breeding and Acclimatization Institute (IHAR), Blonie, Warsaw, Poland

Germplasm Utilization and Enhancement

Breeding Objectives

The breeding objectives in radish are as follows:

1. To develop high-yielding varieties capable of an increased marketable yield of white, long/stumpy, non pithy roots (longer standing period in field without becoming pithy) with thin tap root and non-branching habit.
2. To breed early-rooting short duration varieties fitting in sequence cropping system.
3. To breed varieties with characteristic cylindrical root of appropriate length for mechanical harvesting.
4. To develop varieties and F_1 hybrids having wider adaptability to different agro-climatic conditions.
5. To develop F_1 hybrids utilizing self-incompatible lines.
6. To evolve varieties tolerant to abiotic stresses, particularly tolerance to heat, low temperature, drought, excessive rain and saline, alkaline and acid soils.
7. To develop varieties having superior edible quality such as desired pungency, low nitrate content, less pore formation, etc. For varieties eaten fresh as salad increased sweetness and the root colour are important quality parameters.
8. To breed varieties and F_1 hybrids with multiple resistance to diseases and insect pests such as alternaria blight, white rust, yellows, clubroot, soft rot, radish mosaic virus and aphids.
9. To develop varieties with slow bolting and late-flowering habit.

The major effort in breeding of small-rooted European radishes is associated with physiological adaptations to adapt the crop to different growing seasons (George and Evans, 1981). Resistance to disease is less important perhaps because of very short duration of crop (Crisp, 1995). The major breeding objectives are early rooting under different conditions during the year, non-pithy roots, production of edible roots under low light regimes in winter and slow bolting under high temperatures in summer (Crisp, 1995). In contrast, much of the breeding effort in Asiatic radishes is detected towards resistance to diseases (*Fusarium, Albugo candida, Peronospora parasitica, Alternaria brassicae, Erwinia carotovora*, mosaic virus). Indeed, the Chinese and Japanese radishes are proved rich sources of resistance genes (Williams and Pound, 1963; Shimizu *et al.*, 1963, Hida and Ashizawa, 1985 and Bonnet and Blancard, 1987).

In Japan, autumn radish is the most abundantly cultivated. The characteristic cylindrical root of appropriate length is also required for mechanical harvesting. In winter radish, breeding objectives are cold-hardiness, low temperature plumpness, delayed bolting and less pore formation. Summer radish is mainly utilized for eating fresh as grated radish, and many varieties for this season are derived from 'Minowase' strains that have promising characteristics like heat tolerance, drought resistance and virus disease resistance. 'Minowase' is inferior in eating quality to other strains, and this undesirable trait is derived from 'Ninengo Daikon' (Kumazawa, 1965). It is a prerequisite to clear early bolting for spring radish production. 'Tokinashi Daikon' is used as a predominant parent in hybrid spring radish production. This strain is being replaced by 'Miyashige' and 'Harumaki Minowase", since 'Tokinashi' has also undesirable qualities derived from 'Ninengo Daikon' strain.

Conventional Breeding

At the Indian Agricultural Research Institute, New Delhi, high yielding hybrids of radish were obtained which yielded 30-60 per cent more than the better parent (Pal and Sikka, 1956). Kumar *et al.* (2002) made 56 crosses in radish, and identified the best F_1 hybrids *viz.* Mino Early White x Pusa Himani, Japanese White x DPR-1 (Palam Hirday), and Chinese Pink x Nadauni, manifesting maximum heterosis for root yield and its component traits. Kutty and Sirohi (2003) identified a superior hybrid Acc. No. 30205 x Acc. No. 192 for commercial exploitation. Lal and Srivastava (1975) crossed Kalianpur No. 2 with long root and small fruit and rat-tail radish (*R. sativus* var. *caudatus*) with small root and found that all characters except root thickness gave high estimates of heritability. High values for genetic advance were recorded for cortex thickness and fruit length. Improvement in root or fruit length can be achieved through selection breeding. Murali *et al.* (1998) reported that root yield had a significant positive correlation with root girth and root dry weight at the phenotypic and genotypic levels. A significant correlation of root weight was noted both at genotypic and phenotypic levels, with root length, leaves/plant, average leaf size, leaf weight/plant and plant weight, indicating selection on the basis of these characters will be effective (Danu and Lal, 1998). Kumar and Chandel (2003) observed higher SCA estimates than GCA for root yield, dry matter, ascorbic acid and nitrate content, revealing preponderance of non-additive gene action. The crosses Mino Early White x Pusa Himani and Japanese White x DPR-1 exhibited high heterosis over Japanese White and Chinese Pink with significant SCA effects. Sirohi and kutty (2000) reported overdominance for root weight, shoot weight, leaf length, leaf width and yield/plant, whereas dominance was observed for petiole length and root diameter.

Selection of cultivars suitable for early production led to small size, high growth rate and annual life cycle. Raising growth rate to promote earliness may increase sensitivity to pithiness. Many pithiness-resistant cultivars are developed, the first being A.R. Zawan's Cherry Belle, which spread all over the world. Cultivar Arka Nishant released from Indian Institute of Horticultural Research, Bangalore is also resistant to pithiness.

Radish breeding in Europe concentrated mainly on Red Globe, the Oval Red and White types, which are bred by means of mass selection or combined mass and pedigree selection methods.

Japanese radish now comprises many varieties bred by means of pedigree selection and natural or artificial hybridization made on the original base genotypes. The genetic variability of Japanese radish was brought about by the valuable introduction and breeding practices (Kaneko and Matsuzawa, 1993). Frost (1923) was the first to develop hybrid seeds in radish. Hybrid breeding exploiting the incompatibility system started in Japan and China in the early 1960s, and Japanese seed companies exported seeds of hybrid cultivars, notably the long white 'mouli' type, to Western growers (Crisp, 1995). A wide range of F_1 and other hybrid cultivars of the large-rooted types are now being bred using CMS in Japan and China. Moreover, the Ogura CMS was bred into other radish types including the European small-rooted radishes, and F_1 cultivars of these appeared (Crisp, 1995). This monoculture in radish varieties has essentially eradicated many traditional varieties which are reservoirs of useful genes.

Indian Cultivars

A large number of cultivars, indigenous and introduced, are cultivated in different regions of India. The indigenous radish cultivars are usually white with conical shape, attaining 25-40 cm in length and are more pungent than the introduced European types. Bombay Red, Jaunpuri *mooli* and Contai are local varieties. 'Baramasi' is another indigenous type that can be grown throughout the year. A type cultivated in and around Jaunpur in Uttar Pradesh, known as 'Newari' or 'Jaunpuri Giant', reaches an enormous size with a length up to 75-90 cm and a girth of 50-60 cm and weighs upto 5-15 kg or even more. 'Pusa Desi' is an improved indigenous type.

Important characteristics, breeding method, and pedigree of some of the recommended cultivars in India are given below.

CMS in Radish

Cytoplasmic male sterility (CMS) is a maternally-inherited trait, characterized by inability of a plant to produce functional pollen. Male-sterile cytoplasms are of interest because of their use in hybrid seed production and also as an invaluable material to study mitochondrial function and interaction between mitochondrial and nuclear genes. The CMS trait is mitochondrially encoded and always associated with mitochondrial DNA (mtDNA) rearrangements. The Ogura cytoplasm, first found in a Japanese radish (*Raphanus sativus* L.) by Ogura (1968) near the Kagoshima prefecture of Japan, is one of the most extensively studied male-sterile cytoplasms to date (Yamagishi and Terachi, 1996). In Ogura radish, abortion of male gametogenesis occurs at early microspore stage due to degeneration of the tapetal tissues (Ogura 1968). The mitochondrial genome of Ogura radish is highly rearranged relative to its normal (fertile) counterpart, and contains sequences not present in the normal radish genome (Makaroff and Palmer, 1988). An open reading frame, designated *orf138,* specific to Ogura cytoplasm (Bonhomme *et al.,* 1992; Krishnasamy and Makaroff, 1993), is the actual determinant of cytoplasmic male sterility (CMS) for Ogura male sterility (Grelon *et al.,* 1994 and Krishnasamy and Makaroff, 1994). The *orf138* gene is one of the most intensively studied "CMS genes" (Makaroff, 1995) and is found both in wild and cultivated radishes (Yamagishi and Terachi 1994a,b, 1996). The kos CMS isolated from a population of the Japanese radish cultivar Kosena (Ikegaya, 1986) is related to Ogura CMS (Iwabuchi *et al.,* 1999).

Several studies indicate the origin of Ogura-type male-sterile cytoplasm in Japanese wild radishes (Yamagishi and Terachi, 1994a,b, 1996, 2001). In contrast, Ogura-type cytoplasm is found in only a few cultivars of Japanese cultivated radishes and has *orf138* and its variant, *orf125* (Yamagishi and Terachi, 2001).

Recently, Nahm *et al.* (2005) characterized a new CMS, NWB CMS, obtained from a novel male sterile radish line, collected from South Korea. NWB CMS is unique and has more profound degree of sterility than any of the other CMS types, including Ogura CMS. It induced complete sterility in progenies, when crossed with any of the tested breeding lines. The CMS phenomenon of the NWB CMS line may be induced by cytoplasmic factor(s). NWB CMS differs from other CMS types with regard to its degree of sterility and mitochondrial genome arrangement. The Ogura CMS specific factor, *orf138* is absent from NWB CMS line. A specific 2 kbp DNA fragment confers male sterility in the NWB CMS line. The "WonBaek" CMS line from China is different from Ogura CMS. The *orf138* gene is not present in the WonBaek CMS. The crucial factors that determine male sterility for the Ogura and Kosena CMS, NWB CMS, and WonBaek CMS lines are different (Nahm *et al.,* 2005).

Hawlader *et al.* (1997) crossed Ogura radish, a cytoplasmically genetic male sterile line, with four local and three Japanese cultivars to identify maintainer lines. Out of seven F_1 families, one cross involving a local cultivar, Aushi, produced 100 per cent male sterile (MS) progeny. The crosses involving the other two local cultivars, Tangail Local and Kuni, produced about 90 per cent MS progeny indicating presence of maintainer gene (s) for male sterility. All three exotic cultivars (Japanese) possessed chromosomal gene (s) for controlling the male sterility. The successful transfer of Ogura CMS to three cultivars of radish (Japanese White, Pusa Chetki and Pusa Desi) is reported by

Table 6.2: Promising Cultivars of Radish from India and their Salient Features

Crop/ Variety	*Breeding Method (Pedigree/ Germplasm Used)*	*Year of Release/Year of Notification*	*Institute*	*Areas of Adaptation*	*Salient Features*
			Tropical/Asiatic types		
Japanese White	An introduction from Japan. Originally named as Shiroaguri-Kyo	1973	IARI Regional Station, Katrain	Jammu and Kashmir, Himachal Pradesh, Punjab, Uttaranchal, Uttar Pradesh, Bihar, Jharkhand, Madhya Pradesh, Maharashtra, Karnataka, Kerala, Tamil Nadu	Roots are thick, 20-30 cm long, 5 cm in diameter, plumpy, cylindrical, blunt at tip, mild pungent, smooth and white with green shoulder. Tops medium with deeply cut leaves. Flesh snow-white, crispy, solid and mildly pungent. Suitable for sowing between October to December in plains and July-September in hills. Yield 15-30 t/ha in 60-65 days of cropping duration.
Pusa Chetki	Selfing and massing from a collection from Denmark	1975	IARI, New Delhi	Jammu and Kashmir, Himachal Pradesh, Uttaranchal, West Bengal, Assam, Punjab, Uttar Pradesh, Bihar, Jharkhand, Chhattisgarh, Orissa, Andhra Pradesh, Madhya Pradesh, Maharashtra	Roots are white, smooth, soft in texture, less pungent, 15-20 cm long, thick and stumpy. Leaves are dark green, entire, slightly lobed and upright. Suitable for sowing in hotter months *i.e.* between middle March to middle of August in northern plains. Yield 25 t/ha in 45 days of crop duration.
Pusa Desi	Selection from local material	1965	IARI, New Delhi	Punjab, Uttar Pradesh, Bihar, Jharkhand, Rajasthan, Gujarat, Haryana, Delhi	Roots are white with green shoulder, 30-35 cm long, medium thick, tapering and pungent. Tops have dark green cut leaves. Suitable for sowing from mid August to September in northern plains. Yield 30 t/ha in 50-55 days of crop duration.
Pusa Reshmi	Selection from local material	1971	IARI, New Delhi	Jammu and Kashmir, Himachal Pradesh, Uttaranchal, Punjab, Uttar Pradesh, Bihar, Jharkhand, Rajasthan, Gujarat, Haryana, Delhi	Roots are white with green shoulder, 30-35 cm long, medium thick, uniformly smooth and mildly pungent. Tops medium and leaves light green. Tolerant to slightly higher temperature. Suitable for sowing from mid September to October in northern plains. Yield 35 t/ha in 55-60 days of crop duration.

Contd..

Table 6.2–Contd..

Crop/ Variety	Breeding Method (Pedigree/ Germplasm Used)	Year of Release/Year of Notification	Institute	Areas of Adaptation	Salient Features
Kashi Sweta (IIVR-1)	Selection from Pusa Chetki	2001	IIVR, Varanasi	Punjab, Uttar Pradesh, Bihar, Jharkhand	Roots are 25-30 cm long, 3.3-4.0 cm in diameter, straight, tapering with pointed tip. Yield 45 t/ha in 30-35 days of crop duration. Suitable for growing in spring, summer, autumn and winter seasons.
Kashi Ganga (IIVR-2)	Selection from local collection	2001	IIVR, Varanasi	Punjab, Uttar Pradesh, Bihar, Jharkhand	Foliage is soft and smooth. Roots are straight, 35-40 cm long, 3.5-4.2 cm in diameter. Suitable for planting from September to February. Harvesting within 40-50 days after sowing. Yield 45 t/ha. Due to late pithiness, it can stand the field up to 10-15 days after ready for harvesting.
Kalianpur No. 1	Selection from local material	1982		Punjab, Uttar Pradesh, Bihar, Jharkhand	Roots are 22-23 cm long, smooth, white, crispy, thick, tapering with green shoulder. Top is heavy, leaves and stem green, long, broad and less lobed. It is free from mustard sawfly, aphids and white rust.
Arka Nishant	Mass selection from IHR 72, a material from Singapore	1980	IIHR, Bangalore	Karnataka, Tamil Nadu, Kerala	Roots are long, marble-white, crisp in texture with mild pungency. Matures in 45-55 days. Roots are medium sized (25 cm x 3-4 cm), pleasant aroma. Sets seeds freely in plains. Resistant to pithiness, root branching and forking. Yield 20-30 t/ha.
CO 1	Selection from RS 44-2-1	1981	TNAU, Coimbatore	Karnataka, Tamil Nadu, Kerala	Plants vigourous with green leaves and wavy margins. Roots milky white, 23 cm long, cylindrical and less pungent. Sets seeds freely in plains. Yield 20 t/ha in a crop duration of 45 days. Susceptible to rust and aphids.

Contd..

Table 6.2–Contd..

Crop/ Variety	*Breeding Method (Pedigree/ Germplasm Used)*	*Year of Release/Year of Notification*	*Institute*	*Areas of Adaptation*	*Salient Features*
Punjab Safed	A derivative of the cross White 5 x Japanese White	2002	Punjab Agricultural University, Ludhiana	Punjab, Uttar Pradesh, Bihar, Jharkhand, Rajasthan, Gujarat, Haryana, Delhi	Foliage light green. Roots 33-40 cm long, 3-5 cm thick, smooth and white. Root formation in 50-60 days. Free of forking. A quick growing type with roots remain edible for 10 days after attaining full size.
Punjab Ageti	Selection from local material	1973	Punjab Agricultural University, Ludhiana	Punjab, Uttar Pradesh, Bihar, Jharkhand, Rajasthan, Gujarat, Haryana, Delhi	Roots are long, thin with lower half white and upper red. Average yield 30t/ha. Suitable for April to August sowing.
Punjab Pasand	Selection from Pusa Chetki	1998	Punjab Agricultural University, Ludhiana	Punjab, Uttar Pradesh, Bihar, Jharkhand	Roots are long, white and semi-stumped. Average yield 50 t/ha. Suitable for main season.
Chinese Pink	An introduction		Dr. Y.S. Parmar University of Horticulture and Forestry, Nauni, Solan	Jammu and Kashmir, Himachal Pradesh, Uttaranchal, Punjab, Uttar Pradesh, Bihar, Jharkhand	Roots are 12-15 cm long, pink with white towards tip and semi-stumpy to stumpy. Root formation in 50-55 days.
Nadauni	Selection from local material			Jammu and Kashmir, Himachal Pradesh, Uttaranchal	Very popular in Himachal Pradesh. Tops dark green with cut leaves. Roots long, tapering, smooth and light pink in colour.
IIHR-1-1			IIHR, Bangalore	Karnataka, Tamil Nadu, Kerala	Roots are thin, 30 cm long, avg wt 300g, mildly pungent. Root surface is smooth and shining white. Highly resistant to white rust and free from pre-mature bolting, pithiness, splitting and forking. It takes 45 days to reach edible maturity.

Contd..

Table 6.2–Contd..

Crop/ Variety	Breeding Method (Pedigree/ Germplasm Used)	Year of Release/Year of Notification	Institute	Areas of Adaptation	Salient Features
			Temperate Types		
Pusa Himani	Hybridization between Black and Japanese White	1971	IARI Regional Station, Katrain	Jammu and Kashmir, Himachal Pradesh, Uttaranchal, West Bengal, Assam, Punjab, Uttar Pradesh, Bihar, Jharkhand, Chhattisgarh, Orissa, Andhra Pradesh, Madhya Pradesh, Maharashtra	Roots are white, medium thick, 30-35 cm long, semi-stumpy, tapering with light green shoulders. Flesh is white, crisp, sweet flavoured with mild pungency. Tops short with cut semi-erect leaves. Suitable for sowing from December to February in northern plains and throughout the year except winter months of November-January in hills. Yield 32 t/ha in 60 days of crop duration. Tolerant to high temperature and humid conditions.
Rapid Red White Tipped	Introduction	1976	IARI Regional Station, Katrain	Jammu and Kashmir, Himachal Pradesh, Uttaranchal, Punjab, Uttar Pradesh, Bihar, Jharkhand	A table extra early type. Roots small, round or globular, upper portion red, lower portion white. Flesh white and sweet in taste. Suitable for sowing from mid October to January in northern plains. Yield 35-40 t/ha in 21-25 days of crop duration.
White Icicle	Introduction	1976	IARI Regional Station, Katrain	Jammu and Kashmir, Himachal Pradesh, Uttaranchal, Punjab, Uttar Pradesh, Bihar, Jharkhand	Roots pure white, thin, tender, 12-15 cm long, 2-3 cm in diameter, icicle shaped, solid, straight and tapered. Flesh is icy white, crispy, mildly pungent and flavoured. Plant top is medium short. Suitable for sowing from October to January in northern plains. Yield 45 t/ha in 30 days of crop duration.
Scarlet Globe	Introduction			Jammu and Kashmir, Himachal Pradesh, Uttaranchal	Foliage height 10-15 cm. Roots round, small, 2 cm in diameter and bight red. Flesh is crisp and white. Root formation in 25-30 days.
Scarlet Long	Introduction			Jammu and Kashmir, Himachal Pradesh, Uttaranchal	An early variety. Light green foliage, 15-20 cm high. Roots long, tapering to a point with red skin. Flesh is white and crisp.

Contd..

Table 6.2–Contd..

Crop/ Variety	*Breeding Method (Pedigree/ Germplasm Used)*	*Year of Release/Year of Notification*	*Institute*	*Areas of Adaptation*	*Salient Features*
Palam Hriday	Introduction followed by selection from an exotic source	2003	HPAU, Palampur	Jammu and Kashmir, Himachal Pradesh, Uttaranchal	Strap-like entire margin leaves. Roots are oblong having green shoulders and cream globe with 6-7 cm diameter and small top. Flesh is attractive pink, crisp, juicy. Late bolting and free from pithiness. Roots reach edible maturity in 50-60 days. Suitable for sowing from October to November in north Indian plains and September in mid hills, March-May and June-July in dry temperate areas. High TSS 7.7 per cent.

Sodhi *et al.* (1993). A CMS line of broccoli with radish cytoplasm, BC7-19, was successfully developed by backcross method (Zhu *et al.*, 2001).

Fertility Restoration

Many mitochondrial genes that determine CMS can be suppressed or counteracted by the products of one or more nuclear genes known as fertility restorer genes (*Rf*). CMS-*Rf* systems are used for commercial-scale F_1 hybrid seed production in many crops. Besides, the study of CMS and *Rf* genes help in increasing our understanding of nuclear-cytoplasmic interactions.

Early study on male sterility in Japanese radish suggested that sterility in Ogura CMS could be overcome by a single dominant restorer gene (Ogura, 1968). Nieuwhof (1990) proposed that up to two dominant and one recessive restorer genes, acting independently, were involved in determining the male sterility in Ogura CMS. Yamagishi and Terachi (1994b) reported two dominant restorer genes controlled the Ogura CMS in radish. Koizuka *et al.* (2000) demonstrated that two dominant restorer genes (*Rf1* and *Rf2*) were necessary for fertility restoration in kos CMS radish. The kosena CMS-*Rf* system is genetically the same as that of the Ogura CMS-*Rf* system (Koizuka *et al.*, 2000).

The fertility restorer genes for Ogura CMS are widely distributed in the *Raphanus raphanistrum* and Japanese wild radish (Hamadaikon). Among the cultivated radishes, the restorer gene is frequently found in the European and Chinese varieties, but not in Japanese varieties (Yamagishi, 1998). It is suggested that the Ogura CMS restorer gene(s) alters the expression of *orf138* at a translational or post-translational level and in an organ-specific manner (Krishnasamy and Makaroff, 1994 and Bellaoui *et al.*, 1999). In Kosena CMS radish, the fertility restorer gene *orf687* has recently been cloned by positional cloning (Koizuka *et al.*, 2003). *Orf687* functions either directly or indirectly to lower the levels of Kosena CMS-associated gene, *orf125*, thereby restoring the fertility (Koizuka *et al.*, 2003). Genomic structures of Kosena CMS-associated gene *orf125* and Ogura CMS-associated gene *orf138* are almost identical (Bonhomme *et al.*, 1992 and Iwabuchi *et al.*, 1999).

Various molecular markers linked to the *Rf* gene are explored. AFLP markers linked to the *Rf* gene are identified in radish (Murayama *et al.*, 2003), which can be further utilized for identification of QTLs using molecular maps. The molecular mechanism underlying the restoration of fertility still remains unclear. Currently there is great interest in the map-based cloning of the *Rf* gene.

Self-Incompatibility

Radish exhibits sporophytic self-incompatibility with multiple S-alleles located at a single locus. The papillae on the stigma surface form the barrier to pollen tube penetration. By disrupting the stigma surface or by applying pollen immediately to the conducting tissue of the style or bud pollination, it is possible to get normal seed set by selfing. Self-incompatibility is being employed by radish breeders to produce F_1 hybrids. In this context, identification of S-genotypes of plants before flowering assumes significance. Recently the PCR-RFLP method h proved useful for identification of Salleles in inbred lines and for listing S haplotypes in *R. sativus* (Niikura and Matsuura, 2001 and Lim *et al.*, 2002). Hawlader and Mian (1997) investigated self-incompatibility mechanism in 10 local radish lines in Bangladesh through seed set analysis and pollen tube growth behavior. Based on a low seed set ratio as an estimate for self-incompatibility, cultivars Tangail Local, Tasaki, Kuni, Aushi and Indian Aushi were identified as self-compatible. Nikornpun *et al.* (2004) bred F_1 hybrids exhibiting good horticultural characteristics (such as uniformity, dark green leaf without hair, cylindrical shape, smooth skin, firm root texture and 360 to 400 g root weight), utilizing self-incompatible inbred lines of Chinese radish (*R. sativus* var. *longipinnatus*).

Interspecific and Intergeneric Hybridization

When a trait of interest is not available in the primary or secondary pool, the breeders look to tertiary gene pools to transfer valuable genes controlling traits from different wild species into related crop species. Gene transfer from a tertiary gene pool can only occur if there is sufficient homoeology between chromosome segments in the donor and recipient genomes. Of late, with the advent of modern tissue culture techniques, wider crosses have become possible for interspecies gene transfer.

Karpechenko (1924) crossed *R. sativus* (R) with *B. oleracea* (C) to produce amphidiploid *Raphanobrassica* (RRCC). Since then interspecific or intergeneric crosses involving *Raphanus* are used for transfer of beet cyst nematode resistance (Lelivelt *et al.*, 1993), clubroot resistance (Hagimori *et al.*, 1992), restorer gene (Heyn, 1976), shattering resistance to rapeseed (Agnihotri *et al.*, 1990) and production of monosomic chromosome addition lines (Kaneko *et al.*, 2001). Now, forage varieties of *Raphanobrassica* (RRCC) are also developed for commercial production. Choudhary *et al.* (2000) synthesized Amphidiploid *Raphanofortii* by colchicinization of F_1 hybrid *Brassica tournefortii* (TT, 2n=20) x *Raphanus caudatus* (RR, 2n=18) which has great potential as a new commercial crop, and a bridge species for transfer of economically important attributes of both the species to other *Brassica* species.

Interspecific crosses or somatic cell fusions can be used to create alloplasmic plants, in which the cytoplasm of one plant species is present in combination with nuclear genome of a different species. These cytoplasmic substitutions often result in expression of mitochondrial genes determining CMS (Chase, 2006). CMS cybrids derived from protoplast fusion are useful tools for identification of CMS-related regions because of frequent recombinations among parental mtDNAs cause rearrangement of the mt genomes in the fusion products. CMS was transferred from *Raphanus sativus* cv. Kosena (CMS line) to *B. napus* rape cv. Westar through cytoplast-protoplast fusion (Sakai and Imamura, 1990). Motegi *et al.* (2003) obtained Ogura CMS plants by asymmetrical protoplast fusion between red cabbage (fertile) and normal radish (fertile).

Breeding for Disease Resistance

The promising germplasm identified for various diseases in Japanese radish, Chinese radish and Indian radish are presented below.

Table 6.3: Promising Germplasm for Various Traits in Japanese Radish

Traits	*Genotypes*
	Resistance/Tolerance to Biotic Stress/Disease
Mosaic virus	Minowase group: Bansei-Mino, Motohashi-Mino (resistant) Nerima group: Aki-wase, Takakura, Tosai, Nishimachi-riso (resistant) Awa-Bansei group: Choko-hatsusyu (resistant)
Yellows (*Fusarium oxysporum* f. sp. *conglutinans*)	Hatsuka group: Red Prince (resistant) Minowase group: Otabe (resistant) Motohashi-Taibyo-Minowase, Natsutomi (Moderately resistant) Nerima group: Tosai (moderately resistant) Miyashige group: Utsugi-Gensuke (moderately resistant)
Soft rot (*Erwinia carotovora* subsp. *carotovora*)	Shiroagari group: Yokokadokei (highly resistant) Nerima group: Takakura, Miura, Tosai (resistant) Miyashige group: Asahi-shirokubi, Taibyo-miyashige (moderately resistant) Ninengo group: Ninengo (moderately resistant)
Plasmodiophora brassicae (race 2) and *Erwinia carotovora*	Waincha, Kuro Daikon A, Yokomonkei (moderately resistant)

Kaneko and Matsuzawa, 1993; Bunin *et al.*, 1991.

The disease resistance traits from *Raphanus* are also utilized by the *Brassica* breeders, *viz.* resistance to turnip mosaic virus (Fujisawa, 1990), clubroot (Hagimori *et al.*, 1992) and beet cyst nematode (Lelivelt *et al.*, 1993).

Table 6.4: Promising Germplasm for Various Traits in Chinese Radish

Traits	*Genotypes*
	Resistance/Tolerance to Biotic Stress/Disease
Radish mosaic virus	Rose China Winter, Zheltyi (Resistant)
Yellows (*Fusarium oxysporum* f. sp. *conglutinans*)	China Rose, Kinmon-Aka (Resistant)
Soft rot (*Erwinia carotovora* subsp. *carotovora*)	North Chinese radish group: Pekin Shinribi (Highly resistant) Kiriba-Matsumoto (Moderately resistant)

(Mavlyanova, 1986; Kaneko and Matsuzawa, 1993)

Table 6.5: Promising Germplasm for Various Traits in Indian Radish

Traits	*Genotypes*
	Resistance/Tolerance to Disease
Alternaria blight (*Alternaria brassicae*)	Accessions 7401, 6802, 8801, 8803, 8805, Pusa Desi and Jaunpuri (Resistant)
	Accessions 7212, 7110, 7108, 7208, 7218, 8804, S-271 and IHR-1 (Moderately resistant)
White rust (*Albugo candida*)	IIHR-1-1 (Resistant)

(Kumar and Singh, 2003)

Breeding for Quality Traits

In radish, quality is related to pungency, sugar and water content and pore extent, degree of bolting and also external appearance like root length, shape, colour, gloss, smoothness of skin, etc. Indian varieties Arka Nishant and IIHR 1-1 are free from premature bolting, pithiness, splitting and forking.

Pungency

In Europe and USA, radishes are eaten fresh, hence low pungency level is preferred. In Japan, radish is eaten in cooked or salted vegetable or as 'soba' soups, where pungent radishes are used.

Pore Extent

Excessive root development causes pore development in radish, thus adversely affecting its commercial value. Breeding work is now focused for breeding varieties with early root enlargement and late pore development

In China, genetic resources of superior radish types for pod are 'Beijing Heart Beauty' and 'Tianjin Green'; for cooking purposes 'Beijing Red Robe' and 'Late Long White of Taihu'; for processing 'Yan-zhong', 'Dutch Egg Head' and 'One Cut Variety'; for early spring production 'March Radish'; for early harvest varieties with prolonged sowing date 'Short Leaf No. 13' and for bigger root 'Lasa

Winter' (Depel, 1989). A high quality, non-pungent variety 'Yidaozhong Luobo' with a low water content, tolerance to cold and high fertilizer doses, and disease resistance was identified in China. The preserved radishes are bright yellow and crisp (Zhou, 1991). The salad types of Chinese radish (Xin Li Mei, with red flesh; and Tianjin, with green flesh), possess visual appeal and eating qualities. Within Europe, old types like the Italian Treviso Red, long red and white skinned cultivars (*e.g.* cv. French Breakfast), and German Rettish types are marketed as high-value exotics (Crisp, 1995).

Genomics and Molecular Mapping

Rapid developments in DNA technologies opened the door for marker-assisted breeding of complex traits in several crops, including radish. In radish, the RAPD, AFLP, ISSR and RFLP molecular markers are used for cultivar identification (Pradhan *et al.*, 2004a), phylogenetic (Kong *et al.*, 2005) and genetic diversity analysis (Muminovic *et al.*, 2005) and construction of genetic maps (Bett and Lydiate, 2003 and Tsuro *et al.*, 2005). Pradhan *et al.* (2004b) identified several RAPD markers associated with high or low seed weight, germination proportion, seedling length and fresh weight in radish, which can be further used for marker-assisted breeding and selection of improved radish cultivars.

Bett and Lydiate (2003) developed the first genetic map of *Raphanus* genome, based on meiosis in a hybrid between *R. sativus* (cultivated radish) and *R. raphanistrum* (wild radish). The genetic markers and reference map of *Raphanus* are of considerable value for mapping quantitative trait loci (QTL) controlling important traits in radish like root shape and size, skin colour, nutritional value, quantity of glucosinolate, development time, self incompatibility, restoration of CMS and resistance to various diseases.

By comparative mapping, radish breeders can easily and cheaply access genetic information from related species. The comparative maps of *Raphanus* and *Brassica* genomes can be utilized for efficient interspecies transfer of *Raphanus* genes to *Brassica* crops. In future, comparative mapping with *Arabidopsis thaliana* will greatly facilitate map-based gene cloning, candidate gene identification and help in understanding the evolutionary relationships between *Raphanus* and its genetically well-characterized relatives.

Genetic Transformation

Radish is one of the most recalcitrant crop plants in culture and has a very low shoot regeneration frequency (Curtis, 2003). Since an efficient tissue culture system does not exist for production of transgenic radish in culture (Curtis, 2003), the first transgenic radish (*Raphanus sativus* var. *longipinnatus*) was produced in Korean cultivar 'Jin Ju Dae Pyong', using an alternative method known as 'floral-dip' (Curtis and Nam, 2001). Floral-dip method is currently the only procedure of gene transfer available for producing transgenic radish (Curtis, 2003).

Korean ecotypes of radish are cold sensitive, hence readily bolt at 5-6^{0}C during autumn, making production of high quality roots not possible (Curtis, 2003). Transfer of late-flowering genes into radish is important. *GIGANTEA* (*GI*), a late-flowering gene in *Arabidopsis*, was cloned recently. The 'floral-dipping' technique is used successfully to delay both bolting and flowering in radish by co-suppression of the photoperiodic gene, *GIGANTEA* and a late-flowering transgenic radish produced (Curtis *et al.*, 2002). The production of late-flowering germplasms of radish allows cultivation of cold-sensitive long-day radish over an extended period. Development of an efficient gene transfer system in radish enables radish breeders to transfer agronomically important traits, and manipulate the pharmaceutically valued endogenous chemicals for production of novel germplasms for benefit of mankind (Curtis, 2003).

Future Perspectives

There is emphasis to collect the existing variability available in *Raphanus*, particularly from Europe, China, Japan, Korea and Southeast Asia. Adequate description of the diversity of these collections should result in much greater emphasis on both conservation and exploitation of this variation.

It is suggested to preserve the ancient species, wild forms, landraces and primitive cultivars of *Raphanus sativus* found throughout the world for use in future breeding programmes. Landraces and wild relatives need to be augmented. It is essential to investigate and preserve the economically useful traits in the wild forms of *Raphanus*.

There is need to develop more promising varieties and hybrids of different types of radishes that can resist diseases and insects, and can meet a more varied range of dietary demands. Breeding of restorer lines with good agronomic properties is needed for use in hybrid radish seed production.

Raphanus genome appears to be extensively duplicated. There is need to study *Raphanus* genome organization to develop a better understanding of these gross chromosomal rearrangements, leading to an increased understanding of evolution of *Raphanus*. The molecular mechanism by which the *Rf* genes restore male fertility in CMS plants is not clearly understood, thus necessitates a better understanding of the genetics of restoration in *Raphanus*.

Somatic hybridization using protoplast fusion offers the opportunity to transfer germplasm from more wild radishes into cultivated radishes. The variability created through interspecific hybridization needs to be fully exploited. Marker-assisted selection (MAS) can greatly accelerate the breeding process and has been started to be used in radish breeding programmes. Development of dense genetic maps in radish using molecular markers would facilitate identification and tagging of agronomically important QTLs, map-based gene cloning, candidate gene identification and marker assisted selection. Further, comparative mapping of *Raphanus* with the other crucifers will elucidate the evolutionary relationships. Genetic engineering offers a direct access to the vast gene pool of radish for the introgression of economically useful genes.

References

Agnihotri, A., Shivanna, K. R., Raina, S. N., Lakshmikumaran, M., Prakash, S. and Jagannathan, V. 1990. Production of *Brassica napus* X *Raphanobrassica* hybrids by embryo rescue: an attempt to introduce shattering resistance into *B. napus*. *Plant Breed*. 105(4): 292-299.

Aoba, T. 1967. On wild radish, Nora-daikon (*Raphanus* spp.). *J. Yamagata Agr. Forest Soc.*, 24: 7-12 (in Japanese).

Aoba, T. 1979. Varieties and breeding in Japanese radish. *The iden* 33 (12): 55.

Aoba, T. 1983. Root crops: 'Radish'. In: Japanese Vegetables, Greens and Roots. Yasakashobou, Tokyo. 106.

Aoba, T. 1989. The diversity and a genealogy of Japanese wild radishes. *Noko Gijutu* (Arts in Agriculture), 12: 92-114 (in Japanese).

Aoba, T. 1993. Vegetables in Japan. Yasaka Shobo, Tokyo, 311 p. (in Japanese).

Badiger, S., Sivakumar, K. C., Reddy, M. A. N. and Mathew, D. 2001. Evaluation of exotic radish hybrids under Bangalore condition. *Current Res. Univ. Agric. Sci.–Bangalore*. 30(11-12): 201-202.

Bailey, L. H. and Bailey, E. Z. 1956. Hortus second–A concise Dictionary of Gardening and General Horticulture. Macmillan Co., New York.

Banga, O., 1976 . Radish. (In) *Evolution of Crop Plants*, (Ed.) N.W. Simmonds, Longman, London and New York, p. 60-62.

Barillari, J., Rollin, P. and Lori, R. 2002. Isolation of dietary glucosinolates and their derived isothiocyanates involved in cancer prevention. *NutraCos.* 1(6): 6-9.

Becker, C. 1962. Rettish und Radies (*Raphanus sativus*). *Handbuch der Pflanzenzuchtung*, 6: 23-78.

Bellaoui, M., Grelon, M., Pelletier, G. and Budar, F. 1999. The restorer *Rfo* gene acts post-translationally on the stability of the *ORF138 Ogura* CMS-associated protein in reproductive tissues of rapeseed cybrids. *Plant Mol. Biol.* 40: 893–902.

Bett, K. E. and Lydiate, D. J. 2003. Genetic analysis and genome mapping in *Raphanus*. *Genome* 46(3): 423-430.

Bonnet, A. and Blancard, D. 1987. Resistance of radish (*Raphanus sativus* L.) to downy mildew, *Peronospora parasitica*. *Eucarpia Cruciferae Newsletter*, 12: 98.

Bonhomme, S., Budar, F., Lancelin, D., Small, I., Defrance, M. C. and Pelletier, G. 1992. Sequence and transcript analysis of the *Nco2.5 Ogura* specific fragment correlated with cytoplasmic male-sterility in *Brassica* cybrids. *Mol. Gen. Genet*. 235: 340-348.

Bunin, M. S., Esikava, Kh. and Yoshikawa, H. 1991. Selection of sources of resistance to the main diseases of cruciferous crops from a Japanese germplasm collection of Daikon radish. *Doklady Vsesoyuznoi Ordena Lenina-i-Ordena Trudovogo Krasnogo Znameni Akademii Sel'skokhozyaistvennykh Nauk-im. V.I. Lenina* 11: 32-38.

Chase, C. D. 2006. Cytoplasmic male sterility: a window to the world of plant mitochondrial–nuclear interactions. *Trends Genet.* 23(2): 81-90.

Choudhary, B. R., Joshi, P. and Singh, K. 2000. Synthesis, morphology and cytogenetics of *Raphanofortii* (TTRR, 2n=38): a new amphidiploid of hybrid *Brassica tournefortii* (TT, 2n=20) x *Raphanus caudatus* (RR, 2n=18). *Theor. Appl. Genet.*, 101(5-6): 990-999.

Clapham, A. R., Tutin, T. G., Warburg, E. F. 1962. Flora of the British Isles. Cambridge University Press, Cambridge.

Crisp, P. 1995. Radish. (In) *Evolution of Crop Plants*, 2nd edn, (Eds.) J. Smartt and N.W. Simmonds. Longman Scientific and Technical, Harlow, Essex, England. p. 86-89.

Curtis, I. S. 2003. The noble radish: past, present and future. *Trends Plant Sci.* 8(7): 305-307.

Curtis, I. S. and Nam, H. G. 2001. Transgenic radish (*Raphanus sativus* L. *longipinnatus* Bailey) by floral-dip method– plant development and surfactant are important in optimizing transformation efficiency. *Transgenic Res.* 10(4): 363-371.

Curtis, I. S., Nam, H. G., Yun, J. Y. and Seo, K. H. 2003. Expression of an antisense GIGANTEA (GI) gene fragment in transgenic radish causes delayed bolting and flowering. *Transgenic Res.* 11(3): 249-256.

Danu, N. S. and Lal, S. D. 1998. Correlation studies in radish (*Raphanus sativus* L.). *Progressive Hort.* 30(3-4): 135-138.

De Candolle, A. 1883. Origin of Cultivated Plants. D.Appleton and Co., New York.

Depel, Z. 1989. The genetic resources of radish in China. Intl. Symp. Hortic. Germ. Cult. and Wild. Part II. Vegetables.

Dhar, S. 1998. Evaluation of radish cultivars in South Andaman. *Progressive Hort.* 30(3-4): 217-220.

Du, W. F. 1995. Radish germplasm resources in Hanzhong district, Shaanxi province. *Crop Genetic Res.* 1: 14-15, 35.

Frost, H.B. 1923. Heterosis and dominance of size factors in *Raphanus*. *Genetics* 8: 116-153.

Fujisawa, I. 1990. Turnip mosaic virus strains in cruciferous crops in Japan. *Japan. Agr. Res. Quarterly* 23: 289-293.

Furusato, K. and Miyazawa, A. 1958. Varieties of Japanese radish viewed from horticulture. (In) Japanese radish. Ed. I. Nishiyama. p 138.

George, R. A. T. 1986. Radish: *Raphanus sativus* L. (In) Vegetable Seed Production. Longman, London.

George, R. A. T. and Evans, D. R. 1981. A classification of winter radish cultivars. *Euphytica* 30: 483.

Gill, H. S. 1993. Improvement of root crops. (In) Advances in Horticulture, Vegetable Crops–Vol. 5. (Ed). Chadha, K.L. and Kalloo, G. Malhotra Publishing House, New Delhi, India, pp. 201-215.

Grelon, M., Budar, F., Bonhomme, S. and Pelletier, G. 1994. Ogura cytoplasmic male-sterility (CMS)-associated *orf138* is translated into a mitochondrial membrane polypeptide in male-sterile *Brassica* cybrids. *Mol. Gen. Genet.* 243: 540-547.

Hagimori, M., Nagaoka, M., Kato, N. and Yoshikawa, H. 1992. Production and characterization of somatic hybrids between the Japanese radish and cauliflower. *Theor. Appl. Genet.* 82: 819-824.

Hawlader, M. S. H. and Mian, M. A. K. 1997. Self-incompatibility studies in local cultivars of radish (*Raphanus sativus* L.) grown in Bangladesh *Euphytica*, 96: 311-315.

Hawlader, M. S. H., Mian, M. A. K and Ali, M. A. 1997. Identification of male sterility maintainer lines for Ogura radish (*Raphanus sativus* L.). *Euphytica*, 96: 297-300.

Hegi, G. 1935. Illustrierte Flora von Mittel-Europa. Band IV, Haffte IJR Lehmann's Velarg, Munchen.

Helm, J. 1957. Ueber den typus der Art *Raphanus sativus*, deren Gliederung und Synonymie. *Kulturpflanze* 5: 41-54.

Henslow, G. 1898. The history of the radish. *Gardeners Chronicke* 23: 389.

Heyn, F. W. 1976. Transfer of restorer genes from *Raphanus* to cytoplasmic male sterile *Brassica napus*. *Cruciferae Newsl.* 1: 15–16.

Hida, K. 1990. Root crops: 'Radish'. (In) *Collection of Plant Genetic Resources*. (Ed) T. Matsuo. p. 823.

Hida, K. and Ashizawa, M. 1985. Breeding of radishes for Fusarium resistance. *Japan Agric. Res. Quarterly* 19: 190-195.

Hinata, K. and Nishio, T. 1980. Self-incompatibility in Crucifers. (In) Brassica Crops and Wild Allies, (Ed) S. Tsunoda, K. Hinata and C. Gomez-Campo Jap. Sci. Soc. Press, Tokyo, p. 223-234.

Huh, M. K. 1995. Morphological variation and genetic structure in Korean populations of *Raphanus sativus* L. Ph.D. Thesis, Pusan National University, Pusan, 145 p.

Huh, M. K. and Ohnishi, O. 2001. Allozyme diversity and population structure of Japanese and Korean populations of wild radish. *Raphanus sativus* var. *hortensis* f. *raphanistroides* (Brassicaceae). *Genes Genet. Syst.* 76: 15-23.

Huh, M. K. and Ohnishi, O. 2002. Genetic diversity and genetic relationships of East Asian natural populations of wild radish revealed by AFLP. *Breed. Sci.* 52(2): 79-88.

Ikegaya, Y. 1986. Frequent appearance of cytoplasmic male sterile plants in a radish cultivar Kosena (in Japanese). *Jpn. J. Breed.* 36 [Suppl 2]: 106-107.

Iwasa, S. 1980. Root crops: 'Radish'. (In) *Vegetables in Tropics*. Youkendo, Tokyo, p. 293.

Iwabuchi, M., Koizuka, N., Fujimoto, H., Sakai, T. and Imamura, J. 1999. Identification and expression of the kosena radish (*Raphanus sativus* cv. Kosena) homologue of the Ogura radish CMS-associated gene-*orf138*. *Plant Mol. Biol*. 39: 183-188.

Kaneko, Y. and Matsuzawa, Y. 1993. Radish *Raphanus sativus* L. (In) Genetic Improvement of Vegetable Crops, (Ed). G. Kalloo and B.O. Bergh. Pergamon Press, New York. pp. 487-510.

Kaneko, Y., Yano, H., Bang, S. W. and Matsuzawa, Y. 2001. Production and characterization of *Raphanus sativus-Brassica rapa* monosomic chromosome addition lines. *Plant Breed*. 120(2): 163-168.

Karpechenko, G. D. 1924. Hybrids of *Raphanus sativus* L. x *Brassica oleracea* L. *J. Genet*. 14: 375-396.

Karpechenko, G.D. 1927 Polyploid hybrids of *Raphanus sativus* L. x *Brassica oleracea* L. *Bull. Appl. Bot. Genet. Pl. Br.* 17: 305-410.

Kirtikar, K.R. and Basu, B.D. 1935. Indian Medicinal Plants. Published by Lalit Mohan Basu, Allahabad.

Kitamura, S. 1958. Radish, taxonomy and history. (In) *Radish in Japan*, (Ed), Nishiyama, I. Japan Scientific Society Press, Tokyo, p. 1-19 (in Japanese).

Kitamura, S. and Murata, G. 1957. Colored Illustrations of Herbaceous Plants of Japan (Choripetalae). Hoikusha Publ. Co., Ltd., Osaka, Japan.

Kobabe, G. 1959. Nuturliche Einkreuzung von Hedrich (*Raphanus raphanistrum* L.) in Radies (*Raphanus sativus* var. *radicula* DC) and das Verhalten von Knollenform and Farbe in den nachfogenden F-und-R-Generationen. *Zeitschrift fur Pflanzenzuchtung* 41: 1-10 (in German).

Koizuka, N., Imai, R., Fujimoto, F., Hayakawa, T., Kimura, Y., Kahno- Murase, J. K., Sakai, T., Kawasaki, S. and Imamura, J. 2003. Genetic characterization of a pentatricopeptide repeat protein gene, orf687, that restores fertility in the cytoplasmic male sterile Kosena radish. *Plant J*. 34(4): 407–415.

Koizuka, N., Imai, R., Iwabuchi, M., Sakai, T. and Imamura, J. 2000. Genetic analysis of fertility restoration and accumulation of ORF125 mitochondrial protein in the Kosena radish (*Raphanus sativus* L. cv. Kosena) and a *Brassica napus* restorer line. *Theor. Appl. Genet*. 100(6): 949–955.

Kong, Q. S., XiXiang, L., Xiang, C. P., Yang, Q and Shen, D. 2005. Phylogenetic analysis of cultivated radish (*Raphanus sativus* L.) germplasm with AFLP. *Scientia Agricultura Sinica* 38(5): 1017-1023.

Krishnasamy, S. and Makaroff, C. A. 1993. Characterization of the radish mitochondrial *orfB* locus:possible relationship with male-sterility in Ogura radish. *Curr. Genet*. 24: 156-163.

Krishnasamy, S. and Makaroff, C. A. 1994. Ogura-specific reduction in the abundance of a mitochondrial protein accompanies fertility restoration in cytoplasmic male-sterile radish. *Plant Mol. Biol*. 26: 935-946.

Kumar V. and Chandel, K. S. 2003. Genetics of root yield and quality traits in radish (*Raphanus sativus* L.). *Crop Improv*. 30(1): 33-38.

Kumar, V., Chandel, K. S. and Pathania, N. K. 2002. Heterosis in Asiatic and European cultivars in radish (*Raphanus sativus* L.). *Veg. Sci*. 29(1): 34-36.

Kumar, P. and Singh, D.V. 2003. Screening of radish germplasm against Alternaria blight. *Farm Sci. J*. 12(1): 61-62.

Kumazawa, S. 1956. Sosaiengei Kakuron. Yokendo, Tokyo, 637 pp. (in Japanese).

Kumazawa, S. 1961. Horticulture, Detailed Discussion of Crops. Yokendo, Tokyo, 637 p. (in Japanese).

Kumazawa, S. 1965. Radish. (In) *Vegetable Gardening*. Yokendo, Tokyo. p. 295.

Kutty, C. N. and Sirohi, P. S. 2003. Exploitation of hybrid vigour in radish (*Raphanus sativus* L.). *Indian J. Hort.* 60(1): 64-68.

Lal, S. and Srivastava, J. P. 1975. Heritability and genetic advance for root and fruit characters in radish. *Indian J. Hort.*, 32: 91-93.

Langthasa, S.and Barah, P. 2000. Varietal performance of radish under hill conditions of Assam. *Indian J. Hill Frmg.* 13(1-2): 85-86.

Lee, S. K. 2003. Preparation of kimchi and other preserved vegetables. *Acta-Horticulturae*, 620: 127-133.

Lee, W. S., Kim, S. B. and Yamaguchi, H. 1992. Studies on distribution and variability of wild radish (*Raphanus sativus* var. *raphanistroides*) in Korea–1. Distribution and leaf shape variation. *Korean J. Hort. Sci. Tech.* 10: 62-63 (in Korean).

Lelivelt, C.L.C., Lange, W. and Dolstra, O. 1993. Intergeneric crosses for the transfer of resistance to the beet cyst nematode from *Raphanus sativus* to *Brassica napus*. *Euphytica* 68: 111-120.

Lewis-Jones, L. J., Thorpe J. P., and Wallis, G. P. 1982. Genetic divergence in four species of the genus *Raphanus*: implications for the ancestry of the domestic radish *R. sativus*. *Biol. J. Linnean Soc.* 18: 35–48.

Li Shuxuan 1989. The origin and resources of vegetable crops in China. *International Symposium on Horticultural Germplasm, Cultivated and Wild*, Beijing, China, Sept. 1988. Chinese Society for Horticultural Science. International Academic Publishers, Beijing, pp. 197-202.

Lim, S. H., Cho, H. J., Lee, S. J., Cho, Y. H. and Kim, B. D. 2002. Identification and classification of S haplotypes in *Raphanus sativus* by PCR-RFLP of the S locus glycoprotein (SLG) gene and the S locus receptor kinase (SRK) gene. *Theor. Appl. Genet.* 104(8): 1253-1262.

Makaroff, C. A. 1995. Cytoplasmic male sterility in *Brassica* species. (In) *The Molecular Biology of Plant Mitochondria*. Levings, C. S. and Vasil, I. K. (Ed). Kluwer Academic Publ, Dordrecht, The Netherlands, pp. 515–555.

Makaroff, C. A. and Palmer, J. D. 1988. Mitochondrial DNA rearrangements and transcriptional alterations in the male-sterile cytoplasm of Ogura radish. *Mol. Cell Biol.* 8: 1473-1480.

Makino, T. 1909. Observation on the flora of Japan. *Bot. Mag.* 23: 59-75.

Mansfield, R. 1959. Die Kulturpflanzen baikeft. Akademic Velag, Berlin, pp. 1-659.

Mavlyanova, R.F. 1986. Breeding exploitation of the diversity of radish in the foothills zone of Uzbekistan. *Nauchno Tekhnicheskii Byulleten' Vsesoyuznogo Ordena Lenina-i-Ordena Druzhby Narodov Nauchno Issledovatel'skogo Instituta-Rastenievodstva-imeni N. I. Vavilova* 161: 61-63.

Motegi, T., Nou, I., Zhou, J., Kanno, A., Kameya,T., Hirata, Y., Nou, I. S. and Zhou, J. M. 2003. Obtaining an ogura-type CMS line from asymmetrical protoplast fusion between cabbage (fertile) and radish (fertile). *Euphytica* 129(3): 319-323.

Muminovic, J., Merz, A., Melchinger, A. E. and Lubberstedt, T. 2005. Genetic structure and diversity among radish varieties as inferred from AFLP and ISSR analyses. *J. Amer. Soc. Hortl. Sci.* 130(1): 79-87.

Murali, K., Reddy, N. S., Anjanappa, M. and Krishnappa, K. S. 1998. Correlation studies in radish (*Raphanus sativus* L.). *Karnataka J. Agric. Sci.*, 11(4): 1140-1141.

Murayama, S., Habuchi, T., Yamagishi, H. and Terachi, T. 2003. Identification of a sequence-tagged site (STS) marker linked to a restorer gene for Ogura cytoplasmic male sterility in radish (*Raphanus sativus* L.) by non-radioactive AFLP analysis. *Euphytica* 129(1): 61-68.

Nahm, S. H., Lee, H. J., Lee, S. W., Joo, G. Y., Harn, C. H., Yang, S. G. and Min, B. W. 2005. Development of a molecular marker specific to a novel CMS line in radish (*Raphanus sativus* L.). *Theor. Appl. Genet.* 111: 1191–1200.

Nakagawa, Y., Murayama, S. and Yamagishi, H. 1998. DNA sequence polymorphism in the Ogura CMS gene (*orf138*) in wild and cultivated radishes. *Breed. Sci.* 48 (Supplement 1): 200 (in Japanese).

Nakamura, Y., Iwahashi, T., Tanaka, A., Koutani, J., Matsuo, T., Okamoto, S., Sato, K. and Ohtsuki, K. 2001. 4-(Methylthio)-3-butenyl isothiocyanate, a principal antimutagen in daikon (*Raphanus sativus*; Japanese white radish). *J. Agric. Food Chem.* 49(12): 5755-5760.

Nieuwhof, M. 1990. Cytoplasmic-genetic male stedity in radish (*Raphanus sativus* L.). Identification of maintainers, inheritance of male sterility and effect of environmental factors. *Euphytica* 47: 171-177.

Niikura, S. and Matsuura, S. 2001. Genetic variation of the self-incompatibility alleles (S-alleles) in the cultivated radish (*Raphanus sativus* L.) by the PCR-RFLP method. *Acta Horticulturae* 546: 359-365.

Nikornpun, M., Putivoranat, M. and Zhong, L. 2004. Uses of self-incompatibility genes in Chinese radish. *Southwest-China J. Agril. Sci.* 17(1): 78-83.

Ogura, H. 1968. Studies on the new male sterility in Japanese radish with special reference to utilization of this sterility toward the practical raising of hybrid seeds. *Mem. Agric. Kagoshima Univ.* 6: 39-78.

Pal, B. P. and Sikka, S. M. 1956. Exploitation of hybrid vigour in the improvement of crop plants, fruits and vegetables. *Indian J. Genet.*, 16: 98-104.

Panetsos, C. A. and Baker, H. G. 1967. The origin of variation in "wild" *Raphanus sativus* (Cruciferae) in California. *Genetica* 38: 243-274.

Pistrick, K. 1987. Untersuchungen zur Systematik der Gattung *Raphanus* L. *Kulturpflanze* 35: 225-321.

Pradhan, A., Yan, G. and Plummer, J. A. 2004a. Development of DNA fingerprinting keys for the identification of radish cultivars. *Australian J. Exptl. Agric.* 44(1): 95-102.

Pradhan, A., Yan, G. and Plummer, J. A. 2004b. Correlation of morphological traits with molecular markers in radish (*Raphanus sativus*). *Australian J. Exptl. Agric.* 44(8): 813-819.

Rabbani, M.A., Murakami, Y., Kuginuki, Y. and Takayanagi, K. 1998. Genetic variation in radish (*Raphanus sativus* L.) germplasm from Pakistan using morphological traits and RAPDs. *Genetic Res. Crop Evol.* 45(4): 307-316.

Rana, R. S. 1993. Genetic Resources in Vegetable Crops. (In) *Advances in Horticulture, Vegetable Crops–* Vol. 5. (Ed) Chadha, K.L. and Kalloo, G. Malhotra Publishing House, New Delhi, India, pp. 27-43.

Richharia, R. H. 1973. Cytological investigation of *Raphanus sativus*, *Brassica oleracea* and their F_1 and F_2 hybrids. *J. Genet.* 34: 19-44.

Rubatzky, V.E. and Yamaguchi, Y. 1997. *World Vegetables*. Chapman and Hall, New York, USA, p. 843.

Sakai, T. and Imamura, J. 1990. Intergeneric transfer of cytoplasmic male sterility between *Raphanus sativus* (cms line) and *Brassica napus* through cytoplast–protoplast fusion. *Theor. Appl. Genet.* 80(3): 421-427.

Schulz, O.E. 1919. (In) A. Engler (Ed.), *Cruciferae-Brassiceae,* Part I: Brassicinae and Raphaninae.

Sharma, M. and Dhaliwal, D.S. 1997. *Raphanus raphanistrum* Linn. (Cruciferae)–an alien weed established in Kullu District (Himachal Pradesh). *Indian Forester,* 123(9): 870-872.

Sharma, V. K., Chandel, K. S., Kalia, P. and Pathania, N. K. 2002. Performance of different cultivars of radish (*Raphanus sativus* L.) under the midhill conditions of Himachal Pradesh. *Himachal J. Agric. Res.* 28(1-2): 26-29.

Shimizu, S., Kanazawa, K., Kono, H. and Yokota, Y. 1963. Studies on breeding radish for resistance to virus. *Bull. Hort. Resista: Series A* No. 1: 83-106.

Sidhu, A. S. 2002. Current Status of Vegetables Research in India, Tropical Seed, Com; Publication and Research: Articles.

Sinskaia, E. N. 1928. The oleiferous plants and root crops of the family Cruciferae. *Bull. Appl. Bot. Genet. Pl. Br.* 19(3): 1-648.

Sinskaia, E. N. 1931. The wild radish from the sea coast of Japan in connection with the problem of the origin of the cultivated forms belonging to the genus *Raphanus. Turdy Prikladnoi Bot., Genet. Seletskii* (Leningrad) 26: 1-58 (in Russian).

Sirks, M. J. 1957. Japanese genetica. *Genen en Phaenen* 2: 2-10.

Sirohi, P. S. and Kutty, C. N. 2000. Genetic analysis of yield and its components in radish. *Veg. Sci.* 27(2): 142-144.

Sodhi, Y. S., Pradhan, A. K., Mukhopadhyay, A. and Pental, D. 1993. Stability of 'Ogura' cytoplasmic male-sterility in three Indian cultivars of radish (*Raphanus sativus*). *Indian J. Agric. Sci.,* 63(7): 421-422.

Stewart, A. V. and Moorhead, A. J. 2004. The development of a fodder radish suitable for multiple grazing. *Agron. New-Zealand* 34: 1-7.

Thellung, M.A. 1927. L' origine de la carotte et du radish cultives Revue de Botanique appliqué et d' Agriculture tropicale 7, 666-671 (in French).

Toxopeus, H. and van Sloten, D.H. 1981. The genetic resources of cruciferous crops. *Plant Genetic Res. Newsl.* 47: 3-8.

Trouard-Riolle, Y. 1914. Recherches morphologiques sur les radis cultives. *Ann. Sci. Agron. Fr. Etrang* 31: 295-322, 346-550.

Tsuro, M., Suwabe, K., Kubo, N., Matsumoto, S. and Hirai, M. 2005. Construction of a molecular linkage map of radish (*Raphanus sativus* L.), based on AFLP and *Brassica*-SSR markers. *Breed. Sci.* 55(1): 107-111.

Werth, E. 1937. Abstammung und Heimat des Rettichs. *Angew. Bot.,* 19: 194-205.

Williams, P.H. and Pound, G.S. 1963. Nature and inheritance of resistance to *Albugo candida* in radish. *Phytopathology* 53: 1150-1154.

Yamagishi, H. 1998. Distribution and allelism of restorer genes for Ogura cytoplasmic male sterility in wild and cultivated radishes. *Genes Genet. Syst.,* 73(2): 79-83.

Yamagishi, H. 2001. Genetic relationships between cultivated radish and wild radish from the viewpoint of the cytoplasmic male sterility gene. (In) *Natural History of Cultivated Plants*, Yamaguchi, H. and Shimamoto, Y. (eds.), Hokkaido Univ. Press, Sapporo, p. 96-107 (in Japanese).

Yamagishi, H. and Sasaki, J. 2004. Differentiation of the cytoplasm of Japanese cultivated radishes (*Raphanus sativus*) into four types in the 3' region of mitochondrial Atp6. *Breed. Sci.* 54(2): 189-195.

Yamagishi, H., and Terachi, T. 1994a. Molecular and biological studies on male-sterile cytoplasm in the Cruciferae. I. The origin and distribution of Ogura male sterile cytoplasm in Japanese wild radishes (*Raphanus sativus* L.) revealed by PCR-aided assay of their mitochondrial DNAs. *Theor Appl Genet* 87(8): 996-1000.

Yamagishi, H., and Terachi, T. 1994b. Molecular and biological studies on male-sterile cytoplasm in Cruciferae. II. The origin of Ogura male sterile cytoplasm inferred from the segregation pattern of male sterility in the F_1s progeny of wild and cultivated radishes (*Raphanus sativus* L.). *Euphytica* 80: 201-206.

Yamagishi, H. and Terachi, T. 1996. Molecular and biological studies on male-sterile cytoplasm in the Cruciferae. III. Distribution of Ogura-type cytoplasm among Japanese wild radishes and Asian radish cultivars. *Theor Appl Genet* 93(3): 325-332.

Yamagishi, H. and Terachi, T. 2001. Intra- and inter-specific variations in the mitochondrial gene orf138 of Ogura-type male-sterile cytoplasm from *Raphanus sativus* and *Raphanus raphanistrum*. T*Theor. Appl. Genet.* 103(5): 725-732.

Yamagishi, H. and Terachi, T. 2003. Multiple origins of cultivated radishes as evidenced by a comparison of the structural variations in mitochondrial DNA of *Raphanus*. *Genome* 46: 89–94.

Yamagishi, H., Tateishi M., Terachi T. and Murayama S. 1998. Genetic relationships among Japanese wild radishes (*Raphanus sativus* f. *raphanistroides* Makino), cultivated radishes and *R. raphanistrum* revealed by RAPD analysis. *J. Jpn. Soc. Hort. Sci.* 67: 526-531.

Yamaguchi, H. 1987. Latitudinal cline and intrapopulational differentiation in leaf shape of wild radish in Japan. *Jap. J. Breed.*, 37(1): 54-65.

Yamaguchi, H. and Okamoto, M. 1997. Traditional seed production in landraces of daikon (*Raphanus sativus*) in Kyushu, Japan. *Euphytica* 95(2): 141-147.

Yamane, K., Lu, N. and Ohnishi, O. 2005. Chloroplast DNA variations of cultivated radish and its wild relatives. *Plant Sci.* 168(3): 627-634.

Yuan, H. 1995. Radish germplasm and the elite cultivars of Anhui Province. *Crop Genetic Res.* 4: 33-34.

Zhao, D. 1989. The genetic resources of radishes in China. *International Symposium on Horticultural Germplasm, Cultivated and Wild*, Beijing, China, Sept. 1988. Chinese Society for Horticultural Science. International Academic Publishers, Beijing, pp. 388-393.

Zhou, M.X. 1991. Special vegetable germplasm resources in Zhejiang province. *Crop Genetic Resources* 3: 41-42.

Zhou, C.J. and Chen, H.M. 1991. Study on the distribution of Chinese radish and its centre of origin. *Acta Agric. Universitatis Pekin.* 17(4): 47-53.

Zhu, Y. Y., Wu, X. G., Gong, J., Jin, H. J., Yao, W. Y., Li, S. J. and Wu, Q. S. 2001. Breeding and utilization of a male sterile line with radish cytoplasm in broccoli. *J. Nanjing Agril. Univ.*, 24: 3, 15-18.

Biodiversity in Horticultural Crops Vol. 2 *Pages* **139–174**
Editor: **Professor K.V. Peter**
Published by: **DAYA PUBLISHING HOUSE**

Chapter 7

Tropical Cauliflower

Pritam Kalia
Division of Vegetable Crops, IARI, New Delhi – 110 012

Cauliflower (*Brassica oleracea* L var. *botrytis* L; 2n=2x=18) is an important cole vegetable belonging to Brassicaceae family. It grows at latitudes from 45^0 in New Zealand to 65^0 N in Scandinavian countries. It is grown in all continents of the world, of which Asia is the leading one followed by Europe and covers small acreages in North America, South America, Africa, Australia and New Zealand. India and China together have about 60 per cent of the total cauliflower area of the world and 38.81 per cent of it is in India alone. India is the second largest producer of cauliflower in the world after China. Presently, cauliflower occupies 238.2 million ha in India with a production of 4507.9 million t. (FAOSTAST, 2007).

Cauliflower is used as cooked vegetable in India either singly or in combination with other vegetables, especially potato, peas etc. Surface of the cauliflower curd is a dome of tissues made up of a mass of proliferated floral meristems at harvest. It is also relished in curry and sambhar and often used in pickles along with carrot. Recently, the frozen cauliflowers are available in the Indian markets.

History, Evolution and Adaptation of Indian Cauliflower

North Indian *Kharif* and autumn season cauliflowers are distinct from European types. The European snowball is grown as the late variety in winter-spring season of North India. Swarup and Chatterjee (1972) demonstrated that the tropical (or Indian) cauliflower has close morphological affinity with different West European types like Cornish, Roscoff, Italian, Northern, Angiers and Snowball or Erfurt, though not exactly the same. As per the climatic conditions of North India, the typical 'Indian' cauliflowers are categorised into Groups I and II (maturing at average daily temperatures above 20 ^{0}C). The observations of Indian cauliflower showed that long stalk, open growth habit, exposed yellowish to creamy and uneven cruds, which loosen up easily and their strong flavour, are some of the characters typical of Cornish while some leaf and curd characters can also be Roscoff and Italians like. Late maturing types of December-January maturity show similarities with Italian (autumn) and Snowball or Erfurt types (European Summer Group).

Tracing the chronology of cauliflower varieties introduced into India, Swarup and Chatterjee (1972) pointed out that a Botanist from Kew Gardens London, introduced cauliflower seeds in Uttar Pradesh (then called United Provinces) during 1822 during the East India Company days. The Royal Agri-Horticultural Society Calcutta also introduced seeds of temperate (then called 'English') vegetables. Seeds were imported from South Africa and England and were distributed to farmers in North India. During a period of one hundred years (1822-1929), introduced cauliflower varieties underwent involuntary selection by local growers when they attempted seed production in North Indian plains. Selections were made for early maturity and wide adaptability to hot and humid weather. It was reported from Saharanpur that adapted varieties selected by local growers were good for early sowing and in later season they could be followed by the imported varieties. The first four Indian varieties listed by Sutton and Sons during 1929 were "Early and Main crop Patna" and "Early and Main crop Benares".

Swarup and Chatterjee (1972) further elaborated that self-incompatibility status (especially in Group-I), high degree of heterosis and resistance to black rot disease, testified the greater genetic diversity not reported in Snowball varieties. It is likely that "Cornish"-perhaps the first one to be introduced to India and their contribution of many genes to present day Indian varieties–might have gone out of cultivation in Europe. When the Cornish types were brought from temperate environment to Indian, they adapted slowly to higher temperature and humid conditions.

Other tropical regions of the world also supported this fascinating account of Indian cauliflower where similar varieties like 'Pua Kea' in Hawaii, 'Caminus' in Brazil and 'Extra Early' in Taiwan etc. were developed from the Indian materials. Incidentally, the Indian cauliflower varieties marketed by Sutton and Sons Calcutta, performed well in Ceylon (Sri Lanka), West Indies (British), the Philippines and Florida (USA). Thus, the 'aristocrat' character of European cauliflower transformed to a 'cosmopolitan' status showing flexibility of adaptation to regions from New Zealand to Scandinavian Countries.

Indian seedsmen and growers contributed significantly to the development of 'Indian' cauliflower varieties. The seedsmen of Hajipur (Bihar) specialize on the varieties of groups I and II. Foreign seedsmen like Sutton and Sons (which later became Indian) played a significant role in this venture. Ihe cauliflower has undergone such a 'quick' diversification within a short period of around two centuries and conscious utilization of this phenomenon in the breeding programmes, started in India during the last three or four decades only. Out of the total area under cauliflower in India, about 40 per cent goes to hot weather tropical cauliflowers (Seshadhri and Chatterjee, 1996).

Although the early Indian cauliflower originated from European types, they were characteristically different from cauliflower grown in Europe and other parts of the Western world. The parental variety, which contributed most to the early-maturing Indian types tolerant to high temperature and rainfall conditions, is the Cornish type. The morphological characteristics, namely stalk length, open plant habit, exposed yellow uneven curds and strong flavor are a few common attributes of both Cornish and Indian cauliflower, particularly the early maturing in September-November. They are, perhaps, the earliest-maturing types in the world, hot weather cauliflower (forming curds above 20^{0}C) and they are widely used to develop many improved varieties adapted to the tropics and the subtropics in various countries, namely, Pua Kea in Hawaii (Gilbert 1953, 1955), Campinas in Brazil (Camargo, 1956), Improved Japanese and D-96 in Israel (Feldner, 1956) and Extra Early in Taiwan. The Indian varieties released by Sutton and Sons Co. were also promising in other tropical areas like Sri Lanka, the West Indies, the Philippines, Florida, etc. (Paul, 1933; Rodrigo *et al.*, 1935; Wood and Jones, 1936; Mortensen and Bullard, 1968).

Broad Grouping

In addition to the two annual types, namely tropical Indian cauliflower and snowball types, there are other developments, which took place, in this crop independently in different regions around the world. They remained genetically isolated for a long period (except for the Italians or originals) and, thus, maintained their characteristic features. It would, therefore, be worthwhile to classify these broad groups so that a proper understanding and relationship of the present-day cultivars are possible (Swarup and Chatterjee, 1972).

Table 7.1: Broad Groups of Cauliflower

Cauliflower Types	*Centre of Origin*	*Probable Period of First Cultivation*	*Characters*
Italians or Original	Mediterranean	16th Century	Plants short; leaves erect broad with rounded tips, bluish green; curds good not protected by leaves
Cornish	England	Early 19th Century	Plants vigorous; long stalked; leaves loosely arranged, broadly wavy; curds flat, irregular, loose, not protected, yellow, highly flavoured
Northerns	England	19th Century	Leaves petiolate, broad, very wavy, serrated; curds good, well protected
Roscoff	France	19th Century	Plants short; leaves long erect, slightly wavy with pointed tip, midrib prominent, bluish green; curds white or creamy, hemispherical, well protected
Angers	France	19th Century	Leaves very wavy, serrated, greyish green; curds solid, white, well protected
Erfurt and Snowball	Germany and Netherlands	18th Century	Plants dwarf; leaves short, erect, glaucous green; curds solid, well protected
Indian cauliflower	India	Late 19th Century	Plants short, long stalked; leaves loosely arranged, broadly wavy; curds flat, somewhat loose, yellow to creamy, not protected and highly flavoured

Adapted from Sharma *et al.*, 2004.

The classification of cauliflowers according to their phyllogeny was proposed by Crisp (1982).

Table 7.2: Grouping of Cauliflower According to Phyllogeny

Groups	*Chief Characteristics*	*Common types*
Italian	Very diverse, include both annuals and biennials and curds with peculiar conformations and colours	Jezi, Naples (Autumn Giant), Romanesco, Flora Blanca
North-West European biennials	Derived within the last 300 years from Italian material	Old English, Walcheran, Roscoff, Angers, St. Malo
North European annuals	Developed in northern Europe for atleast 400 years. Origin unknown, perhaps Italian or Eastern Mediterranean	Lecerf, Alpha, Mechelse, Erfurt, Danish
Asian	Recombinants of European annuals and biennials developed within 250 years, adapted to tropics	Four maturity groups are recognised by Swarup and Chatterjee (1972)
Australian	Recombinants of European annuals and biennials and perhaps Italian stock, developed during the last 200 years	Not yet been categorized

Adapted from Sharma *et al.* (2004).

Further studies were recommended for separate grouping of the North European annual and Australian types (Chatterjee, 1993).

Genetic Diversity

The greatest genetic diversity of cauliflower is in the Mediterranean gene centre including Greece, Syria, Cyprus, Sicily, Italy, Spain and Portugal. Wild species or subspecies and improved heterogeneous cultivars are the primary genetic resources for various characters available to the breeders. Introgression from wild species in primitive form of *B. oleracea* in the Mediterranean area and Northwestern Europe resulted in great genetic diversity in cauliflower. The annual types of cauliflower occur in Italy and surrounding areas. Introduction of improved cultivars and hybrids into Europe replaced many of the landraces, primitive cultivars and traditional varieties or types. There is extensive genetic erosion but fortunately many genetic resources are still available in the Mediterranean gene center.

In addition to landraces in Italy, genetic diversity of cauliflower is found in France, U.K., Sweden, Denmark and the Netherlands. Many types of cauliflower were developed in Europe in 16th century, like Originals or Italians (Jezi, Naples, Romanesco and Flora Blanca), Erfurt, Alpha and Snowball in Germany and the Netherlands, Cornish and Northerns in England and Roscoff and Angers in France. The European biennials include Old English, Walcheran, Roscoff, Angers and St. Malo and the annuals like Alpha, Erfurt, Danish, Lecerf, and Mechelse. The two other groups of cauliflower suggested are the Asian (or Indian) and the Australian arising as a result of recombination of European annuals and biennials. Although cauliflower is grown the world over, the leading producers are India, China, France, Italy, United Kingdom, Spain, USA, Poland, Germany, Pakistan, Japan.

Cauliflower Research Centres in India

Research on various aspects of cauliflower is carried out at several research centres in India. The Indian Agricultural Research Institute, New Delhi is the main centre working on varietal improvement of tropical cauliflowers. Other centres are G.B. Pant University of Agriculture and Technology, Pantnagar, Punjab Agricultural University, Ludhiana and BCKV, Kalyani. Recently, IIVR, Varanasi, IIHR, Bangalore and N.D.University of Agriculture and Technology, Faizabad have also initiated tropical cauliflower improvement work.

The varietal improvement work on snowball type is mainly carried out at IARI Regional Station, Katrain. The other centres working on late group are Dr. Y.S. Parmar University of Horticulture and Forestry, Nauni-Solan, TNAU Research Station, Ootacommund, Nilgiri, Hill Campus of GB Pant University Agriculture and Technology, Ranichauri.Research on production and protection aspects are undertaken at several centres located in and around cauliflower growing areas.

An All India Coordinated Vegetable Improvement Project, AICRP (VC), was started in 1971 with its headquarters at IARI, New Delhi which was later(during 1992) shifted to Indian Institute of Vegetable Research, Varanasi. The project comprises 8 main centres, 17 subcentres and a number of volunteer centres located in ICAR Institutes and State Agricultural Universities throughout the country in different agro-climatic regions. The technology developed on cauliflower production is tested at all the centres located in the regions for which it is developed before its final approval.

Thermoclassification of Indian Cauliflowers

Based on their thermo sensitibity, Indian cauliflower is classified into four groups *viz.* early [extra early, early (Ia), early (Ib)], mid early, mid late and late or snowball types. The first three types include typical Indian or tropical types and the selections made for this purpose perform well, which produce quality curds even during May-June making it possible to grow cauliflower almost all the year round (Chatterjee 1993; Singh and Sharma, 2001).

Table 7.3: Cauliflower Cultivars for Different Times of the Year*

Maturity Group	*Cultivar*	*Sowing Time*	*Mean Temperature Required for Curd Initiation and Development*	*Availability Period*
Extra Early	Pusa Meghna, DC 23 Pusa Kartik Sankar	End of February	24-30°C	May-June
Early I (a)	Pusa Early Synthetic, Pusa Meghna, Pusa Kartik Sankar	End of May	22-27°C	End of August-October
Early I (b)	Pusa Deepali Pusa Katki	June	20-25°C	Oct.-Nov.
Mid Early	Pusa Sharad, Pusa Hybrid-2, Improve Japanese	End of July-August	16-20°C	Nov.-Dec.
Mid Late	Pusa Synthetic Pusa Himjyoti Pusa Shubhra	End of August-Sept.	12-16°C	Dec.-Jan.
Late	Pusa Snowball K-1, Pusa Snowball 1, Pusa Snowball K-25	Sept.-Nov.	10-16°C	Jan.-March

*: Under North Indian plains.

Character Divergence

Maturity of Curd

Chaterjee and Swarup (1972) classified Indian cauliflower into three maturity groups:

Maturity group I–Curds harvested from September to early November;

Maturity group II–Curds harvested from mid-November to early December; and

Maturity group III–Curds harvested from mid-December to mid-January.

Plant Type

On the basis of growth habit, four plant types were broadly identified:

Plant type No. 1–curds completely exposed with flat leaves

Plant type No. 4–curds well-covered with erect leaves; and

Plant types No. 2 and 3–intermediates, the former being close to No. 1 and the latter approaching No. 4.

Plant type No. 3 was the best as it had long, erect leaves with or without blanching habit and medium-sized curds. It corresponds to the plant type of Snowball. Plant type No. 2 is the most common among Indian cauliflower.

Stem Length

There is large variation in stem length of different cauliflower germplasm. Chatterjee and Swarup (1972) classified Indian cauliflower into three groups based on stem length: shoot (15 cm and below); medium (16-20 cm) and long (21 cm and above).

Medium stem length was predominant in maturity groups 1 and 2 while maturity group III had mostly short stem.

Stem Pigmentation

The stem may be green or pigmented as in Snowball. Intensity of pigmentation varies in different types of cauliflower. Stem pigmentation is important as a dominant marker gene character.

Leaf Characters

The leaves may be long and narrow, long and broad, and short and broad. The leaf margins are straight or broadly wavy. The leaf color varies from bluish green and wavy green to glossy green. The glossy leaf is governed by a recessive gene and can be used as a marker in seedling stage. Number of leaves ranges 18 to 50 during curding stage. Generally, there are more leaves in early-maturing types than in late-maturing types.

Self-blanching Character

When the inner leaf whorls are folded to cover the curd, it is known as self-blanching type. The cauliflower types greatly vary in this character. Generally, the early-maturing types lack in this character. The late-maturing types usually are self-blanching.

Curd Shape

Shape of cauliflower curd may be flat with even surface or hemispherical with uneven surface. In some cases, the curds have slightly conical shape. The hemispherical and conical curds are usually more compact than the flat type.

Curd Size and Weight

Cauliflower germplasm has large genetic variation for size and weight of curds.

Curd Colour

Color ranges from yellow to bright white. The bright white curds, as in Snowball, have wide market preference. The early-maturing hot weather cauliflowers are mostly yellowish to somewhat creamy white. The curd color is influenced by blanching habit of variety and growing temperature.

Male Sterility

Two types of male sterility are observed in Indian cauliflower: (a) anthers are not fully developed and the pollens are not fertile; and (b) the stigma protrudes from the bud and does not get any chance for self-pollination, although the pollen grains are fertile. Swarup and Chatterjee (1972) isolated a line combining both types of male sterility, *i.e.*, the protruding stigma and poorly developed stamens with sterile pollen. The recessive gene governs it.

Self-incompatibility and Heterosis

Based on the results of morphological and genetical studies conducted on a wide collection of indigenous varieties and inbred lines that had been developed from them, Swarup and Chatterjee (1972) revealed that:

1. The early-maturing Indian cauliflowers are highly self-incompatible while the late maturing types are more self-compatible.
2. Appreciable extent of heterosis for almost all economic characters was observed in Indian cauliflower. Erfurt or Snowball types did not show pronounced heterosis, mainly due to the fact that these types did not have much genetic diversity.

On the basis of genetic stocks studied in Indian cauliflowers, the ideotype of cauliflower should possess following attributes:

Stem length	:	12-25 cm
Plant type	:	No. 3
Frame (spread)	:	35-45 cm
Leaf number	:	18-22
Leaf length	:	50-55 cm
Curd shape	:	Hemispherical
Curd diameter	:	15-18 cm
Curd weight	:	750-1000 g
Curd colour	:	Retentive white
Resistance to	:	Black rot, curd and inflorescence blight

Gene Action Studies

In cauliflower, transformation from vegetative phase to curding is governed by temperature; varieties differ in optimum temperature requirement for curd formation (Haine, 1959; Kato, 1964; Sadik, 1967). Any study on this character must take into consideration suitability of prevailing temperatures for curding of varieties. Watts (1964) made two series of diallel crosses between (a) eight varieties of autumn cauliflower (intervarietal) and (b) six inbreds of a single variety of early-summer type. He did not find any F_1 (in either series), earlier than the early parent. Further, in autumn types, he noted additive effects and those varieties with early mid-maturity possessed dominant polygenes while those with late maturity possessed recessive polygenes. In early-summer type (intravarietal diallel), some interaction was noted. There was association between early curding and low leaf number and between later curding and high leaf number. Swarup and Pal (1966), in a similar study of late-maturing cauliflower, found that dominance and epistasis contributed most towards inheritance of curd maturity. Heterosis was manifested in earliness.

Nieuwhof and Garetsen (1961) reported that the compactness or solidity of curd is controlled by polygenic factors. They observed negative correlation between firmness of curd and seed yield. The pioneer Indian researchers on cauliflower, Swarup, V. and Chatterjee, S.S., started a comprehensive breeding programme for improvement of Indian cauliflower in 1965 at the Indian Agricultural Research Institue, New Delhi. Studies on genetic architecture and breeding were pursued for nearly two decades by a number of research scholars. Three hundred true breeding inbred lines for different characters developed from about 200 accessions collected from India were used to study genetic variability, combining ability, correlations, heterosis, inheritance pattern of characters, and gene effects for plant growth and curd attributes. Combining ability analysis of seven inbred lines of curd maturity group III (mid-December to mid-January) indicated that nonadditive effects were more important in the expression of plant height, plant spread, curd maturity, curd weight and curd size index (Lal *et al.*,

1978). Studying nature of gene effects using six generation means (P_1, P_2, F_1, F_2, B_1, B_2) of 36 cross combinations obtained by crossing six inbred lines of maturity group II with six inbreds of maturity group III, Lal *et al.* (1979) concluded that dominance and epistasis were quite high in expressions of curd weight and curd size indexes; the crosses showing high performance for these characters may be utilized for heterosis breeding.A few of the crosses also revealed significant additive component of variation indicating possibility of improvement in these characters by selection.

High heritability and genetic advance were observed for net curd weight, total plant weight, harvest index, curd size index, curd diameter, stalk length and leaf length in Indian cauliflower (Dubey *et al.*, 2003).

The genetics of qualitative and quantitative traits, genetic advance, heritability and combining ability are reported by several workers in cauliflower. Inheritance of qualitative characters was studied in detail in Indian cauliflower by Ahluwalia *et al.* (1977) and gene symbols for different traits were assigned as: stalk length–long *St*, short *st*; leaf apex–round *Ro*, pointed *ro*; habit–errect *E*, branching *e*; curd colour–yellow *Y*, white *y*; flower stalk length–long *F*, short *f*; flower stalk colour-varigated *V*, green *v*; siliqua length–long *SL*, short *sl* (Ahluwalia *et al.*, 1977)

Table 7.4: Genetics of Quantitative Characters

Character	*Nature of Gene Action*	*References*
Curd weight	Dominance and epistasis	(Swarup and Pal, 1966)
	Pronounced overdominance and epistasis	(Singh *et al.*, 1975)
	Additive and dominance gene action	(Singh *et al.*, 1976a; Jyoti and Vashistha, 1986; Gangopadhayay *et al.*,1997; Sharma *et al.*,1988)
Curd to plant ratio	Partial dominance	(Kale *et al.*, 1979)
Curd diameter	Predominance of dominance gene action	(Lal *et al*, 1979)
Curd size index	Pronounced overdominance and epistasis	(Singh *et al.*, 1975);
	Dominance and epistasis	(Swarup and Pal 1966, Lal *et al.*, 1979)
	Additive dominant gene action	(Singh e*t al.*,1976,Sharma *et al.*,1988);
	Partial dominance	(Kale *et al.*, 1979)
Curd angle	Pronounced additive gene action	(Lal *et al.*, 1979)
	Additive and dominant gene action	(Dadlani, 1977; Chand, 1980)
Curd compactness	Polygenic	(Nieuwhof and Garretson, 1961)
	Dominance and additive gene action	(Lal *et al.*, 1979);
	Additive	(Vashistha *et al.*, 1985).
Maturity		
(i) Earliness	Partially dominant gene action	(Watts, 1964)
	Dominance and epistasis	(Swarup and Pal, 1966)
	Predominance of additive gene action	(Singh *et al.*, 1975; 1976b; Lal *et al.*, 1979; Mahajan *et al.*, 1996; Gangopadhayay *et al.*, 1997)
	Additive gene action	(Kale *et al.*, 1979)
	Additive and dominant gene action	(Sandhu and Singh,1977; Sharma *et al.*1988).
(ii) Lateness	Recessive polygenes	(Watts, 1963)

Adapted from Sharma *et al.* (2004).

Combining ability was studied to select parents for hybridization and good cross-combinations for production of hybrids. The inbred 103 exhibited the best *GCA* for all characters and cross-combination 105 x 108 showed maximum *SCA* for yield potential in early Indian cauliflower (Lal *et al.*, 1977). Further, Lal *et al.* (1978) found that parents 308,303,302 had high GCA for curd weight and curd size index in mid Indian cauliflower. In Snowball group, Lawyana was the best general combiner for curd weight, curd size and leaf size. Sel.12 and Pyramis were the best general combiners for early maturity and gross weight respectively (Sharma *et al*, 1988). Lines IHR3, IHR4, IHR9 and IHR36 were good combiners for most of the characters. The selection of parents on the basis of *per se* performance and general combining ability was effective (Pandey and Naik, 1986). Days to curd initiation, curd weight, number of leaves and plant height and diameter and curd weight influenced the plant weight (Pandey and Naik, 1985). According to Gangopadhaya *et al.* (1997), in early cauliflower, self-incompatible lines cc-13 and vv (351) were the best general combiners for earliness, curd colour, compactness and yield contributing characters, respectively.

Genetics

Several plant characters like plant type, leaf characters (petiolate, leaf apex, margin and arrangement and glossy), stalk length, curd colour and flower colour have simple inheritance. The important quantitative traits governed by polygenes in cauliflower include curd diameter, compactness, maturity, weight, depth, size index, shape and yield. The loose, bracted and ricey defects in cauliflower curds are perhaps polygenic characters greatly influenced by environment. Perhaps looseness and riceyness of curds are highly heritable and it is difficult to eliminate them completely.

Table 7.5: Important Simple Inherited Characters in Cauliflower

Plant Characters	*Inheritance Pattern*
Plant type	Single gene, erect dominant (EE)
	Three major genes-additive, dominance and epistasis
Leaf characters	
Petiolate	Single gene, dominant (PET)
Leaf apex	Single gene, dominant for round apex (RO)
Leaf margin	Single gene, dominant for wavy (WY)
Leaf arrangement around curd	Two genes, dominant for semi-blanched (Bl_1, Bl_2)
Glossy leaf	Single gene, recessive (gl)
	Two genes, inhibitory (IG)
Stalk length	Single gene, long stalk dominant
Curd colour	
Orange	Single gene
Green	Two genes
Retentive White	Three genes for bleaching, controlling peroxidase activity
Flower colour (white, yellow, cream)	Two independent genes, dominance, epistatic interaction with few modifiers

Adapted from Swarup (2006).

Table 7.6: Inheritance of important Quantitative Traits in Cauliflower

Character	Inheritance
Curd maturity	Polygenic, predominance of additive gene action; earliness partially dominance gene action additive and dominant gene action; lateness controlled by recessive polygenes, dominance towards earliness
Curd diameter	Polygenic, predominance of dominance gene action; additive and dominance gene action; low heritability
Curd depth	Polygenic, additive genetic variance; dominance and additive gene action
Curd size index (diameter/depth)	Polygenic, epistasis; over dominance and epistasis; dominance and epistasis; partial dominance and additive gene action; partial dominance; highly complex inheritance
Curd compactness	Polygenic, dominance and additive gene action; two major genes (Co_1, Co_2) in which Co_1 is epistatic to Co_2; semi-compactness controlled by both recessive genes
Curd weight	Polygenic, dominance and epistasis in Snowball group epistasis; partial dominance and additive gene action; over dominance and epistasis; additive and dominance gene action, in tropical cauliflower
Curd shape	Polygenic, partial dominance for smooth curds in Italian types.

Adapted from Swarup (2006).

Self-incompatibility

Most of the heat tolerant tropical Indian cauliflower varieties have self-incompatibility. Self-incompatibility system in cole crops is sporophytic and governed by multiallelic single genes (S gene) with dominance relationship. In self-incompatible plants, the incompatibility is effective two to three days before anthesis and hence bud pollination procedure is adopted for self-pollination. The S-alleles are influenced by environment, especially temperature, and may not be stable in all cases. The breeders in India developed stable self-incompatible lines. Gangopadhyay *et al.* (1995) studied levels of self incompatibility in five lines of maturity group I and three lines of maturity group II Indian cauliflower on the basis of seed set data and fluorescence microcopy. All the four lines, aa (395), vv (351), CC 12 and CC 14 of maturity group I showed a high level of self-incompatibility. Lines aa (395) and CC 13 were promising. Intermediate to weak level of SI was found in all the three lines of maturity group II namely Improved Japanese, 443-6 and 443-8.Sharma *et al.* (2000) reported variety 74-6C as strongly self-incompatible. They also found one plant (P4) as self incompatible and stable in N-9102 and two plants (P14 and P15) in 74-6C.High level of self-incompatibility was reported in early cauliflower cultivars Agahani Small Leaf, IIVR-50 and Agahani-JBT-23/60(Singh *et al.*, 2002).

Cytoplasmic Male Sterility (CMS)

The *Ogura* cytoplasm of *Raphanus* is the most important source of sterile cytoplasm used in cauliflower. Cybrids were utilized to transfer CMS to cauliflower from *Ogura* radish.It had the problems of temperature sensitivity and chlorosis in hybrid plants. Now this problem is overcome by protoplast fusion. The CMS male sterility, especially *Ogura* is established in different groups of cauliflower and is being explored for F_1 hybrid development at IARI. The transfer of sterile 'Anand' cytoplasm from *B. rapa,* originally derived from wild species, *B. tournefortii* via *Brassica napus* into cauliflower is also explored as a new source for facilitating F_1hybrid breeding. A recessive *ms* gene in cauliflower is tagged using RAPD and RFLP markers to accelerate breeding of hybrids.

The transgenes, *Barnase, Bar* and *Barstar* are being utilized to develop transgenic cauliflower hybrids. The male sterility transgene *barnase* is also *Bacillus amyloliquifaciens,* in which ribonucleases destroy the tapetum layer in pollen and thus produce stable male sterile plants. The male sterility *ms* gene is linked to herbicide resistance gene, *bar.* The restorer line for the *barnase* sterile lines can be obtained by expressing the gene coding for *barstar.*

Local Cultivars

There are several local cultivars in India of varying maturities, commonly named after the season of curd maturity like Kunwari, Kataki, Agahani, Poosi, Maghi, etc. Mainly private seed companies of Hajipur (Bihar) and Faizabad (UP) market seeds of these local types. Many of these local cultivars are grown in Bihar, UP, Punjab, Haryana, Rajasthan, MP, Maharashtra and Gujarat.

The earliest selected local cauliflower varieties, Early and Main crop Patna and Early and Main Crop Benaras were from M/s Sutton and Sons, India, 1929.

Crop Improvement

There are two constituencies, the grower and the consumer the world over, cauliflower breeders need to satisfy. From growers point of view, the breeding objectives in crop improvement are: a) increased crop density; b) reduction in crosses due to diseases, insect pests and physiological disorders; c) heat tolerant; d) improved partition from curd weight/plant weight point of view and uniformity *i.e.* F_1 hybrid development. From consumers' point of view,quality is the main concern and for this whiteness, structure (density, surface texture, suitability for floreting), freedom from diseases, nutrition and sensorial value are the main breeding goals.

Previously, lack of useful self-incompatibility alleles in cauliflower limited development of F_1 hybrids. Identification of more useful alleles and the advent of genetic and cytoplasmic male sterility stimulated more cauliflower F_1 hybrid development in the last decade or so.

In India, main emphasis in cauliflower improvement is on development of cultivars and hybrids (SI and CMS system based) to heat and resistance to diseases (black rot, downy mildew, alternaria leaf spot, sclerotinia rot) and insect pests (diamondback moth and cabbage butterfly). Earlier emphases were on breeding heat tolerant early maturing varieties and hybrids. As a result, improved varieties Pusa Meghna, Pusa Early Synthetic and F_1 hybrid Pusa Kartik Sankar (CC x Sel 41-5) were developed and released for cultivation. Pusa hybrid 2 (CC x Sel 1-3-18-19) was the first hybrid developed in cauliflower in India using SI mechanism in November maturity group (Singh *et al.*, 1994). Recent emphasis on resistance breeding led to isolation of multiple resistance sources *viz.*, Kn-81 (DM, SR and Alt), BR-2 (DM, BR and DMB), Lawyana (BR, SR and DBM) and Armel (BR, SR and DM). These were involved in hybridization with commercially consumer acceptable land varieties/lines of each maturity group to transfer resistance. The improved elite material has already reached an advanced F_5/F_6 generation stage, being evaluated for resistance to respective disease/insect pest; yield potential and other horticultural traits.

In cauliflower, besides yield, special emphasis is needed to improve its quality characters, including nutritive value and insect-pest and disease resistance.Improvement in yield can be achieved through its component characters, which have direct or indirect effects. The important components are curd size, weight, depth, compactness and colour of curd besides their uniformity in size and maturity. The plant type including frame, stem length, harvest index and resistance to common biotic and abiotic stresses also need attention of the breeder. Gill *et al.* (1987) isolated a pure line, called Pusa Himjyoti, having retentive white curd from MGS 2-4 for July to October cultivation in hilly areas of Himachal Pradesh. Sel.12, a black rot resistant line is developed, using a black rot resistant tropical type line Sn-445 and Pusa Snowball-1 at IARI, Katrain (Gill *et al.*, 1983).

The systematic work on crop improvement started in forties at the IARI and later on at State Agricultural Universities (SAUs) located at Ludhiana, Pantnagar, Faizabad, Rahuri, Hisar and IIHR, Bangalore. Pusa Katki developed at IARI was released in 1954. It is suitable for October maturity. It has open type, yellowish-white and medium-sized compact curds. Two more introductions, Improved Japanese and D 96 from Israel belong to mid-season group. Improved Japanese has creamy white, compact curd and bluish-green, up-right leaves. It matures during November to mid-December. D96 could not become popular due to its susceptibility to several diseases. More cultivars are added in each group on the basis of work done at different institutions.

Table 7.7: Varieties of Indian Cauliflower Developed and Released in Different Maturity Groups

Name	Source	Year	Maturity	Yield (t/ha)	Remarks
Pusa Early Synthetic	IARI, New Delhi	1990	September-October	11	Helped in spreading cauliflower in non-traditional areas
Pusa Deepali	IARI, New Delhi	1975	October	12	Retentive white curds
Early Kunwari	PAU, Ludhiana	1960	September-October	8	Grown in limited areas only
Pant Gobhi 3	GBPUAT, Pantnagar	1993	October	10	
Pusa Hybrid 2	IARI, New Delhi	1993	November-December	27	First Public Sector hybrid, resistant to downy mildew
Pant Gobhi 2	PAU, Ludhiana	1993	November-December	12	
PG 26	PAU, Ludhiana			17	
PG 35	CCSHAU, Hisar		November-December	17	
Hisar 1	CCSHAU, Hisar		November-December	12	
IIHR 101	IIHR, Bangalore		October-November	10	
IIHR 105	IIHR, Bangalore		October-November	10	
Pant Shubhra	GBPUAT, Pantnagar	1981	December-January	20	A synthetic variety
Pusa Shubhra	IARI, New Delhi	1985	December-January	21	Resistant to black rot and curd-inflorescence blight
Pusa Synthetic	IARI, New Delhi	1975	December-January	22.5	Resistant to curd blight
Pusa Himjyoti	RS, IARI, Katrain	1985	December-January	22	Retentive white curd, suitable for summer sowing in hills above 1,000 m sea-level
Palam Uphar	HPKV, Palampur	2006	December-January	23	Retentive white compact curds
Pusa Sharad	IARI, New Delhi	1999	November-December	24	Retentive white compact curds
Pusa Meghna	IARI, New Delhi	2002	September-October	13	Suitable for May-June sowing in north Indian plains
Pusa Kartik Sankar (F_1)	IARI, New Delhi	2003	September-October	15	Suitable for May-June sowing and resistant to downy mildew.

Adapted from Gill and Sharma (1996)

Besides these cultivars, a number of varieties developed by private sector are marketed under different names. The improvement work on the Snowball or late types is being done at Regional Station, IARI, Katrain, Dr. YSPUHF, Solan, HPKV, Palampur and on a limited scale at Research Station, BCKVV, Kalimpong. Initially recommended varieties were Snowball or China Moti marketed by seed companies by importing their seeds. Availability period of this group is January-February, which later got extended to March with the development of new cultivars. The temperature requirement of this group is 10-16^{0}C. The seed of this group was commercially produced in 1956 for the first time at Katrain. Since 1956 the Katrain station developed and recommended a number of varieties like EC 12013 (Snowball 16 Holland), which was later replaced by Pusa Snowball 1 and Pusa Snowball 2 in 1975. Pusa Snowball 2 was dropped because of its poor seeding ability. Pusa Snowball K-1 released in 1983 is the most popular variety. It has field resistance to black-rot disease of cauliflower. It produces 30 t. ha^{-1} of retentive snow-white curds. Variety Dania in this group was developed at Kalimpong and is suitable for growing in eastern parts of the country. It provides white, compact curds during January to mid-February.

Development of F_1 hybrids is an important area of research in cauliflower. Self-incompatibility level is the highest in-group I, followed by group II and group III and the lowest in Snowball types. A number of self-incompatible alleles were isolated and maintained at the IARI. In the Indian cauliflower SB, SD, SV alleles are very strong and dominant, whereas SA and SE are very weak; SM is intermediate and more susceptible to breakdown. Using this self-incompatibility mechanism, Pusa Hybrid 2 and Pusa Kartik Sankar were released.

Cauliflower is also susceptible to a number of diseases and insect pests. Sources of resistance against major diseases are identified. In Indian cauliflower, indigenous collections, Sn 445, MGS 2-3 and Pua Kea of Israel possess resistance to black rot (*Xanthomonas campestris*). Variety Pusa Shubhra was released in 1985 as resistant to black rot using MGS 2-3 as resistance source. Resistance to black rot was dominant and polygenically inherited. Similarly, a heading broccoli (EC 103576) was resistant to black rot and sclerotinia rot (*Sclerotinia sclerotiorum*). Five more heading broccoli, *viz.* EC 131592, Janavon, Early Winter Adam's White Head, EC 177283 and KN 81, possess resistance to sclerotinia rot. Selection 12, resistant to black rot was evolved through hybridization of Sn 445 (tropical) with EC 12012 (temperate) types. Similarly, KT 25 developed by crossing EC 103576 (heading broccoli) and Pusa Snowball K 1, has high resistance to sclerotinia rot. Resistance to sclerotinia was recessive and polygenically controlled. An artificial inoculation technique was developed which involved partial colonization of detached petals with either the ascospore or mycelium for initiating the infection in cauliflower seedling. Under optimum environmental conditions, typical symptoms appeared 6 days after placement of partially colonized petals on the host surface. With this technique a large amount of breeding material can be screened at seedling stage. Downy mildew (*Peronospora parasitica*), is another serious disease of cauliflower, and can attack the plant at any stage of its growth. Resistance is found in breeding lines 3-5-1-1, 1-6-1-2, BR 2 and CC at IARI. Pusa Hybrid 2 having resistance to downy mildew was released in 1993. Resistance to downy mildew is dominant and polygenic. The pedigree and characteristic features of important varieties, synthetics and hybrids developed in India are given below.

Table 7.8: Important Varieties, Synthetics and Hybrids of Cauliflower Developed in India

Variety/Synth/Hybrid	*Pedigree*	*Chief Characteristics*	*Sources*
		September maturity	
Early Kunwari	Selection from local cultivar	Curds yellowish, mature in mid Sept-mid Oct., leaves bluish green, wavy with waxy bloom	PAU, Ludhiana
Pusa Early Synthetic	Synthetic variety	Curds yellowish white, medium size, moderately compact	IARI, New Delhi
Pant Gobhi-3	Synthetic variety of 8 inbred lines	Stem long, semi-erect leaves, curds creamy white and compact	GBPUAT, Pantnagar
Pusa Meghna	Selection	Semi-dome shaped, white, compact, medium sized curd weighing about 400g. Maturity 80-90 days	IARI, New Delhi
Pusa Kartik Sankar	F_1 hybrid (ccx Sel. 41-5)	Compact white curds, weighing about 450 g, slightly vigorous growing plant, maturity days 85-90	IARI, New Delhi
		October maturity	
Pusa Kataki	Selection from local cultivar	Plants medium-sized, leaves bluish green, wavy, curd creamy white	IARI, New Delhi
Pusa Deepali	Selection from local cultivar	Plants erect, leaves waxy green, short, rounded tips, curds compact, self-blanching	IARI, New Delhi
RC-JOB-1	Selection	High yielding, med. early maturing, resistant to stress conditions and resistant against caterpillars	SKN College of Agriculture Jobner, Rajasthan
		November maturity	
Pant Gobhi-4	Recurrent selection in a local Agahani cultivar	Medium-long stem, leaves semi-erect, curds creamy white, medium compact	GBPUAT, Pantnagar
Punjab Giant-26	Selection	Plants tall and spreading, curds big, yellowish white, compact	PAU, Ludhiana
Punjab Giant-35	Selection	Curds cream-coloured, compact	PAU, Ludhiana
Hisar 1	Selection	Curds medium to large sized, white	HAU, Hisar
Pusa Sharad	Selection	Plants spreading, curds white, very compact	IARI, New Delhi
Pusa Hybrid-2	F_1 hybrid cc x Sel-1-3-18-19	Plants medium, semi-erect, curds white, medium sized, compact, field resistance to downy mildew	IARI, New Delhi
		December maturity	
Pusa Shubhra	Selection from a triple cross, MGS-2-3, 15-1-1 and D96	Plants erect, stalk long, leaves bluish green, curds compact, white, highly resistant to black rot and curd and inflorescence blight	IARI, New Delhi
Pusa Synthetic	Synthetic, involving 7 inbred lines	Plants erect, plant frame narrow to medium, curds creamy white to white, tolerant to curd and inflorescence blight	IARI, New Delhi
Pusa Himjyoti	Selection	Plants erect, leaves bluish green, curds retentive white, self-blanched, compact, suitable for hills	IARI, New Delhi
Pant Shubhra	Recurrent selection	Curds compact, slightly conical, creamy white	GBPUAT Pantnagar

Contd..

Table 7.8–Contd..

Variety/Synth/Hybrid	*Pedigree*	*Chief Characteristics*	*Sources*
		Late maturity	
Pusa Snowball-1	Selection from a Snowball type cross between EC 1203 and EC 1202	Leaves upright, inner leaves tightly cover the curd, curds white, medium size, very compact	IARI regional station, Katrain
Pusa Snowball K-1	Selection from an exotic Snowball variety	Curds very white, very compact, leaves puckered, serrated, light green, tolerant to black rot, one week later than Pusa Snowball-1	IARI regional station, Katrain
Pusa Snowball KT 25	Selection from a cross between winter cauliflower EC 103576 and Pusa Snowball-1	Semi-erect dark green upright growing foliage and retentive white, very compact and medium sized curds. It has resistance to sclerotinia rot and black rot	IARI regional station, Katrain
Ooty-1	Selection	White compact curds	TNAU, Coimbatore

Adapted from Swarup (2006).

Breeding Open Pollinated Varieties

Crop breeding is a breeder's activity, picking up useful characteristics and putting up them together to develop a variety having desirable traits. Population improvement method is commonly followed for improvement of cole crops. In India mass selection is widely used for improvement of cauliflower. Inbreds are used in the breeding of hybrids, synthetics and open pollinated varieties or in intervarietal hybridization programme. In recent past, a variety, 'Pusa Sharad' in mid maturity group of Indian cauliflower was developed using recurrent selection method (Sharma *et al.*1999). The cross combinations involving inbreds, with low inbreeding depression and high heterotic residual effects due to additive gene action, which show better response to selection, are used in the breeding of composite varieties. Back cross method is commonly followed to transfer resistance from donor to recurrent parent. This method is followed to develop 'Pusa Shubhra' a cauliflower variety resistant to black rot, curd blight and riceyness (Singh *et al.*1993).

Pusa Early Synthetic and Pusa Synthetic cauliflower varieties were developed by synthesizing 6 and 7 parents, respectively and were released in India for commercial cultivation (Singh *et al.*, 1997 and Gill, 1993).

Intervarietal Crosses for Yield Improvement

Selection in F_2 generation of intervarietal crosses can be practiced to recombine and fix the favourable yield genes. A well-pronounced additive variance for yield or other characters in elite F_2 generation offers great potential for improvement in such characters through appropriate selection procedures. The occurrence of transgressive segregants (*i.e.*, individual plants, the performance of which is better than that of both the parents of the cross) towards favourable directions for many desirable economic characters in the F_2 generation of intervarietal crosses of cauliflower was reported by Swarup and Pal (1966). Such segregants can be selected to develop superior lines/varieties.

Heterosis Breeding

In cauliflower, heterosis was first reported by Jones (1932). Swarup and Pal, and Pal and Swarup (1966) found appreciable heterosis in snowball types for earliness (5-7days), curd weight (24.5–28.2 per cent), curd size index (22.54-34.85 per cent) over better parent. In different maturity groups of Indian cauliflower, many workers reported appreciable heterosis for economical characters. Kumaran (1971) made a study of three maturity groups of Indian cauliflower and recorded appreciable amount of heterosis for curd weight, plot yield, curd size index, maturity of curd, stalk length and leaf size index. Maximum heterosis(per cent) observed by him was 22.85, 12.07, 24.64, 21.53, 49.76, and 32.46, respectively. Swarup and Chatterjee (1972, 1974) investigated heterosis in Indian cauliflower and reported better manifestation of it in first maturity group, compared to other groups. Similarly, maturity group second was superior over maturity group third. In this respect, they recorded heterosis of 41.23, 16.00, and 23.41 percent for yield, curd size and curd weight, respectively in maturity group second, while in maturity group third, it was 22.22 and 27.00 per cent for curd size and curd weight, respectively. Appreciable amount of heterosis was reported in Indian cauliflower for different characters in different groups (Deshpande, 1975, Singh *et al.*1975, and Sandhu *et al.*, 1977).

Verma (1979) using male sterile lines as female parent, observed sufficient heterosis over better parent. The observed heterosis(per cent) was 93.3, 91.1, 55.6, 44.4, 20.0 and 2.2 for curd weight, plant height, curd size index, curd to plant ratio, early maturity and number of leaves, respectively. Hoser-Krauze *et al.* (1982) reported heterosis for earliness, curd diameter, curd weight and quality in 2 and 3 way reciprocal F_1 hybrids, which were made using three self-incompatible Indian cauliflower and three temperate self-incompatible lines. Pandey and Naik (1985) studied heterosis in hybrids of different sub-species of *Brassica*. Brazilian Broccoli line 137 was crossed with two cauliflower lines 138 and 149 from the USA and 10 Indian cauliflower cultivars. Heterosis for tallness, number of leaves and leaf area index was positive, except for the cross, Superfine Maghi x 137, which showed negative heterosis for leaf number. They also concluded that lines with high heterosis could be selected at the seedling stage itself.

Gangopadhayay *et al.* (1997) using four self-incompatible lines in maturity group first as female parent with 11male parents in the early and 7 in mid group reported 31.2, 25.3, 34.5, 25.0, 16.6, 25.7, 53.5, 87.0, 49.5, 71.8 and 82.3 per cent heterosis in early and 19.4, 17.7, 39.3, 42.9, 13.5, 20.8, 38.3, 20.6, 49.6, 51.9 and 63.2 per cent heterosis in mid group for days to maturity, days to 50 per cent maturity, curd compactness, colour, number of leaves, curd diameter, curd depth, curd size index, gross weight, marketable weight and curd weight, respectively.

Synthetic Varieties

About four to six inbred lines are used in a synthetic. The inbred lines after testing their general combining ability in a diallel cross, polycross or in a top cross can be synthesized to form a synthetic variety. Exploiting the pronounced additive genetic variance can develop a synthetic variety. A variety produced in this manner has a three-fold advantage: (1) its seeds can be easily produced from open-pollination and maintained by the farmer; (2) it is useful particularly in places where the commercial seed industry is not well-developed; and (3) it is broad-based and, therefore, can adapt better to changing growing environments.

Bhatia *et al.* (1978) reported that S_1, S_2 and S_3 generations of early maturing synthetic cauliflower showed 72.0, 37.0, and 20.8 per cent increase in curd weight, respectively, over that of the standard variety, Improved Japanese. This indicated a high-yield potential of synthetics evolved from Indian cauliflower.

Hybrid Seed Production

The advantage of using F_1 hybrids for commercial production in cole crops including cauliflower is well documented. Besides superiority for yield contributing characters, earliness, quality parameters, wider adoptability, resistance to some insect-pest and diseases; hybrids are advantageous by having uniformity in maturity, size and shape of the curds in cauliflower. Indian cauliflowers, especially those types belonging to early maturity groups are grown under adverse climatic conditions need immediate attention of breeders to develop commercial F_1hybrids. Lot of variability within the commercial varieties was observed in this group in size, shape and maturity time. Recently, an F_1 hybrid Pusa Kartik Sankar (DCH-541) developed by Indian Agricultural Research Institute was identified and released for cultivation under North Indian conditions through All India Coordinated Research Project on Vegetable Crops (Sharma, 2003) in early maturity group of cauliflower. Commercial acceptance of hybrids, in addition to their superiority for desirable traits also depends upon reasonable cost of F_1hybrid seed production. Cost of seed should be within the reach of common growers. Hand emasculation and pollination are not economically effective in cole crops, since the seed quantity produced/pollination is very low. Cauliflower is a naturally cross-pollinated crop and to substantiate this has inbuilt genetic mechanisms like self-incompatibility and male sterility. The self-incompatibility and male sterility are used in the production of F_1hybrid seed in most of the cole crops to avoid hand emasculation and to reduce cost of F_1seeds.

Since bud pollination is labour intensive, time consuming and expensive, spraying of 5 per cent Sodium Chloride (Nacl) solution at the time of full bloom to overcome self-incompatibility in Indian cauliflower as an alternative for maintenance of homozygous selfincompatible lines was advocated from IARI.

Marker Assisted Improvement of Quantitative Traits

Until recently, most analysis of quantitative trait loci (QTLs) in *Brassica* focused on locating QTLs, respectively for disease resistance (Camargo *et al.*, 1995; Voorrips *et al.*, 1998). Early inheritance studies focused on morphological traits, although complex inheritance was often observed, suggesting that many genes controlled the traits. Lan and Paterson (2000) attempted to resolve these complex inheritance patterns *i.e.* a series of QTLS associated with cauliflower curd traits using three F_2 populations. Identification of QTLs for a particular trait contributes to crop improvement strategies through marker-assisted selection (MAS), especially where the trait is of high value. Segregating populations of F_1 derived doubled haploid (DH) lines from divergent parents were used to construct a detailed linkage map of the *Brassica oleracea* genome based on RFLP, AFLP and simple sequence repeats (SSR) markers and identification of QTLs controlling developmental characteristics (Sebastian *et al.*, 1999, 2002). Genetic linkage maps of *Brassica oleracea* created at HRI, Wellesbourne (UK), comprise over 600 molecular markers (King, 2004).

Breeding for Improved Quality

The extrinsic quality characters in cauliflower are curd colour, solidity, freeness from bracting, pinking and riceyness. The cauliflower curd consists of a mass of short peduncles bearing many thousands of apical meristems. These meristems normally develop to give flowers, but a large proportion of them (usually over 90 per cent) abort before or occasionally during floral phase. The curd is, thus, a precociously developed floral button, and its appearance at marketable stage is affected by the normal ontogeny of flowering and by death of excess floral material.

The quality of the curd, in terms of cosmetic appeal to consumers is largely determined by the timing of morphological changes associated with flowering in relation to curd reaching a marketable size. The distinct features characteristic of poor cosmetic quality are disfiguring defects like elongation of peduncles, the precocious development of apical meristems into flower buds, growth of bracts and leaves from peduncles through surface of curd, and development of pink or purple colourations in the curd. The first two of these defects, known respectively as looseness and riceyness, are essential parts of the flowering process, and selection against them is largely a matter of ensuring that they appear as late as possible after the curd reached a marketable size. The later two, namely bracting and pinking are essential for flowering and therefore selection against them is for their complete elimination.

The environment affects the expression of bracting, pinking and riceyness. This may, for example, be completely absent in some seasons but a serious problem in others. On comparison of the morphology of curds when aseptically cultured in nutrient solution with the effects of their genotype on bracting and pinking as shown by a progeny test (Crisp, 1975; Crisp, Gray and Jewell, 1975; Crisp, Jewell and Gray, 1975) and found that the formation of bract-like structures in culture was reflection of the genotypic tendency to do this in the field. Assessment in culture and in the field increased the selection pressure against bracting. The results with pinking are even more useful because whilst the assessments of purple colourations in field and culture were phenotypically independent, both were genotypically related to appearance of defect in the field, possibly owing to different genes governing the same phenotype. With respect to riceyness, having inherent association with endogenous synthesis of auxins, screening of genotypes can be done at an early stage by growing young cotyledons aseptically on auxin free nutrient medium where ricey susceptible genotypes will strike roots within a week (Kalia *et al.*, 1994).

This suggests that a two-tier system be applied to selecting against bracting, pinking and riceyness in cauliflower to produce potential new cultivars with bract free, ricey free non-pink curds. If high proportions of plants selected from the field are free from these defects, their defect development in culture should allow more stringent selection against the defects. In addition, assessment of bracts in culture may be carried out as a complement to screening for anthocyanin production (Crisp *et al.*, 1975), to the vegetative propagation of selected plants (Crisp and Walkey, 1974) and to elimination of viruses from infected plants (Walkey *et al*, 1975), any of which may be necessary in a breeding programme with cauliflower. Graham Teakle (2004) reported two MADS-box regulatory genes *viz.* BoAP1-a and BoCAL-a that are present at loci having key roles in determining formation of cauliflower. A number of cDNA for different MADS-box genes are isolated from cauliflowers, which include both floral promoters and repressors. The expression pattern and genetic map position of these genes will help predict their potential relationship with cauliflower quality *i.e.* bracting and riceyness.

The intrinsic quality traits include nutrition related parameters like vitamins, minerals, protein, carbohydrates, fats, flavour etc. Inherently, cauliflower is not rich in nutritional traits, but since 'Indian cauliflowers' is a new group with wide variability for horticultural traits, exploited for varietal improvement. Variability for nutritional traits needs to be investigated and exploited in the present scenario of nutritional security.

Cauliflower flavour is very delicate which increases its popularity, therefore breeding for flavour will also draw attention of breeders in future programmes. Cauliflowers like other crucifers contain glucosinolates. These are hydrolyzed by myrosinase to give volatile products, mainly nitriles and isothiocyanates, which contribute characteristic flavour to the curd. A varietal difference in both individual and total glucosinolate content was reported in cauliflower (Sones *et al.*, 1984). Breeding for different glucosinolate contents might, therefore, be possible.

The work of Li and Garvin (2003) revealed that in cauliflower *Or* gene showed promise for unraveling the molecular control of carotenoid accumulation and for creating a potentially new avenue for enhancing provitamin A content in food crops. Their efforts focused on a map-based cloning strategy for isolating the *Or* gene, with the establishment of a high resolution map of the *Or* locus and the construction of a cauliflower BAC library. A BAC contig encompassing the *Or* gene was built, physical mapping delimited the *Or* locus to a 50 kb genomic fragment within a single BAC and sequencing of this BAC revealed 15 putative ORFS. From this single ORF as the *Or* candidate gene was identified. On cloning *or* gene, it may be a very useful tool for significantly enhancing the carotenoid content of plant foods normally low in β-carotene.

Breeding for Biotic Resistance

With the evolution of a large number of cultivars for different seasons/climates, cauliflower and other cole crops are grown round the year. All these resulted in the continuous build-up of disease inoculum and insect population. The common diseases of Indian cauliflower are black rot, bacterial soft rot, sclerotinia rot, downy mildew and black spot (*Alternaria sp.*). The important insect pests are Diamondback moth, cabbage butterfly caterpillar, aphids, cabbage head borer and Bihar hairy caterpillar (*Spilosoma oblique*). Management of these diseases and pests using chemical pesticides is not only cumbersome and costly but also poses health hazards. It is imperative to have resistant varieties for which identified resistant sources and knowledge of genetics of resistance for a particular disease and pest are imperative. The related information for economically important biotic stresses of cauliflower are discussed hereafter:

Diseases

Downy Mildew

It is caused by an obligate parasite, *Peronospora parasitica* that can infect the crop at any stage of growth. It is systemic in nature and infection observed at seedling stage can reappear at curd and marketing stage (Crute and Gordon, 1987). In Indian cauliflower, BR-2, CC and 3-5-1-1; EC177283, Ec191150, EC191157, Kibigiant, Merogiant, EC191140, EC191190, EC191179 and Noveimbrina are resistant (Singh *et al.*, 1987, Mahajan *et al.*, 1991). MGS2-3, 1-6-1-4,1-6-1-2 and 12C (Chatterjee, 1993); KT-9 (Sharma *et al.*, 1991) Early Winter Adam's White Head (Sharma *et al.*, 1995); CC-13, KT-8, xx, 3-5-1-1,CC (Trivedi *et al.*, 2000); Perfection, K1079, K102, 9311 F1 and 9306 F1 (Jensen *et al.*, 1999); Kunwari-7, Kunwari-8, Kunwari-4 and First Early Luxmi (Pandey *et al.*, 2001) were resistant to moderately resistant. Pusa Hybrid-2 (Singh *et al.*, 1994) of Indian cauliflower and Pusa snowball K-25 of Snowball type having resistance to downy mildew were released for commercial cultivation in India. Resistance to downy mildew is ascribed to a single gene with dominant effect (Sharma *et al.*, 1991; Mahajan *et al.*, 1995; Jensen *et al.*, 1999), single gene with recessive effects (Hoser-Krauze *et al.*, 1984; Mahajan *et al.*, 1995) or several genes (Hoser-Krauze *et al.*, 1995).

Sclerotinia Rot

This disease is caused by *Sclerotinia sclerotiorum*. It has a very wide host range and infects most of dicot crops, but is more severe in seed crop of cauliflower, though, it may attack the crop at an early stage of its growth also. Moderate resistance to this pathogen was reported in EC131592, Janavon, EC103576, Kn-81, Early Winter Adam's White Head, EC162587, EC177283 (Kapoor, 1986: Baswana *et al.*, 1991, Singh and Kalda; 1995, Sharma *et al.*, 1995 and Sharma *et al.*, 1997). Resistance is polygenically controled and recessive in nature (Baswana *et al.*, 1993 and Sharma *et al.*, 1997). Pusa Snowball K-25 developed using EC103576 as resistant source with Pusa Snowball-1, is released for commercial cultivation, which possesses field resistance against this disease.

Pandey *et al.* (2003) reported moderately resistant lines of early cauliflower to sclerotinia rot, namely Kataki-6, Kataki-13, Patna Kataki, Deep Malika, Suryamukhi, Pusa Himkaran, Early Laxmi and PDVR early. Kataki-13 and Kataki-6 showed high degree of tolerance.

Black Rot

The disease is caused by a bacterium *Xanthomonas campestris* (Pam) Dawson. Yellowing of leaves starts from leaf margin and extend in the direction of midrib, followed by blackening of veins (vascular bundles). Cauliflower lines reported as resistant sources are Sn 445, Pua kea and MGS2-3 (Sharma *et al.*, 1972); RBS-1, EC162587 and Lawyana (Sharma *et al.*, 1995); Sel-12 (Gill *et al.*, 1983); Sel-6-1-2-1 and Sel-1-6-1-4 (Chatterjee, 1993) and Avans and Igloory (Dua *et al.*, 1978).

Some of the above sources are used in the development of resistant varieties. Pusa Shubhra was developed, using Pua kea and MGS2-3 lines and recommended for commercial cultivation (Singh *et al.*, 1993). Pusa Snowball K-1 is field resistant to black rot (Gill *et al.*, 1983).

The resistance was dominant and governed by polygenes and dominance components of variation were more pronounced than additive (Sharma *et al*, 1972). Jamwal and Sharma (1986) reported that a single gene governs dominant resistance.

Of the 54 accessions wound inoculated with four isolates of X_{CC} race 4 at the juvenile stage, A 19182 and A 19183 exhibited no symptoms, and the accessions PI 199947, PI 199949 and PI 194256 segregated for resistance to X_{CC} (Tunguc and Griffths, 2004).

Three segregating F_2 populations were developed by self-pollinating 3 black rot resistant F_1 plants, derived from a cross between black rot resistant parent line 11B-1-12 and the susceptible cauliflower cultivar 'Snow Ball'. Plants were wound inoculated using 4 isolates of *Xanthomonas compestris* pv *campestris* (X_{CC}) race 4, and disease severity ratings of F_2 plants from the three populations were scored. A total of 860 arbitrary oligonucleotide primers were used to amplify DNA from black rot resistant and susceptible F_2 plants and bulks. Eight RAPD markers amplified fragments associated with completely disease free plants following black rot inoculation, segregated in frequencies far lower than expected. Segregation of markers with black rot resistance indicates that a single, dominant major gene controls black rot resistance in these plants. Stability of this black rot resistance gene in populations derived from 11B-1-12 may complicate introgression into *B. oleracea* genotypes for hybrid production (Tonguc *et al.*, 2003).

Black Leaf Spot (*Alternaria* sp.)

In cole crops, the black leaf spot disease is caused by *Alternaria brassicae* or *Alternaria brassicicola*. Brown to black, small to elongated spots appear on leaves, stems of older leaves. In younger plants, it causes symptoms like *Rhizoctonia solani*. When the fungus infects the curd, especially in seed crop, the disease is called as inflorescence blight. Resistance was found in Indian cauliflower lines, MGS2-3, Pua Kea and 246-4 (Sharma *et al.*, 1972); 23-7, 466, MS98, 210-21, Sel-9, 443-7(Trivedi *et al.*, 2000); IIHR142 and IIHR217 (Pandey *et al.*, 1995) and Snowball KT-9 (Sharma *et al.*, 1991). Resistance to curd blight is dominant in nature, polygenically inherited and in general additive effects were more pronounced than dominant one (Sharma *et al.*, 1975). Pusa Shubhra having resistance to curd blight is released for commercial cultivation (Singh *et al.*, 1993). Both additive and dominant gene action played a role in resistance but partial dominance is more important (King and Dickson, 1994). They found no linkage between leaf colour (red or green) and leaf spot resistance.

Insect Pests

Many workers reported varietal differences in susceptibility to various pests in cauliflower under field conditions. The glossyness of leaves though not a undesirable character, but is associated with small sized plants, having poor curd quality. Recombinants have not yielded encouraging results in breeding programme. Dickson *et al.* (1986) reported that PI 234599, a glossy-leaved cauliflower from Australia, was highly resistant in the field to diamondback moth (*Plutella xylostella*) but the resistance could not be separated from glossy leaf character because the two traits are closely linked.

Resistance to cabbage head borer (*Hellula undalis* L. *Fabricius*) was reported in cauliflower, ES-97, ES-96, Katiki (J.B), KW-5, KW-8, KW-10, Kunwari (RB), Kathmandu Local, Early Patna, EMS-30 and PSK-16 (Lal *et al.,* 1991). Lal *et al.* (1994) also found resistance under field conditions in Indian cauliflower F_1 hybrids like aa X ES102, aa x Kataki (JB), aa x First Early, aa x First Crop, aa x Sel.100, aa x Sel.41 and aa x 824 have resistance to Bihar hairy caterpillar (*Spilosoma oblique*). Aphids cause major losses to cole crops. The aphid species responsible for economic losses in cauliflower and other cole crops are cabbage aphid (*Brevicoryne brassicae*), green peach aphid (*Myzus persicae*) and turnip aphid (*Lipaphis erysimi*). Resistance to cabbage aphid is reported in NY 13816, NY101181, NYIr 9602 and NYIR 9605 but work on cauliflower is very scanty. Natural occurring compounds like, glucosinolates, pipecolic acid and b-nitroprionic acid in the tissues of *Brassica* plants are responsible for resistance to cabbage looper and imported cabbageworm. Breeding resistant varieties in cauliflower and other cole crops for most of pests and some diseases still remain elusive because hardly any resistant source with desirable degree of resistance is available in the germplasm. Genetic engineering offers a safe and long lasting solution. Cauliflower can be conveniently transformed by using *Agrobacterium* provided suitable genes are available. Introduction of genes coding for *Bacillus thuringiensis* (Bt) insecticidal crystal protein; cowpea and soyabean trypsin inhibitors and cytokinin biosynthesis enzymes provide substantial protection against the insects in the crop (Kumar and Sharma, 1997).

Table 7.9: Insect Pests of Vegetable Brassicas and Potential Resistance Conferring Genes

Common Name	*Scientific Name*	*Useful Genes*
Diamond back moth	*Plutela xylostella*	*Cry 1A* class
Cabbage butterfly	*Pieris brassicae*	*Cry 1A* class
Stem-borer	*Hellula undalis*	*Cry 1A* class
Hairy caterpillar	*Spilosoma oblique*	*Cry 1A* class
Cut-worm		*Cry 1A* class
Aphids	*Agrotis ipsilon*	Snowdrop lectin
Thrips	*Caliothrips indicus*	Cry 2A
Leaf-minor	*Phytomyza hartiocola*	

These Bt genes in genetically transferred plants have proved quite successful. Similarly, there are plant enzymes, which solublize fungal cell walls and cause membrane damage. Some of the antifungal proteins listed below for different fungi are responsible for causing disease in cauliflower and other crops (Kumar and Sharma, 1997). Transgenic plants expressing genes coding for such proteins could be developed for developing resistant varieties.

Table 7.10: Important Fungal Diseases of Vegetable Brassicas and Useful Genes that Confer Fungal Resistance

Disease	*Causative Fungus*	*Useful Genes*
Damping off	*Pythium* spp	Permantis
Downy mildew	*Peronospora parasitica*	Chitinase
Wire-stem	*Rhizoctonia solani*	Glucanase
Leaf-spot	*Alternaria brassicae*	Thionin
Cabbage yellows	*Fusarium oxysporum*	RIP
White rust	*Albugo candida*	Osmotin
Black-rot	*Xanthomonas campestris*	Chitinase
Soft-rot	*Erwinia carotovora*	Chitinase

Economically potential and useful target genes, like trypsin inhibitors, CaMv caspid, antisense CaMv gene VI and anther specific BcP1 genes were also explored in transforming cauliflower crop (Hu and Leonard, 2003).Seedling explants based *Agrobacterium* mediated transformation was proposed for cauliflower genetic engineering. Transferring/incorporating proteinase inhibitor genes are also proposed to develop transgenic cauliflowers resistant to insects.

Improved Growing Practices of Raising Successful Cauliflower Crops

Climatic Requirements

The Indian cauliflowers grow under high temperature and rainfall conditions, forming curds at 12 to 27°C. The late Snowball types form curds at 10 to 16°C. Based on the temperature requirement for curd initiation and development, cauliflower cultivars are divided into four maturity groups. These maturity groups are also known by the name of English or Hindi months in which the curd reaches marketable stage in North Indian plains *viz.*, September or Kunwari; October or 'Kataki'; November or 'Agahani' and December or Poosi and/or Maghi. Snowball types belong to late group.

Fluctuation in temperature at the time of curd initiation and development stage adversely affects the quality of curds. If the temperature were higher or lower than the required, the curds would either be leafy and ricey.

Varietal Selection

Earlier, the cauliflower was cultivated only during winter. Now, it can be cultivated from May to March and curds are available from August end to March under North Indian conditions. Specific cultivars are available with ability to form curd at a particular range of temperature from 10 to 27 °C.

If early cultivars are planted late then, instead of normal curd, they form small buttons and ricey curds. Similarly, if late types were planted in the early season, these would continue to grow vegetatively and form curds only when the required temperature is set on.

Sowing and planting time as suggested for different maturity groups may be adjusted in the different parts of India based on the climatic conditions. Some early cultivars are also found suitable for cultivation in spring summer (March- April) plantings. These are also suitable for cultivation in Kerala, Karnataka, Andhra Pradesh and Orissa. With wide range of varieties available today, cauliflower can be cultivated round the year in one or the other parts of India.

Seed Management

Cauliflower is a cross-pollinated crop and is propagated through seeds. Genetically pure seed is required to raise good crop. For raising a healthy crop, it is desirable that the seeds of a selected variety are procured from a Government agency or reputed traders/company or supplier. It is always advisable to consult local Agriculture Department or Krishi Vigyan Kendra about the reliable source of quality seed.

Early cultivars are planted at a closer spacing than mid and late types. The former requires more number of seedlings/unit area than the later. In general, the early group requires 400 to 500 g and the late group 300 to 400 g seed to raise seedlings to be planted in one hectare.

Hot water treatment of seed at 50^{0}C for 30 minutes controls seed borne diseases. Seed treatment with captan (3 g/kg) or thiram (3 g/kg) or bavistin (2 g/kg) is done before sowing for surface sterilization to check damping off and get healthy seedlings.

Nursery Management

Select a sunny place having proper drainage to prepare the beds. The nursery area should be treated with formaldehyde (1:38) solution. After drenching with formaldehyde, the area is to be covered with polythene or tarpaulin at least for 72 h. Before preparation of beds, the polythene sheet should be removed and left open for at least for 4 to 5 days.

The nursery beds should be of 15 cm height and half-meter width in soil of good tilth. Add farmyard manure (FYM) @ 6-8 kg/10m^{2}with small amount of phosphorus and potash. Excessive amount of nitrogenous fertilizer should be avoided because this results in the development of lanky seedlings. About 50-75 m^2 area is enough to raise nursery for one-hectare land.

For May or early June nursery raising for early maturity group, when the temperature is very high, humidity is low and hot winds are blowing, the nursery beds should be 30 cm wide only so that water percolates up to the centre of beds from the side channels at the time of irrigation. This will help in keeping them wet. After seed germination, the beds are covered with *sirkis* or shading nets to protect the young seedlings from hot winds.

Early and mid season cauliflowers are prone to pre and post emergence "damping off" diseases because, the climatic conditions are very congenial for their development for which nursery management is essential.

Seeds should then be sown at a depth of 1.5 to 2.0 cm in rows spaced at 7 to 8 cm apart. Thick sowing is avoided and about 20 to 30 seeds are sown/30 cm long row. After sowing, seeds are covered with a thin layer of mixture of fine manure and soil and the beds are gently levelled. The seedbed is covered with a thin layer of dry grass. This would help in maintaining uniform moisture and facilitate germination. Sufficient moisture should be nmaintained. The grass cover is removed as soon as the seeds sprout 3-4 days after sowing. Regular hoeing and weeding are done and proper moisture is maintained to get healthy seedlings. Crowded seedlings are spaced by proper thinning for healthy growth.

Transplanting

Transplant 4 to 5 weeks old, healthy, strong and stocky seedlings to get uniform growth and stand in the field. Reject blind and diseased seedlings.

Seedlings are ready for transplanting after 30 to 40 days of sowing. Watering is stopped at least 4 to 5 days before transplanting to harden the seedlings. But, a day before uprooting, the seedling beds are watered to avoid damage to the root system and facilitate uplifting.

Avoid use of more than 6 weeks or less than 4 weeks older seedlings to get good cauliflower crop. Generally, the cultivars of early cauliflower group are transplanted on ridges and that of mid and late groups on flat beds keeping in view the climatic conditions.

Transplanting is done either on a cloudy day or in the evening hours to get good stand and crop followed by light irrigation.

Spacing

The spacing between the plants in early, mid and late crops is kept 35-45, 40-45 and 30-40 cm respectively, depending upon cultivars. The spacing between the rows is kept 45-60 cm for all the groups to attend cultural practices including earthing up without damaging the plants. A very close spacing should be avoided as it results in the development of vigorous vegetative growth and small sized curds.

Land Preparation

If fallow land is available, it should be deep ploughed during summer months of May-June. This would help in minimizing the soil borne diseases, pests and perennial weeds.

Green manuring of fallow land with the onset of first rainfall by sowing *dhaincha, sanai* or *lobiya* in the end of June or early July will not only increase soil fertility but also improve the physical structure and water holding capacity. Plough down 50 to 60 days old green manuring crop and give sufficient time for its decomposition before planting cauliflower.

Soil Type

Sandy or sandy loam soils, though, give an early crop but curds are loose, have poor staying ability and yield realization is also low.

Crop raised in the clay loam soil put slow growth in the initial stage, form compact curds having better keeping quality and gives higher yield.

In general, the sandy loam soils for early maturity group and clay loam soils for mid and late maturity groups are preferred.

Cauliflower crop does well in the neutral soil having P^Haround 6.5 Acidic soil affects the availability of molybdenum and magnesium and alkaline soils of boron. These are very important for proper growth and curd formation of in cauliflower.

Early cauliflower types can be grown in the saline soils having C index within tolerance range (C_2 to C_4) but late types commonly cultivated at lower temperature have high salt tolerance (C_6). The yields of cauliflower are affected considerably when cultivated in saline soils.

Nutrient Management

Cauliflower is a heavy feeder of mineral elements. It removes large amount of macronutrients from soil. Use of biofertilizers in combination with chemical fertilizers and organic manures offers a great opportunity to increase production of cauliflower with less cost. This combination was efficient in yield increase over exclusive application of chemical fertilizers and can be attributed to increase in uptake of nutrients resulting in faster synthesis and translocation of photosynthesis from source (leaves) to sink (curd). The lower yield with RDF could be due to reduction in adequate supply of mineral nutrients because of fixation. Application of VAM a+ any level of P+100 per cent NK, *Azospirillum* and *Azotobacter* with 100 per cent NPK and 75 per cent N + 100 per cent PK recorded significantly higher vitamin C content (59.9 to 60.5 mg/100 g) compared to RDF and other treatments (54.5 to 55.4 mg/100 g). The significant increase in the total N, P and K content of curd was observed with biofertilizers compared to the RDF but the contents were compared among different biofertilizers. There was an over all increase in the available NPK in soil after harvest with the application of biofertilizers.

Among the biofertilizers application of VAM with 100 per cent NPK recorded maximum available NPK (301.5, 31.7 and 340.7 kg ha $^{-1}$, respectively) content in soil. This could be due to non-symbiotic nitrogen fixation and phosphorus mobilization by the VAM.

The application of *Azospirillum* and *Azotobacter* with 75 per cent, N + 100 per cent PK or with 100 per cent NPK and PSB or VAM with 75 per cent P + 100 per cent N K or 100 per cent NPK has reputed in 12.72 to 37.00 per cent increased yields. Significantly higher net returns and benefit cost ratio were obtained with the application of VAM with 100 per cent NPK and 75 per cent P +100 per cent over the rest of the treatments. It is possible to reduce the application of nitrogenous fertilizers by 25 per cent through inoculation with *Azospirillum* (or) *Azotobacter* and 25 per cent P through PSB or VAM.

Application of biofertilizers (*Azospirillum, Azotobacter* PSB and VAM) along with inorganic produced significantly higher yield (18.6 to 22.6 t ha $^{-1}$) compared to application of recommended dose of fertilizers (RDF) (16.5 t ha $^{-1}$). Application VAM + 100 per cent RDF gave significantly higher yield (22.6 t ha $^{-1}$) compared to the application of all the biofertilizers + inorganic fertilizer but comparable with VAM + 75 per cent P and 100 per cent N and K. The use of bio-fertilizers also enhanced the nutrient status of available nitrogen, phosphorus and potassium in soil. Higher benefit cost ratio of 2.96 was observed with application of VAM + 100 per cent RDF followed by VAM + 75 per cent P and 100 per cent NK (2.72) and lowest with RDF application (1.95). Hence, 25 per cent of nitrogen and phosphorus can be saved with the use of biofertilizers (N and P sources), respectively to achieve higher benefit cost ratio (Narayanamma *et al.*, 2005).

Successive increase in nitrogen levels up to 150 kg N ha $^{-1}$ increased plant height, leaf length, leaf width, days to curd initiation, days to curd maturity, curd diameter, curd depth, net curd weight and marketable curd yield. There was no significant difference in curd solidity at 150 kg N $^{-1}$ over 100 kg N ha $^{-1}$. Phosphorus levels up to 120 kg P_2O_5 ha $^{-1}$increased plant height, leaf length, leaves/plant, net curd weight and marketable curd yield significantly over its lower levels. Leaf width, curd diameter, curd depth and curd solidity at 120 kg P_2O_5 ha $^{-1}$was statistically at par with of 80 kg P_2O_5 ha $^{-1}$. Application of 150 kg N ha $^{-1}$ and 80 kg P_2O_5 ha $^{-1}$ recorded the highest value w.r.t. leaves/plant, curd diameter, curd depth, net curd weight, curd solidity and marketable curd yield, gave the highest net returns (Rs. 37285 ha $^{-1}$) and improved the benefit: cost ratio (3.25:1) than other treatment combination (Jana and Mukhopadhyay, 2001).

The maximum average weight of curd (408.5 g), diameters (12.84 cm) and curd yield (64.85q ha $^{-1}$) were found with the application of 90 kg N and 80 kg P_2O_5 ha $^{-1}$ (Swarup *et al.*, 1999).

The full doses of phosphorus and potash, one-third dose of nitrogen be applied and mixed thoroughly in soil just before planting. Remaining quantity of nitrogen should be splitted into two equal doses and applied a month after transplanting and at the time of curd initiation. Foliar spray of 1 to 2 per cent urea helps in boosting the crop growth, minimise the formation of buttons (small curds) and increases the productivity. It should be done before the curd initiation stage.

Micronutrients

Micronutrients are also important for normal growth and curd formation. Molybdenum and boron are very important micronutrient in case of cauliflower. Deficiency of molybdenum occurs in acidic soils resulting in to Whiptail that can be corrected by applying sodium molybdate @0.28kg/ha in 250-1000 liters water. Boron deficiency occurs in alkaline soils leading to browning of curds and hollow stem. This can be corrected by applying sodium borate @ 5.6kg/ha in 250-1000 liters water lowering the pH to 6.5(Fordham and Biggs, 1985).

Irrigation

Depending upon soil type and weather condition, the irrigation schedule should be adjusted in such a way that a continuous supply of moisture for uniform growth and development of curd is maintained. During dry weather and in light soil, the frequency of irrigation is kept higher than in wet weather and heavy soils. Proper moisture should be maintained during curd initiation and development to get quality curds and good yield.

For the optimum growth of cauliflower crop, irrigation should be given when the moisture contents have fallen below 50 per cent of field capacity to get good cauliflower crop.

Mulching

Farm waste or polythene of different types can be used for mulching.

Use of clear plastic mulch helps in controlling the weeds, and conserving the soil moisture, thereby resulting in better growth and higher yield of cauliflower crop. Mulching with clear plastic also hastens the maturity of cauliflower crop.

Mulching of cauliflower crop with farm wastes reduces the temperature, conserve the moisture, activate the microorganisms and add nutrient to the soil. These factors also help in increasing the yield and production of quality curds in cauliflower. Substantial increase in cauliflower yield is obtained by mulching with 10 cm thick paddy husk.

Weed Control

As soon as plants establish after transplanting, weeds also start appearing in the field. The weeds need to be kept under check since these compete with the crop for nutrients, water, light and air.

These also harbour various diseases and insect-pests and become a vehicle for their spread in the main crop.

Weeds may be controlled either manually or by the use of herbicides or blend of both may also be used.

Pre-planting application of 3.3 litre Stomp 30 EC (Pendimethalin) or 2.0-2.25-litres Basalin (Alachlor)/ha in the form of spray in the finally prepared field controls the dicot weeds for more than a month.

Integrated Management Practices

Cultural Practices

1. Do the summer ploughing repeatedly to expose the organisms, which remain in the soil and cause damage to the crop.

2. Use available resistant varieties for cultivation such as Pusa Shubra and Pusa snowball K1 in the areas prone to black rot and curd blight and Pusa Snowball K-25 in the areas prone to black rot, downy mildew, sclerotinia rot.
3. Rotate the cauliflower crop with non-cruciferous crops.
4. Avoid "off season" crop in areas, which have high incidence of diseases and insect-pests during the main season.
5. Keep weeds under check by inter cultivation.
6. Use hot water treated seed to get rid of seed borne diseases.
7. Irrigate the crop lightly in late evening hours.

Mechanical Practices

1. Avoid damage to the curd at the time of harvesting.
2. After harvesting, discard the curds affected by insects, soft rot, black rot and other diseases or have mechanical injury or bruising.

Chemical Practices

1. Give one spray of Ridomil MZ-72 (one g in two litres of water) or 0.2 per cent Mancozeb (Dithane or Indofil M-45), when conditions are conducive for the appearance of downy mildew, black leaf spot or other foliar diseases. If needed, repeat the spray.
2. Drench the nursery with captan or thiram (3 g/litre) solution before sowing @ 5 l/m^2 and also treat the seed with above fungicides before sowing to check "damping-off". If post emergence "damping-off" is observed, apply either of the above-mentioned fungicides (2 g/litre) in between the rows.
3. If weather is conducive for spread of black and/or soft rot, spray the crop with streptocycline or Agromycin (1 g/10 litres of water) or plantomycin (1 g/litre of water).
4. Spray the crop judiciously with endosulfan or malathion (2 ml/litre of water) or pyrethrins (1 ml/litre) to keep the sucking, chewing and biting types of insects under control.

Pre- and Harvest Care

During curd developmental stage, proper moisture should be maintained in the field. The dry spell followed by irrigation during this period result in the development of loose, unattractive and poor quality curds.

Excessive application of nitrogenous fertilizer results in the development of ricey, fuzzy and loose curds.

As soon as curds have attained full size these are harvested immediately.

While harvesting hold the leaves on top of the curd in one hand so that it can be cut at ground level with wrapper leaves. Harvesting is done preferably in the late evening so that the curds remain turgid and fresh at the time of marketing in the following morning. The outer large leaves are then trimmed away leaving only sufficient Jacket leaves to protect the curds from bruising and other injuries. More Jacket leaves are kept when cauliflower is to be transported loose in trucks or carts to minimize the bruising of the curds. These plants are loaded by keeping the curd downward. When it is transported to distant markets, plants are packed in gunny bags having wider mesh in such a way to devoid the curd from bruising. While packing in crates, the Jacket leaves are trimmed leaving a fringe of leaves projecting 2-3 cm above the curd.

Curd Maturity and Grading

The curds are sometimes graded after harvesting. Those having weight above two kilograms are graded as large, between one and two kg as medium and below one as small. However, curds must be compact, creamy to white, free from discoloration, riceyness, and fuzziness and bruising. Diseased, loose and bruised curds are also avoided at the time of grading.

After removing all the leaves completely, the curd is packed in cling wrap film. This helps in maintaining the freshness of the curd and helps in fetching higher price in the market.

In glut season, cauliflower curds are frozen and then marketed during lean period. The cauliflower curds are washed, cut into pieces, packed in poly bags of the desired size and then frozen either singly or with carrot or peas.

For dehydration, segments of cauliflower curds are separated and blanched for 4-5 minutes in boiling water, steeped in 0.5 per cent sulphur dioxide solution for about an hour, then drained and washed. The product is dried at 60^{0}C for about 12 hours. These segments can be marketed as dehydrated cauliflower.

Seed Production

The annual cultivars of cauliflower depend upon varying temperature for curd initiation and development. In India, only annual types are cultivated. These are further divided into tropical annual or Indian cauliflower and temperate annual or snowball or late types.

Seed Production Areas

The seed production of tropical cauliflower was undertaken in India since its evolution in subtropical regions. The seed producing states of this group are Bihar, Uttar Pradesh, parts of Haryana, Punjab and Rajasthan. Seed production technique for late cauliflower was standardized at Katrain for which areas in Kullu and Kashmir were identified. India started producing its own seeds of late cauliflower since early sixties. Besides Kashmir and Kullu valley, its seeds are also being produced in Saproon Valley (H.P.), Kalimpong (W.B.), Nilgiri Hills (T.N.) and hills of Uttarakhand.

Planting Time

The cultivars of different groups of tropical cauliflower for seed production are planted at the same time as for vegetable production. Since curd precedes flowering, so for proper selection there should be normal curd formation. However, in early types (September maturity) the seed yield obtained in normal planting is low. These types bolt, when the temperature is not favourable for seed setting and are affected by disease like *Alternaria etc.* Seed growers plant the early crops slightly late than normal season to get better seed yield. It affects the earliness of crop and hence it is suggested to avoid this practice.

However, in case of snowball types, planting is adjusted in such a way that plants complete vegetative growth before the on set of winter and curds develop starts after winter (March). The crop is planted in early October for seed in hills.

Seed Production with Seed to Seed Method

Seed to seed method is followed since transplanting after curd development makes the plant more prone to *Alternaria*, *Erwinia* and other defects. In Indian cauliflower, it takes about 25-40 days from curding to full bloom depending upon the variety. The optimum temperature for fertilization and seed development is 12-18°C. Cultivars of early group are more susceptible to diseases like *Erwinia* and *Alternaria*, distortion of inflorescence due to prevailing high temperature and humidity. Distortion of inflorescence is more common when the day temperature is more than 30°C. The commercial seed growers delay the transplanting of the early maturity group to get higher seed yield. This should not be encouraged since the selection of seed producing plants is in late sowing form curd at lower temperature. This effects the maturity time, is the main reason for longer harvesting period in early cultivars. Compact solid curd gives low seed yield due to lesser number of bolters. Some plants even do not bolt. Scooping of central portion or longitudinal half cutting or partial marginal cutting of curd is practiced to get more number of bolters. But scooped or half cut curds become more prone to diseases. Spray with 0.1 per cent solution of Aureofungin to check decay. Spraying of the crop with Gibberellic acid 50-100 ppm (0.5-1 g/10 litres) at curd development and bolting stage helps in increasing the number of bolters and total seed yield by loosening the curd after proper development. Another spray of GA_3 (50-100 ppm) at bolting stage 20 days after the first spray gives higher seed yield.

The highest seed yield per plant and per hectare (12.05 g and 6.65 g, respectively) was obtained from 5th September transplanting date of early cauliflower and variety Pusa Deepali recorded highest seed yield of 5.67 q/ha. This was attributed to high values of characters like height of inflorescence stalk, length of siliqua, number of seeds/siliqua and seed yield/plant (Thapa *et al.*, 2002).

Isolation

Cauliflower is a highly crosspollinated crop. It does not only cross with its own cultivars but also with other members of cole group *i.e.* cabbage, broccoli, brussel sprouts, knolkhol, kale, collard and other wild species.Proper isolation is kept to produce quality seed. Isolation of 3000m for breeder seed and 1600m for certified seed is recommended.

Pollinators

Like other cole crops, the honeybees are the usual pollinating agents, though bumblebees and other insects are also being reported for pollination.

Four to Five honeybees' colonies/ha are necessary at the time of flowering for proper pollination and to get good seed yield.

If spraying with insecticide becomes necessary, it is advisable to spray the crop in the late evening hours using safe insecticides like endosulfan, which are less harmful to honey bees.

Roguing

At least four inspections/roguings are required in cauliflower. First roguing is done at vegetative stage to remove off type plants and second at curd initiation. Third rouging is done after curd formation. Plants forming loose, ricey, and fuzzy and buttons are rejected. Blind, deformed and diseased plants are also rejected. Fourth rouging is done after bolting but before flowering, plants with peripheral and uniform bolting are kept for seed production. Early and late bolters are also rejected.

Disease and Insect Pest Management

Sclerotinia rot, black rot, alternaria black leaf and pod spot and soft rots are main diseases encountered after curd development when the crop is kept for seed production. The common insects found during this period are aphids, caterpillars of white butterfly and diamond back moth. To control sclerotinia rot, a serious disease of cauliflower seed crop, a spraying schedule starting from curd initiation to pod development is followed. Crop is sprayed with bavistin (0.1 per cent) followed by dithane M-45 (0.2 per cent) + bavistin (0.1 per cent) at 15-20 days intervals. At least, 4-5 sprays are needed. Since, this is a soil borne disease, rotation of monocot crop (not affected by sclerotinia rot) is very important. Other diseases and insect pests are managed using same pesticides as recommended for vegetable crop production.

Soft rot of curd caused by *Erwinia caratovora* and sclerotinia rot caused by *Sclerotinia sclerotiorum* is most serious disease of seed crop besides black rot and Alternaria black leaf spot. A control schedule starting from curd initiation to pod development is recommended (Kapoor *et al.*, 1988). Crop is sprayed with Bavistin (0.1 per cent) followed by Dithane M-45 (0.2 per cent) + Bavistin (0.1 per cent) at 15-20 days interval. At least 4-5 sprays are needed. This also keeps the crop free from other foliar fungal diseases. The soft rot and black rot are managed by Streptocycline (0.01-0.02 per cent) sprays. Aphids and caterpillars of diamond back moth and *Pieris spp*. are more common. Since honey bees and other flies are the main pollinating agent, so spray the crop late in the evening hours with safer insecticides or biopesticides.

Harvesting and Seed Yield

When the pods attain yellow colour, should be harvested, delayed harvesting results into seed shattering. Hence, 2-3 harvesting are required. About 50-60 days are needed for pod maturity after fertilization. Seeds of early types are ready for harvesting in February-March and mid maturity in March-April months in North Indian plains. The harvest is cured under shade for 3-5 days and then gently thrashed after sun drying for 2-3 days. After cleaning the seed is properly graded using grading machine and before storage should be dried to minimum moisture (about 5 per cent). However, the seed of snowball types are ready for harvesting by June in India. Early types when grown under normal planting give about 250-300 kg/ha seed yield. However, seed yield of Indian cauliflower may vary between 500-600 kg/ha and Snowball types from 300-500kg/ha.

References

Ahluwalia, K.S., Swarup V. and Chatterjee S.S. 1977. Inheritance of qualitative characters in Indian cauliflower. *Veg. Sci.* 4: 67-80.

Baswana, K.S., Rastogi, K.B. and Sharma, P.P.1991. Inheritance of stalk rot resistance in cauliflower (*B.oleracea* var.*botrytis L.*).*Euphytica*, 57:93-96.

Baswana, K.S., Rastogi, K.B. and Sharma, P.P. 1993. Inheritance studies in cauliflower. *Veg.Sci.*20: 56-59.

Bhatia, G.L., Chatterjee, S.S. and Swarup V. 1978. Yield potential of early maturing cauliflower synthetics. *Indian J. Agri. Res.*, 12(1): 9-14.

Camargo, L.E.A., Williams, P.H. and Osborn, T.C. 1995. Mapping of quantitative trait loci controlling resistance of *Brassica oleracea* to *Xanthomonas campestris pv* campestris in the field and glasshouse. *Phytopathology*, 85: 1296-1300.

Chand, J. 1980. Studies on selection for yield and quality in cauliflower. Ph.D. Thesis, PG School, IARI, New Delhi.

Chatterjee, S.S. 1993. Cole crops. In: Vegetable Crops. Bose, T.K., Som, M.G., and Kabir, J. (Eds) Naya Prokash, India pp.125-223.

Chatterjee, S.S. and Swarup, V. 1972. Indian Cauliflower has a still greater future. *Indian Hort.*, 17(3): 18-20.

Chatterjee, S.S. and Swarup, V. 1983. Seeds and Farms. 9: 15-16.

Crisp, P. 1975. Meristem culture and the breeding of cauliflowers. *Proc. Eucarpia meeting on cruciferae,* 1974(A.B. Wills and C. North): SCRI, Dundee, U.K., pp 55-59.

Crisp, P. 1982. The use of an evolutionary scheme for cauliflowers in the screening of genetic resources.*Euphytica* 31:725.

Crisp, P., Cray, A.R. and Jewell, P.A. 1975. Selection against the bracting defect of cauliflower. *Euphytica,* 24: 459-65.

Crisp, P., Jewell, P.A. and Gray, A. R. 1975. Improved selection against the purple colour defect of cauliflower curds. *Euphytica,* 24: 177-80.

Crisp, P., and Walkey, D.G.A. 1974. The use of aseptic meristem culture in cauliflower breeding. *Euphytica,* 23: 305-13. Crute, I.R.and Gordon, P.L. 1987. Downy mildew *rep.natn.veg.res.sta.*for 1986/7:54.

Dadlani, N.K. 1977. Studies on selection for yield and quality in cauliflower. Ph.D. Thesis, PG School, IARI, New Delhi.

Deshpande A.A. 1975. Genetical studies in early Indian cauliflower. M.Sc. Thesis, PG School, IARI, New Delhi.

Dickson, M.H., Eckenrode, C.J. and Lin, J. 1986. Breeding for diamondback moth resistance in *Brassica oleracea.* (In) Proceedings of the First International Workshop on Diamondback Moth Management. (Ed). N.S. Talekar. 137-143.

Dua, I.S., Suman, B.C. and Rao, A.V. 1978. Resistance of cauliflower (Brassica *oleracea* var.*botrytis*) to *Xanthomonas comestris* influenced by endogenous growth substances and relative growth rate. *Indian. J. experimental Biology*.16:488-491.

Dubey, R.K., B.P. Singh and Hari Har Ram 2003. Genetic variability, heritability and genetic advance for quantitative characters in Indian cauliflower (*Brassica oleracea* L. var. *botrytis* L.). *Veg. Sci.,* 30(1): 81-82.

Feldner, H. 1956. New varieties of cauliflower developed locally. Bull. Independent Biol. Labbr. Keffrr Malal, Israel, 12(2): 13-14 (Pl. Breed. Abst. 1961, 31: 3063).

Fordham, R. and Biggs, A.G. 1985. Principles of Vegetable Crop production. Publs. Collins, 8 Grafton Street, London.

Gangopadhyay, K.K., H.S. Gill and S.R. Sharma 1997. Heterosis and combining ability studies in early group of Indian cauliflower involving self-incompatibility lines. *Veg. Sci.,* 24(1): 26-28.

Gangopadhyay, K.K., H.S. Gill and T.A. More 1995. Assay of self-incompatibility in Indian cauliflower (*Brassica oleracea* L. var. *botrytis* L.) with special reference to maturity group II and I. *Veg. Sci.,* 22(1): 42-47.

Gill, H.S. 1993. Improvement of cole crops. Advances in Horticulture–Vegetable Crops, Vol.5 (Eds.) Chadha, K.L. and Kalloo, G., Malhotra Publishing House, New Delhi: 287-303.

Gill, H.S., Lakhanpal, R.D., Sharma, S.R. and Bhagchandani, P.M. 1983. K-1, a valuable addition to "Snowball" group of cauliflower. *Indian Hort.,* 27(4): 23-24.

Gill, H.S., Sharma, S.R. and Kapoor, K.S. 1987. Pusa Himjyoti cauliflower for the hills. Indian Hort. 32(2): 13-16.

Gill, H. S. and S. R. Sharma 1996. Cole Crops. (In): 50 Years of Crop Science Research in India (Paroda, R. S. and Chadha, K. L., Ed). ICAR, New Delhi. pp. 635-647.

Gilbert, J.C. 1953. Improved hot weather cauliflower types. *Circ. Hawaii Agric. Exp. Sta.,* 49: 7.

Gilbert, J.C. 1955. Pua Kea cauliflower for low elevations. *Hawaii Fm. Sci.,* 4(2): 5-8.

Haine 1959. Time of heading and quality of curd in winter cauliflower. *J. Nat. Inst. Agric. Bot.,* 8:667-674.

Hoser-Krauze J., Gabryl J. and Antosik J. 1982. Influence of the cytoplasm of Indian self-incompatible lines on the earliness and quality of F1 cauliflower curds. *Eucarpia cruciferae Newl.* 7 : 12.

Hoser-Krauze J., Lakowaska, Ryk E. and Antosik J. 1995. The inheritance of resistance of some *Brassica oleracea* L. cultivars and lines to downy mildew (*P.parasitica*). *Ex. Fr. Jour. of applied genetics*. 36 (1): 27-33. Instt. of Veg. Crops. Poland.

Hu, Chiang-Yeh and Leonard, Clair, M. 2003. Transgenic cauliflower plant. (*In*) Improvement of Plant Genetic Engineering Vol. 5 (Ed.) R.P. Singh and Jaiwal, P.K., pp. 177-200.

Jamwal, R.S. and Sharma, P.P. 1986. Inheritance of resistance to black rot (*Xanthomonas compestris*) in cauliflower (*B.oleracea* var.*botrytis L.*). *Euphytica,* 35(3): 941-943.

Jana, J.C. and T.P. Mukhopadhyay, 2001. Effect of nitrogen and phosphorus on growth and curd yield of cauliflower. Var. Aghani in Teraizone of West Bengal. *Veg. Sci.,* 28(2): 133-136.

Jensen, B.D., Hockenhull, J. and Munk L. 1999. Seedling and adult plant resistance to downy mildew in cauliflower (*Brassica oleracea* var. *botrytis*). *Pl. Pathology*. 48 (5): 604-612.

Jones, H.A. 1932. Vegetable Breeding at the Univ. of Calfornia. *Proc. Amer. Soc. Hort. Sci.* 29 (5): 572-581.

Jyoti, S. and Vashistha, R.N. 1986. Gene effects studies of curd weight in mid season cauliflower (*Brassica oleracea* var. *botrytis* L.). *Haryana J. Hort. Sci*. 15: 263-266.

Kale, P.N., Chatterjee, S.S. and Swarup, V. 1979. Heterosis and combining ability in Indian cauliflower (*Brassica oleracea* L. var. *botrytis* L.) *Veg Sci.* 6 (1): 11-20.

Kalia, P. 1994. *In Vitro* screening against riceyness in cauliflower. Proc. Inter. Hort. Cong., Kyoto, Japan, August 21-27, 1994.

Kapoor, K.S. 1986. Disease resistance in cauliflower (*Brassica oleracea* var.*botrytis*) against *Sclerotinia sclerotiarum* (lib) De Bari. *Veg.Sci.*13: 285-288.

Kato, T. 1964. On the flower head formation and development in cauliflower. I. Ecological studies on flower head formation and development. *J. Jap. Soc. Hort. Sci.,* 33: 316-326.

King, G.J. 2004. Genomics for crop improvement in *Brassica*. HRI Internet Website.

King, S.R. and Dickson, M.H. 1994. Identification of resistance to *Alternaria brasicicola* in *Brassica oleracea*. *Eucarpia Cruciferae Newl.*16: 126-127.

Kumar, P.A. and Sharma, S.R. 1997. Improving vegetable brassicas biotechnologically. Indian Hort. 41 (4): 55-56.

Kumaran, N.M. 1971. Studies on combining ability, Gene effects and Heterosis in cauliflower. Ph.D. Thesis, PG School, IARI, New Delhi.

Lal, G., Chatterjee, S.S. and Swarup, V. 1978. Studies on combining ability in Indian cauliflower. *Genet. Agr.,* 32: 85-87.

Lal, G., Chatterjee, S.S. and Swarup, V. 1979. Genetics of important characters in Indian cauliflower. *Gent. Agr.*, 33: 181-190.

Lal G., Swarup V. and Chatterjee S.S. 1977. Combining ability studies in early Indian cauliflower. *J. Agri Sci* 89: 169-75.

Lal, O.P., Gill, H.S., Sharma, S.R. and Singh, R 1991. Field evaluation in different cultivars of cauliflower against the head borer (*Hellula undalis* Fabricius) (Lepidoptera: pyralidae). *J.Ent.Res.*15 (4): 277-281.

Lal, O.P., Sharma S.R.and Singh, R 1994. Field resistance in different hybrids of cauliflower against Bihar hairy caterpillar (*Spilosoma obliqua* Walker) (Lepidoptera: arctidae). *J. Ent. Res.18* (1): 45-48.

Lan, T.H. and Paterson, A.H. 2000. Comparative mapping of quantitative trait loci sculpting the curd of *Brassica oleracea. Genetics,* 155: 1927-1954.

Li, L. and Garvin, D.F. 2003. Molecular mapping of *Or*, a gene inducing beta-carotene accumulation in cauliflower (*Brassica oleracea* L. var. *botrytis*). *Genome,* 46: 588-594.

Mahajan V., Gill H.S., Singh R. 1991. Screening of cauliflower germplasm lines against downy mildew. *Cruciferae Newl.* 14/15, 148-149.

Mahajan V., Gill H.S., and Sharma S.R. and Singh R. 1996. Combining ability studies in Indian cauliflower (*Brassica oleracea* var. *botrytis* L.) Group III. Veg. Sci. 23 (2): 166-70.

Mahajan, V., Gill, H.S. and More, T.A. 1995. Inheritance of downy mildew resistance in Indian cauliflower (group III). *Euphytica,* 86:1-3.

Mortensen, E. and Bullard, E.T. 1968. Handbook of tropical and subtropical horticulture. Dept. of State Agency for International Development, Washington, DC.

Narayanamma, M., C.H. Chiranjeevi, I.P. Reddy and Riazuddin Ahmed, S. 2005. Integrated nutrient management in cauliflower (*Brassica oleracea* var. *botrytis* L.). *Veg. Sci.,* 32(1): 62-64.

Nieuwhof, M. and Garretsen, F. 1961. Solidity of cauliflower curd. *Euphytica,* 10: 301.

Pal, A.B. and Swarup, V. 1966. Gene effects and heterosis in cauliflower II. *Indian J. Genet* 26 (3): 282-94.

Pandey, K.K., P.K. Pandey and Singh, B. 2003. Artificial screening against white rot for resistance sources in Asiatic group of cauliflower. *Veg. Sci.,* 30(1): 77-78.

Pandey, S.C. and Naik, G. 1985. Factor analyzing of yields components in cauliflower (*Brassica oleracea* var.*botrytis*). Eucarpia Cruciferae Newl. 10:68-70.

Pandey, S.C. and Naik, G. 1986. Genetic and combining ability studies in cauliflower. *Eucarpia Cruciferae Newl.* 11:36-37.

Pandey S.C., Naik G., Ramkishan and Sridhar T.S. 1995. Breeding resistant varieties in cauliflower and cabbage. In Agroecosystem Management {Mandal *et al.,* Eds}, IIHR, Bangalore.

Paul, W.R.C. 1933. A note on the cultivation of cauliflower in the low country districts of Ceylon. *The Trop. Agriculturist,* 81(2): 91-94.

Rodrigo, P.A., Urbanes, P.G. and Hernais, P.S. 1935. A progress report on the acclimatization of cauliflower. *Philippine J. Agri.,* 6: 115-127.

Sadik, S. 1967. Factors involved in curd and flower formation in cauliflower. *Proc. Amer. Soc. Hort. Sci.,* 90: 252-259.

Sandhu, J.S. and Singh, A.K. 1977. Inheritance of maturity period and curd weight in cauliflower (*Brassica oleracea* var. *botrytis* L.).*Haryana J. Hort. Sci*. 6 (3-4): 161-65.

Sandhu, J.S., Thakur, J.C. and Nandpuri, K.S. 1977. Investigations on hybrid vigour in cauliflower (*Brassica oleracea* L. var. *botrytis* L.). *Indian J. Hort*. 34: 430-34.

Sebastian, R.L., Kearsey, M.J. and King, G.J. 2002. Identification of quantitative trait loci controlling development characteristics of *Brassica oleracea* L. *Theor. Appl. Genet.*, 104: 601-609.

Sebastian, R.L., Howell, E.C., King, G.J., Marshall, D.F. and Kearsey, M.K. 1991. An integrated AFLP and RFLP *Brassica oleracea* linkage map from two morphologically distinct doubled haploid mapping populations. *Theor. Appl. Genet.*, 100: 75-81.

Seshadhri, V. S. and Chatterjee, S. S. 1996. The history and adaptation of some introduced vegetable crops in India. Vegetable Science 23(2): 114-140.

Sharma, S.R., Gill, H.S. and Kapoor, K.S. 1988. Heterosis and combining ability studies in late cauliflower (*Brassica oleracea* var. *botrytis* L.). *Veg. Sci.* 15 (1): 55-63.

Sharma, B.R., Swarup, V.and Chatterjee, S.S. 1972. Inheritance of resistance to black rot in cauliflower. *Can.J.Gen.Cytol.*14:363-370.

Sharma, B.R., Swarup, V.and Chatterjee, S.S. 1975. Inheritance of resistance to drying at bolting and seed formation stage in cauliflower. *Indian.J.Genet*.35 (1): 44-48.

Sharma, B.R.,Dhiman,J.S.,Thakur,J.C.,Singh,Aand Bajaj,K.L.1991. Multiple disease resistance in cauliflower. *Advances in Horticultural Science*. 5(1): 30-34.

Sharma, S.R., Gill, H.S. and Kapoor, K.S. 1997. Inheritance of resistance to white rot in cauliflower. *Indian J.Hort*. 54(1): 86-90.

Sharma, S.R., Kapoor, K.S.and Gill, H.S. 1995. Screening against Sclerotinia rot (*Sclerotinia sclerotiarum*), Downy mildew (*Peronospora parasitica*) and black rot (*Xanthomonas compestris*) in cauliflower (*Brassica oleracea* var.*botrytis* sub var.*cauliflora* DC). *Indian J. Agric. Sci*.65 (12)916-918.

Sharma, S.R., Singh, R.and Gill, H.S. 1999. Cauliflower Pusa Sharad. *Indian Hort.* 44(3):7-8.

Sharma, R., Thakur, J.C. and Khattra, A.S. 2000. Studies on self-incompatibility and its stability in open pollinated varieties of Indian cauliflower. *Veg. Sci.*, 27(2): 152-154.

Sharma, S. R., Singh, P. K. Chable V. and Tripathi, S.K. 2004. A Review in Hybrid cauliflower development.In: Hybrid Vegetable Development (Singh, P.K., Dasgupta, S. K. and Tripathi, S. K., eds., Food Products Press, an Imprint of the Haworth Press, Inc.pp.151-193.

Singh, B., Singh, A., Pal, A.K. and Banerjee, M.K. 2002. Evaluation of self-incompatibility in Indian cauliflower (*Brassica oleracea* L. var. *botrytis* L.). *Veg. Sci.*, 29(2): 142-145.

Singh, D.P., Swarup, V. and Chatterjee, S.S. 1975. Genetical studies in Indian cauliflowers (*Brassica oleracea* L.var. *botrytis* L.). I. Heterosis and combining ability in maturity group I. *Veg. Sci*.2:1-7.

Singh, D.P., Swarup, V. and Chatterjee, S.S. 1976b. Genetical studies in Indian cauliflower III (*Brassica oleracea* L. var. *botrytis* L.). Heterosis and combining ability in maturity group-III. *Veg. Sci.* 3 (1): 47-53.

Singh, R. and Sharma, S.R. 2001. Cole crops. (In) Thamburaj, S. and Singh, N. (Eds.) *Textbook of Vegetables, Tubercrops and Spices*. pp 76-146. Directorate of Information and Publications of Agriculture, Indian Council of Agricultural Research, New Delhi.

Singh, D.P., Swarup, V. and Chatterjee, S.S. 1976a. Genetic studies in Indian cauliflower (*Brassica oleracea* L.var.*botrytis*).Heterosis and combining ability in maturity group II.*Veg. Sci.*3 (1): 41-46.

Singh, H. and Kalda, T.S. 1995. Screening of cauliflower germplasm against sclerotinia rot. Indian J.Genet.55(1):98-102.

Singh, R., Chatterjee, S.S., Swarup, V., Trivedi, B.M. and Sen, B. 1993. Evolution of Pusa Shubhra–A resistant cauliflower to black rot and curd blight diseases. *Indian J. Hort.* 50 (4): 370-372.

Singh, R., Gill, H.S. and Sharma, S.R. 1994.Breeding of Pusa Hybrid–2 cauliflower. *Veg.Sci.*21 (2):129-131.

Singh, R., Trivedi, B.M., Gill, H.S. and Sen, B. 1987. Breeding for resistance to black rot, downy mildew and curd blight in Indian cauliflower. *Eucarpia Cruciferae Newsl.* : 96-97.

Singh, R., Gill, H.S. and Chatterjee, S.S. 1997. Breeding Pusa Early Synthetic cauliflower. *Veg.Sci.* 24(1): 23-25.

Sones, K., Heaney, R. K. and Fenwick, G. R. 1984. Glucosinolates in *Brassica* Vegetables: investigation of twenty-seven cauliflower cultivars (*Brassica Oleracia* var. *botrytis* subvar. *cauliflora* DC.). *Food Agric.* 35, 762.

Swaroop, K., Suryanarayana, M.A. and Sharma, T.V.R.S. 1999. Effect of nitrogen and phosphorus on growth and curd yield of cauliflower (*Brassica oleracea* L. var. *botrytis* L.). *Veg. Sci.*, 26(1): 85-86.

Swarup, V. 2006. Cauliflower. (In) *Vegetable Science and Technology in India*. Kalyani Publishers. pp.359-370.

Swarup, V. and Chatterjee, S.S. 1972. Origin and improvement of Indian cauliflower. *Econ. Bot.*, 26: 381-393.

Swarup, V. and Chatterjee, S.S. 1974. Improvement of Indian cauliflower. *Indian Jour. Genet. Pl. Breed.* 34A; 1300-1304 (SABRAO Proceedings).

Swarup, V. and Pal, A.B. 1966. Gene effects and heterosis in cauliflower. *Indian J. Genet.*, 26(3): 269-281.

Teakle, G. 2004. MADS about cauliflower quality. HRI Internet Website.

Thapa, U., Maity, T.K. and Psati, M.K. 2002. Effect of planting date and variety on seed production of early cauliflower. *Veg. Sci.*, 29(2): 189-190.

Tonguc, M., Earle, E.D. and Griffiths, P.D. 2003. Segregation of distortion of *Brassica carinata* derived black rot resistance in *Brassica oleracea. Euphytica,* 134: 269-276.

Tonguc, M. and Griffths, D. 2004. Evaluation of *Brassica carinata* Accessions for resistance to black rot (*Xanthomonas campestris* pv *campestris*). *HortScience,* 39(5): 952-954.

Trivedi, B.M., Sen, B., Singh, R., Sharma, S.R. and Verma, J.P. 2000. Breeding multiple disease resistance in mid season cauliflower. Proc. Indian Phytopath.Soc.golden jublee International Conference on Interegated plant disease management for sustainable agriculture.Vol.2: 699-700.

Vashistha, R.N., Neog, S.J. and Pandita, M.L. 1985. Gene effects studies of curd compactness in cauliflower. Haryana.Agric.Univ.J.Res.15: 406-409.

Verma, S.S. 1979. Studies on heterosis breeding in cauliflower. Ph.D. Thesis, PG School, IARI, New Delhi.

Voorrips, R.E., Jongenus, M.C. and Kanne, H.J. 1997. Mapping of two genes for resistance to club root (*Plasmodiophora brassicae*) in a population of doubled-haploid lines of *Brassica oleracea* by means of RFLP and AFLP markers. *Theor. Appl. Genet.*, 94: 75-82.

Walkey, D.G.A., Cooper, V.C. and Crisp, P. 1975. The production of virus-free cauliflowers by tissue culture. *J. Hort. Sci.* 49: 273-5.

Watts L.E. 1963. Investigations breeding I. Studies on self-incompatibility. *Euphytica* 12: 330-340.

Watts, L.E. 1964. Studies of maturity in F_1 and F_2 generations of cauliflower from crosses between summer, autumn and winter types. *J. Hort. Sci.*, 39: 84-91.

Wood, R.C. and Jones, H.M. 1936. Cauliflower cultivation in the tropics. *Tropical Agriculture,* 13(8): 218-220.

Biodiversity in Horticultural Crops Vol. 2 *Pages* **175–188**
Editor: **Professor K.V. Peter**
Published by: **DAYA PUBLISHING HOUSE**

Chapter 8

Jackfruit

Rema Menon, R. Keshavachandran and K.V. Peter
Kerala Agricultural University, Vellanikkara, Thrissur – 680 656, Kerala

The jackfruit is indigenous to the rain forests of the Western Ghats and is cultivated throughout the tropical lowlands in South and South east Asia, parts of central and eastern Africa and Brazil. Archeological findings in India have revealed that jackfruit was cultivated in India 3000-6000 years ago. Findings also indicate that Indian Emperor Ashoka the Great (274-237 BC) encouraged arbori-horticulture of various fruits including Jackfruit. One of the earliest descriptions of the jackfruit is found in the 16th century memoirs of the Mughal Emperor Babar. Palaces in Bali and Macassar were built of jackwood and was once reserved for temples in Indochina.

Jackfruit is adapted to humid tropical and near tropical climates. It flourishes in humid hill slopes also up to an elevation of 1500m(Samaddar, 2002). Quality of fruits deteriorates in higher altitudes but grow well in arid and warmer places of S.India. Jackfruit thrives in a wide variety of soil although a rich deep alluvial soil is preferred. Soil drainage is of great importance and are severely damaged by flood, and higher water table. Open textured light soils or laterite soils supplemented with nutrients can also support jackfruit.

Major jackfruit producers are Bangladesh, India, Myanmar, Indonesia, Sri Lanka and Malaysia. It is the national fruit of Bangladesh.

In India, it has wide distribution in Assam, Bihar, Uttar Pradesh, the foothills of the Himalayas and South Indian states of Kerala Tamil Nadu and Karnataka.

In India, jackfruit is known by different names as *Chakka.* (Malayalam), *Palaa* (Tamil), *Halasina hannu* (Kannada), *Katahal* (Hindi), *Phanas* (Marathi), *Panasa* (Telugu) and *Panasam* (Sanskrit).

The English name jackfruit derives from Portuguese *jaca*, which is derived from Malayalam *Chakka.* Southeast Asian names include *Nangka* in Indonesia and Malayasia and *Mit* in Vietnam. *Finesi* (Swahili), *Jaca* (Portuguese), *Jaka* (Spanish), *Boluomi* (China) are other names across the globe.

Nutritional Importance

Jackfruit is a comparatively cheap fruit and a fairly good source of vitamins and minerals. The composition o f the edible portion is detailed in Table 8.1.

Jackfruit is commonly used in South and Southeast Asian cuisines. It can be eaten unripe or ripe and cooked or uncooked. The seeds also find use in certain recipes, young jackfruit has a mild flavour and distinct texture and the cuisines of India, Bangladesh, Sri Lanka, Indonesia and Vietnam use cooked young jackfruit. Mature unripe and ripe fruits are used in a variety of preparations.

1. Chakka Pradaman: Jackfruit pudding from Kerala, India
2. Chakka Varatti: Jackfruit jam from Kerala, India
3. Chakka vattal: Jackfruit chips from Kerala, India
4. Panasa koora: Jackfruit curry from coastal Andhra, India
5. Gudeg: Traditional dish from Central Java, Indonesia
6. Lodeh: Traditional Indonesian vegetable dish
7. Jackfruit salad: Vietnamese dish with boiled young jackfruit

. In South India, the jackfruit is a popular food next to mango and banana. In Kerala, tender fruits of about 60 days old are commonly used as vegetable. Due to its varied culinary uses and its availability in plenty during heavy monsoon period, jackfruit has earned the well deserved name poor man's food (Samaddar, 1990). Jackfruit is cited as an example of a food prized in some areas of the world and allowed to go waste in others.

Seeds are rich sources of starch and forms a popular ingredient in culinary preparations. The seeds are also eaten cooked or baked like beans and has a taste similar to chestnuts. Jackfruit seed is also a good source of protein having high solubility near neutral pH with low water and fat absorption capacities and hence suited for incorporation into savoury products (Rajeshwari *et al.*, 1999).

Apart from table use, ripe bulbs are used for making canned products, nectare, preserves, jam, jelly, squash, fruit bar and candy. The ripe bulbs fermented out distilled produce a potent liquor. Mature unripe bulbs are made into pickles, chips and papad (Bose and Mitra, 1990).Rind and perigones are rich source of pectin.

Other Uses

Jackfruit has been described as a multidimensional feed resource for livestock and poultry (Das *et al.*, 1998). Jackfruit foliage is a potential supplement for goats (Asit Das *et al.*, 2001, Nguyen *et al.*, 2001). Allelopathic influence of jackfruit leaf litter has been demonstrated (Jacob John *et al.*, 1999). Nematicidal property was revealed in pot culture studies (Das and Das, 1999). Jacalin, a major protein from jackfruit seeds has proved useful for the evaluation of immune status of patients infected with HIV-1 and for detection of tumour (Kabir, 1998). Jackfruit latex has bacteriolytic activity equal to that of papaya latex.

The tree has got good timber value as it is rarely attacked by white ants and is fairly resistant to fungal and bacterial decay. Jackfruit resembles mahogany and is superior to teak for furniture, construction and musical instruments like Kerala drum *Mridangam*. From the sawdust of jackwood or chips of the heartwood, boiled with alum a rich yellow dye is derived, commonly used for dyeing silk and cotton robes of Buddist priests. Besides morin, artocarpin is also responsible for the yellow color. Chemical analysis of wood of the jackfruit tree revealed a cellulose content of 56.74 per cent, lignin content of 28.76 per cent, pentosan content 18.64 per cent, yield of charcoal 38.74 per cent, charcoal calorific value 7108.37 cal/g and compression strength of 350kg/cm^3 (Komarayati, 1995)

Chemical Composition

Jackfruit volatiles contained 45 components of which esters represent a high proportion (31.9 per cent) and contribute to the distinct flavor (Wong *et al.,* 1992) Maia *et al.* (2004) studied the aroma volatiles from hard and soft variety of jackfruit. Isopentyl isovalerate (28.4 per cent) and butyl isovalerate (25.6 per cent) were the major components identified in the aroma concentrate of hard type while isopentyl isovalerate (18.3 per cent), butyl acetate (16.5 per cent) ethyl isovalerate (14.4 per cent) dominated in soft type, which corresponded to the morphological variation and the distinct aromas.

Fructose, glucose and sucrose are major sugars in all parts of the fruit except in spiny rind, which is devoid of glucose (Choudhary *et al.,* 1997). Flavones with antibacterial activity against carcinogenic bacteria has been isolated from jack fruit (Sato *et al.,* 1996).

The peel of *A. heterophyllus* showed highest amount of 5-CQA (13000 mg/kg) among different tropical fruits analyzed (Pontes *et al.,* 2002)

Morphology

Jackfruit belongs to the family Moraceae, which includes mulberry (*Morus* spp.) and figs (*Ficus* spp.). The genus *Artocarpus* contains about 50 species mostly native to Asia and 15 produce edible starchy fruits.

Morton (1987) described jackfruit as a handsome and stately tree with evergreen alternate glossy some what leathery leaves. All parts contain a sticky white latex. Short stout flowering twigs emerge from the trunks and large branches or even from the soil covered base of very old trees. Jackfruit is a monecious out breeding plant and wind is the chief pollinating agent.

Morphology of inflorescence and fruit has been described (Joseph and Kumaran, 1995). Joseph and Kumaran (1996) conducted detailed studies on the floral biology of jackfruit. The pattern of male and female catkin production differed, the former occurred from October to February while latter was confined to three months from late November to February.

Female Inflorescence

During flowering, specialized, short, stout laterals develop on the trunk or main branches or even from the soil covered base of very old trees initiating leaves and inflorescences. Female inflorescences occur only on these laterals whereas the males may occur also in the terminal leaf axil and lateral shoots. Male and female inflorescences are similar during early development, but later the female is distinguished by a thicker peduncle and a large annular disc at the base of the spike. The light green elliptic or rounded female inflorescence is composed of papillae which become sticky when receptive 1-2 weeks after exertion, this condition may last for a further 2 weeks.Joseph and Kumaran(1996) observed that sequence of emergence of stigma on female catkin was highly erratic and continued for 3-4 weeks. Complete fading of stigma occurred in 21-35 days. Pollination and fertilization of a spike are complete within 3-6 days after anthesis. Prior to anthesis, perianth segment remain free though tightly packed. Later they enlarge in the centre and fuse but remain free at the lower and upper regions. The fruit is botanically called a sorosis.

Male Inflorescence

The male inflorescence is oblong and light green and grows to a length of 5-10 cm. and is densely covered with minute flowers, each with a two lobed Calyx and with a single stamen. Each stamen consists of four anthers which split open at anthesis exposing the masses of spherical pollen grains. The male perianth segments are shorter than the female and though tightly packed are always free.

Anthesis commences 2-3 weeks after emergence from the stipules and lasts about 2 weeks during which inflorescence appears yellow. Pollen production is abundant and ranged from 1, 50, 00, 000 to 1, 70, 00, 000/catkin. The Pollen fertility (per cent) ranged 89-93 and the pollen diameter ranged from 16-22 mm. The pollen production in jack fruit is abundant with fertility ranging from 89 to 93 per cent and lost viability within 24 hours. No marked difference between Varikka and Koozha type in respect of pollen characters was noted (Joseph and Kumaran, 1996).

Fruit

The rind of the compound fruit is green or yellow when ripe and composed of numerous hard cone like points attached to a thick and rubbery pale yellow or whitish wall. The edible part consists of yellow banana flavored bulbs (fully developed perianths) embedded among narrow strips of thin tough undeveloped perianths(or perigones) and a central pithy core. Fruit set was 83.33 per cent with post set drop ranging from 30-40 per cent in this two types. Fruit development which took 100-138 days followed a sigmoid pattern. Occurrence of distorted fruits was attributed to inadequate pollination (Krishnamurthy and Rao, 1963).

Table 8.1: Yield and Fruit Characters of Exotic Jackfruit Varieties

Cultivar	*Country*	*Yield (Kg)*	*Fruit Size*	*Fruit Weight (kg)*	*Edible Flesh %*	*Rind Color*	*Flesh Color and Texture*	*Flavour*
Canopy Black gold	Queensland	55-90	Medium	6.7	35	Dark green, sharp fleshy spines, do not flatten	Deep orange soft	Strong and sweet
Cheena		50-70	Small long and narrow	2.4	33	Green with blunt spines, slightly flatten	Deep orange, soft and fibrous	Earthy flavour
Cochin	Australia		Small	1.5	35	Smooth with flattened spines	Firm and mild	–
Dung Rasmi		75-125	Medium to large	8	32	Bright green to pale yellow with sharp spines	Deep orange and firm	Sweet flavour and pleasant aroma
Golden Nugget	Australia	60-80	Small, green rounded	3.2	41	Fleshy spines flatten to golden yellow	Deep orange, soft and firm fibreless	Excellent
Golden pillow				3.6 to 5.5	35-40		Thick and crunchy with golden colour	Mild and sweet
J-30	Malaysia	50-60		7.6	38	Dark green, uniform with blunt spines	Deep orange and firm	Rich and sweet
J-31	Malaysia	42-60	Large	12	36	Large and irregular, prominent blunt spine	Deep yellow and firm	Sweet and rich earthy aroma
NS1	Malaysia	90		4.2	34	Dark green and blocky with blunt flattened spines	Dark orange and firm	Rich and sweet
Tabouey	Indonesia	50-70	Long, thin and tapered at stem end. Mishapen	-	-	Bright yellow colour and irregular blunt spines	Pale yellow and firm	Mild and pleasant

Each bulb encloses a smooth oval light brown seed covered by a thin white membrane. When fully ripe, the unopened jackfruit emits a strong disagreeable odour resembling that of decayed onions, while the pulp of the opened fruit smells of pineapple and banana. Jackfruit matures 3-7 months from flowering. On maturity fruit color changes from light green to yellow brown and produces a dull hollow sound when tapped.

Related Species

The Champedak–*A. integer* Merr. (Syns. *A. champeden,* Spreng., *A. polyphena* Pers), is also known as chempedak. The tree is native and grows wild in Malaya up to an altitude of 1300 m and is cultivated in Malaysia, Indonesia, Thailand and the Philippines. It is distinguished from the jackfruit by the long, stiff brown hairs on young branchlets, leaves, buds and peduncles. The fruit is borne on a deciduous tree, reaching about 18 m in cultivation and 30-45 m in the wild. The fruits are cylindrical or irregular, mustard yellow to golden brown, reticulated, warty and highly odoriferous. The rind is thinner than that of jackfruit. The pulp is deep-yellow, tender, slimy, juicy and sweet. The wood is strong and durable and yields a yellow dye, and the bark is rich in tannin.

The Lakoocha, *A. lakoocha* Roxb., is also known as monkey jack or *lakuchi* in India, *tampang* in Malayasia and as *lokhat* in Thailand. The tree grows 6-9 m tall with deciduous, large, leathery leaves, downy on the underside. Orange yellow male flowers and reddish female flowers are borne on the same tree. The fruits are round or irregular, 5-12.5 cm wide, velvety, dull-yellow tinged with pink, with sweet sour pulp which is eaten raw or made into curries and chutney. It is a native of the humid sub-Himalayan region of India up to 1200 m and also Malayasia and Sri Lanka.

The wood sold as lakuch and heavier than that of jackfruit is similar to teak, durable outdoors and under water, used for boats, furniture etc. The bark which contains about 8.5 per cent tannin is chewed like betel nut. The wood and roots yield a dye of richer color than that extracted from jackfruit.

The Kwai Muk, *A. lingnanensis Merr* or *A. hypargyraeus*. The tree is a slow growing, slender, erect ornamental 6-15 m tall and native to China and Hong Kong. Male and female flowers are tiny and yellowish borne on the same tree. Fruits are more or less oblate and irregular with velvety, brownish thin tender skin. Ripe pulp is orange red or red and soft with agreeable sub acid to acid flavor, eaten raw and contains 1-7 small pale seeds

The other *Arocarpus* species which produce edible fruits include *A. anisophyllus, A. elasticus, A. odoratissimus, A. rigidus etc*

The phylogenetic relationship among eleven *Artocarpus* species were estimated by comparing 30 restriction site mutations in an amplified region of chloroplast DNA (cpDNA). Close relationship between Chempedak (*A. integer*) and jack fruit was confirmed. Bread fruits(*A.altilis*) and *A. elasticus* formed amonophyletic group indicating that breadfruits were derived from *A. elasticus* or its close relatives. The results were in conformity with conventional classification of *Artocarpus* sp. except for the position of *chaplasha* (Kanzaki *et al.,* 1997).

Propagation

Seed Propagation

Propagation is usually by seeds and cannot be stored longer than a month. Germination requires 3-8 weeks, but soaking in water, removal of leathery seed coat and treatment with growth regulators resulted in earlier and higher germination and higher vigour index (Prakash, 1998).Singh *et al.* (2002)

reported highest germination (95.13 per cent), tallest plants shortest span of germination (13 days) and fastest germination (3.61 days) when seeds were treated with 100 ppm GA.

The seeds may be sown *in situ* or polythene bags and transplanted to the main field. Seedling population has a long juvenile period and are not true to type.

Vegetative Propagation

Vegetative propagation through different methods has been tried.

In air layering experiments, IBA at 7500 ppm applied to hardwood stem produced the most rooted air layers up to 88 per cent (Sabyasachi and Mishra, 1996) . Treatment consisting of etiolation + 5000 IBA + NAA (1:1) gave highest success of rooted layers (90.00 per cent) and favoured root growth (Sengupta and Thakur, 2001). Epicotyl grafting with mature, plump terminal scion shoot on germinating jackfruit seedling of about 8-10 days by wedge method during April-May gave 50-90 per cent success and the successful epicotyl grafts attained saleable size within ayear (Gunjate *et al.*, 1982). Softwood grafting on six month old rootstock has been a success (Swamy and Melanta, 1994). Pre treatment of scion–ringing and blanching for three days gave maximum success of 60 per cent (Maiti and Sen, 1999). Other successful methods reported are patch budding (Kelaskar *et al.*, 1993), rooted cuttings (Kamaluddin and Bhuiyan, 1996) and mound layering (Alila *et al.*, 1997).

Micro Propagation

Amin and Jaiswal (1993) developed a technique for rapid propagation using apical bud cultures. Successful micro propagation of jack fruit using axillary buds on modified MS supplemented with 18 mg/l BA + 0.2 mg/l IBA (Singh *et al.*, 1998). Roy *et al.* (1993) propagated flood tolerant type of jackfruit using shoot buds as explants.Roy *et al.* (1996) standardized *in vitro* multiplication procedure using shoot tips and nodal segments collected from coppiced shoots on the main trunk of a new jackfruit cultivar bearing fruits twice a year. Adiga *et al.* (1998) studied the in vitro response of Singapore jack. Ganeshan *et al.* (2005) standardized the protocol to conserve *A. heterophyllus in vitro* at 10°C without an intervening subculture for four years.

Variability Observed in Jackfruit for Morphological Characters

Tree

Tree growth habit: Erect, semi erect, spreading

Branching: Sparse, medium, dense

Branching pattern: Erect, opposite, verticillate, horizontal, irregular

Trunk surface: Smooth, rough, very rough

Crown shape: Pyramidal, broadly pyramidal, spherical, oblong, semicircular, elliptical, irregular

Leaf

Leaf blade shape: obovate, elliptic, broadly elliptic, narrowly elliptic, oblong, lyrate

Leaf apex shape: acute, acuminate, retuse obtuse

Leaf base shape: oblique, rounded, cuneate, shortly attenuate

Leaf color: Light green, green, dark green, pinkish green

Leaf pubescence: Glabrous, sparse, intermediate, dense

Inflorescence

Inflorescence color: Light green, light yellow, dark yellow, crimson

Female inflorescence density: Sparse, intermediate, dense

Female inflorescence position: Trunk, trunk and primary branch, trunk, primary and secondary branch, the whole stem

Fruit

Fruit bearing position: Main trunk, primary branch, secondary branch

Fruit clustering habit: Solitary, clusters

Fruit shape: Obloid, spheroid, ellipsoid, clavate, oblong, irregular seed attachment to fruit: Depressed, flattened, inflated

Fruit rind thickness: Thin, medium thick and very thick

Fruit rind color: Green, greenish yellow, yellow, reddish yellow

Fruit surface: Smooth, spiny shape of spine: sharp pointed, intermediate and flat

Spine density: Sparse, dense

Latex exudation: Low, medium, high

Flake

Flake thickness: Thin, medium, thick

Flake shape: Spheroid, cordate, twisted, obovate, rectangular, curved tip, irregular

Flake texture: soft, firm, coarse, fibrous, melting

Pulp

Pulp taste: Insipid, acid, bitter, sweet

Pulp consistency: Slimy, soft, medium, firm

Pulp flavor: Weak, intermediate, strong

Pulp juiciness: Not juicy, juicy, very juicy

Pulp color: Coppery red, deep yellow, yellow, light yellow creamy white, white

Seed

Seed shape: Spheroid, ellipsoid, elongate, oblong, reniform, irregular

Seed surface sliminess: Slightly slimy, intermediate, highly slimy

Seed coat thickness: thin, intermediate, thick

Seed surface pattern: uniform, regular striations, patches

Seed coat color: off white, creamish, dull brown, brown

Adherence of seed coat: easily separable, intermediate, difficult to separate to kernel

Table 8.2: Food Value/100 g of Edible Portion

	Tender	Ripe	Seed
Calories	51	98	43
Moisture (g)	84	77.2	64.5
Carbohydrates (g)	94	25.4	38.4
Proteins (g)	2.6	1.9	6.6
Fat (g)	0.3	0.1	0.4
Fibre (g)	2.6	1.1	1.5
Ash	0.7	1.0	1.25
Calcium (mg)	50	22	21.0
Phosphorus (mg)	97	30	28.0
Iron (mg)	0.5	1.2	0.8
Potassium (mg)	246	409	363
Vitamin A (IU)	30	540	–
Thiamine (mg)	0.3	0.9	0.6
Riboflsvin (mg)	0.1	0.05	0.06
Ascorbic acid (mg)	11	10.0	10.0

(Sadhu and Chattopadhyay, 2001; Morton, 1987).

Genetic Resources: Collection, Characterization and Evaluation

Being cross pollinated and predominantly seed propagated, jackfruit exhibits wide variation in terms of vegetative, flowering, fruiting and biochemical characters. The heterogenous make of the seedling population offers tremendous potential for selection of superior types suited for commercial cultivation.

Based on the firmness of flesh, cultivated types are of two general groups.

Firm Flesh

The pulp is firm and crisp and taste varying from very sweet, sweet acid to insipid. The hard pulp group is locally known as varikka, varcha, kujja, korcha or berka. The highly relished pulp is crisp and flavoured. Juice is scanty and the seeds are comparatively smaller (Singh *et al.*, 1967). The firm fleshed type is known as Varika in South India and in other areas as Burka, Berka or Waraka

Soft Flesh

The pulp is very juicy and soft, sometimes fibrous, taste variable with degree of sweetness. The soft pulp group is locally known as pazham, ghula, vela, koozha, ghila, tsjakepa or koppa. The mature pericarp is comparatively smaller in size. The juice is either their or thick, colour varying from pale yellow to dark or golden yellow. Pulp is generally mushy or soft and taste ranging from sweet to insipid. The seeds are comparatively larger. Koozha types are found in Kerala, Tamil Nadu and Karnataka.

Both firm and soft fleshed types occur in natural population and are identical in many plant and fruit characters and to a considerable extent in flakes till they ripen. No significant difference in the floral characters between Varikka and Koozha type was observed (Joseph and Kumaran, 1996).Rahman *et al.* (1999)observed variation in the carbohydrate composition in the soft and firm types with maturity and climatic conditions.Hard variety could be preserved for a longer period than soft type(Sonde and Naik, 1998).

Selection from seedling population is the only conventional breeding method used and explorations have been undertaken to identify elite trees having good agronomic and economic traits.

Based on collecting mission for jack fruit germplasm undertaken in parts of eastern Uttar Pradesh where it grows wild, eighteen trees exhibiting high variability in growth, yield, fruit and flake characteristics were identified (Singh and Srivastava, 2000).

In extensive surveys to identify and collect superior genotypes in Eastern India, 1800 trees were evaluated. Yield potential varied from 16 to 325 fruits/year. Types with TSS >25°Brix and total sugars >20 per cent were identified as table types. Few types bore thrice a year and were suited for vegetable purpose. Two types with higher juice content had potential for processing. Few drought tolerant types were also identified (Mitra *et al.*, 2000).

Mitra and Maity (2002) carried out detailed study on genetic resources of jackfruit of West Bengal covering 1460 fruits. Thirty five superior clones have been identified and conserved.

To combine desirable characteristics of jack fruit cultivars, a breeding programme with twelve selected cultivars in the Farichild Tropical Garden germplasm was attempted (Wasieleuski and Campbell, 1999).

Agronomic evaluation of 12 genotypes collected from lower Brahmaputra Valley zone of Assam showed significant differences in yield and chemical composition and selections were made for higher fruit yield, TSS and lower acidity (Sarma *et al.*, 1997).

In a study conducted at Kannara centre of AICRP(Tropical Fruits) 211 fruits from three districts of Kerala were characterized and evaluated *in situ*. Wide variation in fruit morphology and quality were observed. Out of the sixteen selections made, nine trees were found best for table purpose, four for making chips and two for culinary purpose. Two multi purpose types were also identified. Based on exploration carried out in Mohanpur, West Bengal, sixteen superior types were identified. Fruit shapes recorded were oval, oblong, cylindrical, round and bulbous. A cylindrical shaped type weighing 12.5 Kg with 263 flakes and TSS of 250 B was identified from Umarpur in Nadia district. Another unique clone from Barasat had red flakes with a TSS of 19.708 (ICAR, 2003/03.). An extra early clone maturing during January weighing 1-1.5 Kg and a high yielding clone with 512 fruits weighing on an average 1-2 kg were identified.

Survey carried out in different locations of Upper Assam led to the selection of five fruit types, Shape Varying from spheroid to ellipsoid. Variability was observed in respect of number of fruits, fruit weight, size and number of flakes (ICAR 03-04).

On the basis of fruit morphology and chemical composition of the pulp, Kumar *et al.* (1996) classified various genotypes into nine categories. Selections with moderate yields, large fruits (> 15 kg) and bulb (about 20 g), small cylinders, high pulp/cylinder ratio and average quality were made.

Muthulakshmy (1996) carried out detailed study on genetic diversity in four different topographical regions of Thrissur district, namely plains, hills, coastal and riverside areas. Wide variability was noticed with regards to growth, flowering and yield parameters (Table 8.2). She observed three types of bearing habit- regular, alternate and irregular. Jackfruit trees were grouped based on bearing season as early bearing (Dec-Feb), Mid season bearing (March-May) late bearing (June-August). Early bearing types were common in the plains (18.75 per cent), while late types were noticed in hills (27.25 per cent). Solitary bearing tendency was pronounced than cluster bearing habit. Predominant fruit shape was broadly ellipsoid (25.62 per cent) followed by irregular (11.43 per cent).

The yield/tree (21.23-1710.0kg) and fruits/tree (3.6-230.6) showed remarkable differences. Wide variation was noticed in terms of biometric characters of fruits-length (18.5-80.30cm), girth (30.12-110.0cm) fruit weight (1.05-22.3kg), number of flakes (1.0-456.0) flake fruit ratio (0.69-4.30). Significant variation was observed in respect of quality attributes–TSS (14.63-33.008) total acidity (0.60-4.95) and sugar acid ratio (2.32-20.81 per cent). Nine firm fleshed selections were made from the four topographical regions surveyed.

Rai *et al.* (2003) evaluated 21 genotype of jackfruit from Bihar, West Bengal and Uttar Pradesh and based on the overall performance with respect to the bearing potential maturity period, fruit and flake characters and identified genotypes suited for table purpose (HPJS-5/8 and HPJ3-3/10) and for culinary purpose (HPJS-4/5 and HPJS-2/6). Genotypes HPJS-3/10, HPJS-5/8, HPJS-8/9, HPJS-9/5 and HPJS-11/9 were selected for pleasant aroma in ripe flakes.

Genetic Diversity Analysis

A study on seed characters of yellow-bulb, light yellow-bulb and orange bulb types showed that light yellow types had the highest seed weight (7.66 g), seed length (3.23 cm), seed breadth (2.10 cm) and average total weight of seeds per fruit (913.21 g). The yellow types had the maximum seeds per fruit (124.6) and the highest pulp to seed ratio (4.24) (Guruprasad and Thimmaraju, 1989).

Muthulakshmy(1996) observed association of tree shape and growth characters. Pyramidal tree shape was associated with erect growth habit and low branching density whereas obovate and oblong tree shapes were associated with spreading growth and high branching density. When oblong fruits were mostly solitary, spheroid fruits were frequently noticed with clustering habit. Path analysis revealed that fruit length, fruit weight, seed length and fruit number had direct effect on yield. Based on non-hierachial eucleadian cluster analysis eighty collections in each topography could be grouped in to ten clusters.

Based on genetic diversity analysis of 26 jack fruit accessions using AFLP, 12 primer pairs could uniquely identify the accessions. Jack fruit accessions with South East Asian origin were grouped in one major cluster, Indian accessions were grouped in a different cluster (Schnell *et al.*, 2001).

Maiti*et al.* (2003) carried out genetic variability studies for various physico-chemical attributes of 44 jackfruit genotypes collected from different agro climatic zones o f West Bengal. Leaf size and yield per tree varied due to genotypic and phenotypic variance. Highest magnitude of genotypic and phenotypic variance and genetic advance were observed for fruit weight. The highest heritability was noticed for leaf length. Highest coefficient of variation was reflected in respect of yield per tree.

Breeding

The wide range of variability has not been exploited through systematic breeding efforts. The Fruit Experimental Station at Kallar began breeding work in 1952 to develop short compact, many branched trees precocious and bearing large good quality fruits. Singapore jack was chosen as the female parent because of its early and late crops. Velipala, a local selection from the forest having large fruits with large carpers of superior quality and bearing regularly. After 25 years of testing one hybrid was rated as outstanding for precocity, fruit size, off season as well as main season production and yield excelling its parents(Morton, 1987).

In an effort to combine desirable characters of jackfruit cultivars, a breeding programme was initiated in 1998 at the Fairchild Tropical Garden at Miami, Florida. Twelve cultivars available in the germplasm collection were crossed in different combinations. The progeny has been field established for evaluation(Wassieleuski, 2000)

Varieties

Jackfruit types like 'Varikka', Koozha, Navarikka, Rudraksha chakka or Thamara chakka and other wild forms have been collected from Wynad plateau in the Western Ghat of Kerala, where jackfruit is believed to have originated (NBPGR, 1986). Similarly Rasdar, Khajwa and Sugandhi are three types identified in Eastern Uttar Pradesh (NBPGR, 1988). Other superior selections include Gulabi(rose scented), Champa(flavor like that of *Michelaa* sp.), Hazari(bearing large number of fruits)

Muttam Varikka is a locally selected variety producing medium sized fruits with sweet flesh. Muttam Varikka produce fruits weighing 7.6 kg with 46 cm length and 23 cm width (Srinivasan, 1970).

Singapore or Ceylon jack is a highly precocious bearer introduced to India from Ceylon producing fruit in 2 ½ to 3 years. Fruits are medium sized with small fibrous and very sweet carpers. In addition to the normal summer crop, it yields a second off season crop during October-December(Morton, 1987)

Three jackfruit selections have been released by Tamil Nadu Agricultural University (Veeraghavadattam *et al.*, 1996; Alagiapillai *et al.*, 1996)

Burliar-1: is a T Nagar selection from Fruit Research Station, Burliar.

Palur-1: is a single plant selection made from Panithankuppam village near Panruthi and developed at TNAU centre, Palur. The tree bears twice in a year and produces about 60-80 fruits each weighing 12 Kg, flesh yellow, crisp and sweet. The special feature is that the fully ripe fruits have a smooth rind.

Pechiparai-1: is a selection from Mulagumoodu local developed at Horticultural Research Station, Pechiparai under TNAU.A bearing tree yields 105 fruits each weighing 17 Kg on an average. There are two bearing seasons, April-June and November–December. The flesh is sweet, crisp tasty with pleasant aroma.

Hybrid jackfruit has been developed at Fruit Research Station, Kallar by crossing Singapore Jack and Velipala. Fruits resemble Singapore Jack with bigger carpels and taste.

Exotic Cultivars

The Fairchild Tropical Garden, Florida introduced and evaluated a collection of 22 superior jackfruit cultivars developed in different countries. Black Gold, Golden Nugget, J-31, Honey Gold and Lemon Gold have been the most precocious. The cultivars vary greatly in their size, shape flesh characteristics and taste Delightful, Dang Rasimi, Golden Nugget and Black Gold have been the most vigorous. All cultivars compared well with NS-1, the most popular cultivar in Florida.(Campbell and McNaughton, 1994).The yield and fruit characters of these cultivars are described below:

The popular cultivars of Bangladesh are Topa, Hazari, Chala, Goal, Koa and Khaja (Samaddar, 2002).

References

Adiga,T.D., Khan, M.M. and Sathyanarayana, B.N.1998. Effect of carbon source,culture vessels, gelling agents and GA3 on *in vitro* shoot proliferation of Singapore jack (*Artocarpus heterophyllus* Lam.) *Karnataka J of Ag. Sciences,* 11(3): 632-636.

Alagiapillai,O.A., Premjoshua, J., Kuttalam, S., Subramanian, V. and Jeyasekhar, M. 1996. PPI-1 Jack: A new high yielding, regular bearing variety for Tamil Nadu. *Madras Ag. J.* 83(5): 310-312.

Alila, P., Sanyal, D. and Kar, P.L. 1997. Propagation of jackfruit by mound layering. *Crop Research-* Hisar, 14(3): 433-436.

Amin, M.N and Jaiswal, V.S. 1993. *In vitro* response of apical bud explants from mature trees of jackfruit (*Artocarpus heteropyyllus* Lam.. *Plant Cell Tissue and Organ Culture,* 33(1): 59-65.

Asit Das, S. K. Ghosh, L.C. De, Dhiman, K.R. and Das, A. 1998. Jackfruit: A multidimensional feed reserve for livestock and poultry. *Pasudhan,* 13: 7.

Campbell, R.J. and McNaughton, R. 1994. New jackfruit cultivars under trial in South Florida. In: *Proc. of the Inter American Society for Tropical Horticulture,* 38: 201-205.

Chowdhury, F.A., Rahman, M.A. and Mian, A.J. 1997. Distribution of free sugars and fatty acids in jackfruit (*Artocarpus heterophyllus*). *Food Chemistry,* 60(1): 25-28.

Das, N. 1999. Nematicidal effect of organic soil amendments of some latex bearing plants on root knot nematode, *Meloidogyne incognita. Crop Research,* Hisar, 18(2): 299-300.

Jacob, John and Nair, A.M. 1999. Preliminary investigation on the allelopathic influence of leaf litter of certain multipurpose trees. *Indian J. Forestry,* 22(1-2): 66-69.

Joseph, T. and Kumaran K. 1995. The morphology of inflorescence and fruit in jack (*Artocarpus heteropyyllus* Lam.). *J. Trop. Ag.* 33(2): 113-119.

Joseph, T. and Kumaran, K. 1996. Fruit set, fruit development and fruit drop in hard and soft flaked types of jack (*Artocarpus heteropyllus* Lam.). *J. Trop. Ag.* 33(1): 21-24.

Kabir, S. 1998. Jacalin: A jackfruit seed derived lectin of versatile applications in immuno biological research methods *J. Immunobiological Research,* 212(2): 193-211.

Kamaluddin, M., Ali, M. and Bhuiyan, M.K. 1996. Effect of auxin on rooting of cuttings of jackfruit(*Artocarpus heteropyllus* Lam.).Chittagong University Studies, *Science,* 20(1): 71-75.

Kanzaki, S., Yonemori, K., Sugiura, A. and Subhrabandhu, S. 1997. Phylogenetic relationships between the jackfruit, the breadfruit and nine other *Artocarpus* sp. from RFLP analysis of an amplified region of CpDna. *Scientia Horticulturae,* 70(1): 57-66.

Kelaskar, A.J., Desai, A.G. and Salir, M.J. 1993. Effect of season on the success and growth of bud grafts of jack (*Artocarpus heteropyllus* Lam.). *J. Trop Agriculture,* 31(1): 112-115.

Kumar, S. and Singh, I.S. 1996. Improvement of jackfruit (*Artocarpus heteropyllus* Lam.) through selection. *Recent Horticulture,* 3(1): 55-59.

Maia, J.G.S., Andrade, E.H. A., Zoghbi, M. and Das, G.B. 2004. Aroma volatiles from two different varieties of jackfruit (*Artocarpus heteropyllus* Lam.). *Food Chemistry,* 85(2): 195-197.

Maity,C.S and Sen, S.K. 1999. Effect of pre treatment of scion on grafting success in jackfruit (*Artocarpus heteropyllus* Lam.). *Horticultural Journal,* 12(2): 97-99.

Maity, C.S., Wangchu, L. and Mitra, S.K. 2003. Genotypic variability for physico chemical attributes in jackfruit (*Artocarpus heteorphyllus* Lam.) genotypes of West Bengal. *Indian Agriculturist*, 47(3/4): 193-199.

Pontes, P.V., Moreira, R.F.A., Trugs, L.C. and Maria, C.A.B. 2002. The content of chlorogenic acids in tropical fruits. *J. Science of Food and Agriculture*, 82(10): 1177-1181.

Prakash, M. 1998. Effect of plant growth regulators and chemicals on germination of jackfruit. *Annals of Plant Physiol.* 12(1): 75-77.

Rahman, M.A., Nahar, N. and Mian, A.J. 1999. Variation of carbohydrate composition of two forms of fruit from jack tree (*Artocarpus heterophyllus* Lam) with maturity and climatic conditions. *Food Chemistry*, 65(1): 91-97.

Rai, M., Nath, V., Das, B., Rai, A. and Kumar, M. 2003. Evaluation of jackfruit genotypes for yield and quality attributes under Eastern Indian conditions. *Orissa J.Hort.*, 31(1): 1-6.

Rajeshwari, H.S. and Prakash, J. Jackfruit seeds-composition, functionality and use in product formulation. *Indian J. Nutrition and Dietetics*, 36(6): 312-319.

Roy, S.K. 1993. Development of flood tolerant *Artocarpus heterophyllus* through *in vitro* culture. *Agricultura Ricerca*, 15: 146.

Sabyasachi, R. and Mishra, S.P. 1996. Effect of growth regulators on varying age of shoots on success of marcotage in jackfruit. *Orissa J. Hort.*, 24(1/2): 79-82.

Sadhu, M.K and Chattopadhyay, P.K. 2001. Jackfruit. In: *Introductary Fruit Crops*, Naya Prakash, Calcutta, p. 306.

Samaddar, H. N., 1990. Jackfruit. In: *Fruits: Tropical and Subtropical,* (Eds.) T.K. Bose and S.K. Mitra. Naya Prakash, Calcutta, pp. 84-90.

Samaddar, H.N., 2002. Jackfruit. In: *Fruits: Tropical and Subtropical*, Vol. 2 (Eds.) T.K. Bose, S.K.Mitra and D. Sanyal. Naya Prakash, Calcutta, pp. 541-564.

Sarma, R., Mohan, N.K.and Nath, J.C. 1997. Evaluation of yield potential and physico chemical composition of some local types of Assam. *Horticultural Journal*, 10(2): 22-30.

Sato, M., Fujiwara, S., Tsuchiya, H., Fujii, T., Linuma, M., Tosa, H. and Ohkawa, Y., 1996. Flavones with antibacterial activity against carcinogenic bacteria. *J. Ethnopharmacology*, 54(2-3): 171-176.

Schnell, R.J., Olano, C.J., Campbell, R.J. and Brown, J.S. 2001. AFLP analysis of genetic diversity within a jackfruit germplasm collection. *Scientia Horticulturae*, 91(3/4): 261-274.

Sengupta, S. and Thakur, S. 2001. Studies on the effect of growth regulators on rooting of air layers of jackfruits (*Artocarpus heterophyllus* Lam.). *Orissa J. Hort.*, 29(1): 63-65.

Singh, S., Krishnamoorthy, S. and Katyal, S.L. 1967. *Fruit Culture in India*. ICAR, New Delhi, p. 27.

Singh, R., Tiwari, J.P. and Singh, R. 1998. *In vitro* clonal propagation of jackfruit (*Artocarpus heterophyllus* Lam.). *Indian J. Hort.*, 55(3): 213-217.

Singh, D.K., Bhattacharya, B. and Mondal, K. 2002. Role of pre sowing seed treatment with different chemicals on germination behaviour and seedling growth of jackfruit (*Artocarpus heterophyllus* Lam.). *Environment and Ecology*, 20(3): 741-743.

Sonde, N. and Naik, R.K. 1998. The effect of processing methods on the acceptability of two jackfruit varieties. *Karnataka J. Ag. Sciences*, 11(2): 471-473.

Srinivasan, K. 1970. Muttom Varikka: A promising jackfruit variety. *Agric. Res. J. Kerala,* 18: 51-52.

Swamy, G.S.K. and Melanta, K.R. 1994. Effect of age of rootstocks on the success of softwood grafting in jackfruit (*Artocarpus heterophyllus* Lam). *Karnataka J. Ag. Sciences,* 7(4): 471-473.

Veeraraghavadattam, D., Jeeva, S. and Rabindran, R., 1996. *Scientific Fruit Culture.* Suri Associates, Coimbatore, p. 129.

Wong, K.C., C.L. Lim and Wong, L.L. 1992. Volatile flavor constituents of Chempedak (*Artocrpus polyphena* Pers) fruit and jackfruit (*Artocarpus heterophyllus* Lam) from Malaysia. *Flavor and Fragrance J.* 7: 307-311.

Biodiversity in Horticultural Crops Vol. 2 *Pages* **189–201**
Editor: **Professor K.V. Peter**
Published by: **DAYA PUBLISHING HOUSE**

Chapter 9

Grape in India

G.S. Karibasappa
National Research Centre For Grapes, Manjri Farm, Pune – 412 307

Grapes are grown in India in 60,000 ha. producing annually 1.6 million t. of fruits (CMIE, 2001). Most of the area is confined to Maharashtra, Karnataka, Andhra Pradesh and parts of Tamil Nadu. Commercially only a few varieties like Thompson Seedless and its mutants (Tas-A-Ganesh, Sonaka and Manik Chaman), Dilkush (mutant of Anab-E-Shahi), Gulabi (Muscat Hamburg), Sharad Seedless (clonal selection from Kishmish Chernyi), Bangalore Blue (Isabella) and recently introduced Flame Seedless and RedGlobe are grown. Wine varieties cover about 7000 ha. in Maharashtra, Karnataka, and Himachal Pradesh, with annual wine production of over ten million litres. Major red wine varieties are Cabernet Sauvignon, Merlot, Shiraz, Zinfandel, Grenache and Pinot Noir and white wine varieties are Sauvignon Blanc, Chenin Blanc, Clairette, Chardonnay, Symphony and Semillon. In fact grapes are processed into raisins and about 55,000 t. of raisins are produced annually *i.e.*, 13.75 per cent of fresh grape production.

Status of *Vitis in-situ* Germplasm in India

The indigenous *Vitis* species were initially recorded through botanical surveys conducted in erstwhile British India by Sir, Hooker and his associates (Lawson,1875; King,1897; Cooke, 1903; Gamble,1918, Kanjilal *et al.*,1936) and later by the BSI at its various zonal centres in the country during post independent period. Following species are described by earlier workers;

V. lanata Roxb.,. *V. parviflora* Roxb., *V. polystachya* Wall., *V. pycnantha* Coll. and Hems., *V. obtecta* Wall. and *V. vinifera* Linn.

Based on above publications and personal surveys, Patil (1993) depicted the *in situ* availability of wild genera of the family *Vitaceae* in India as in Figure 9.1.

Olmo (1970) opined that only a few edible indigenous *Vitis* species are recorded confined to foothills of North-Western Himalayas, located at elevations 1000 m and slightly above. Jindal and

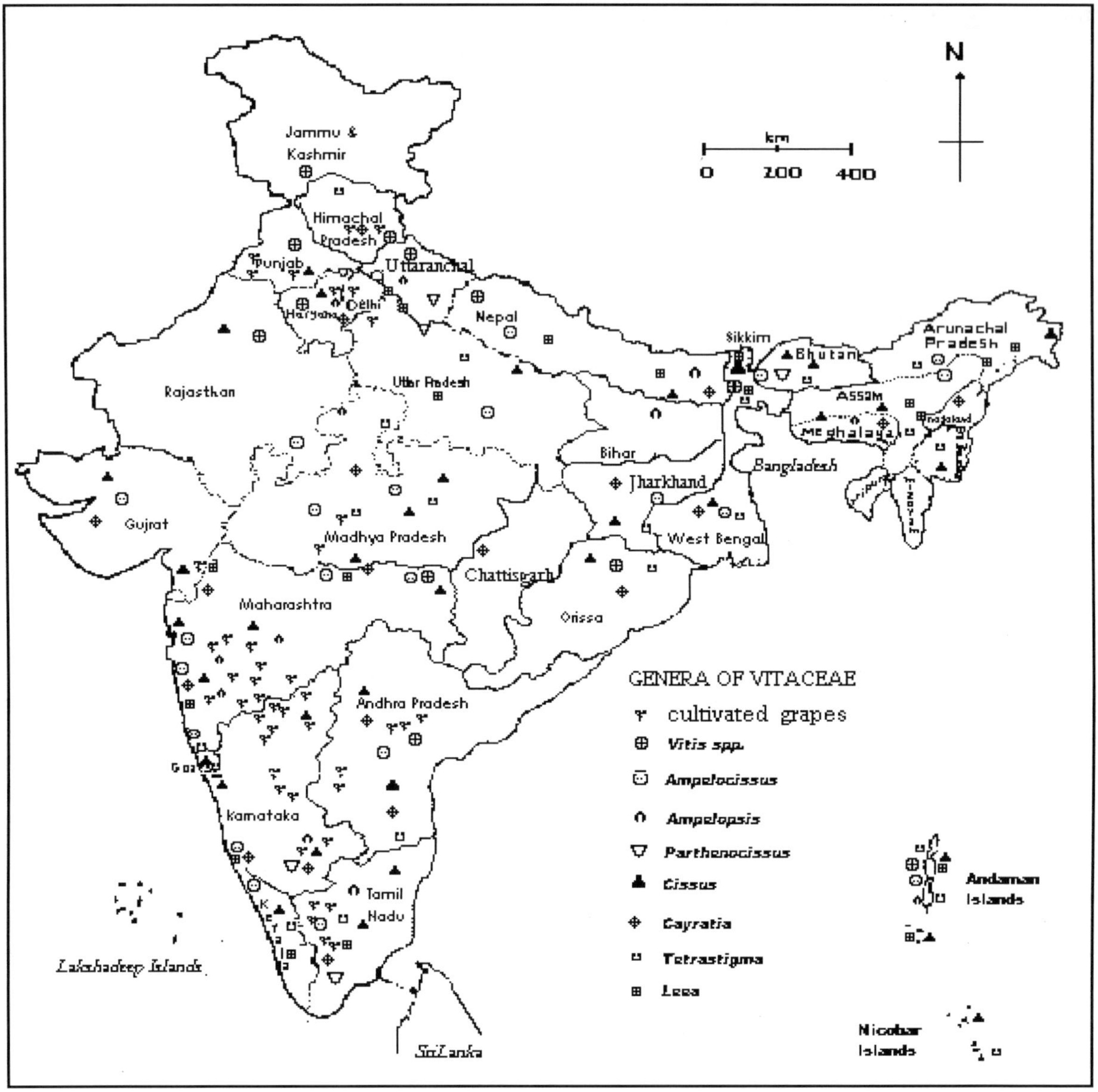

Figure 9.1: *In situ* Germplasm of Wild Relatives of Grapes in India
Source: Patil (1993)

Singh(1993) claimed that none of the species related to *Vitis* in the peninsular India is edible. Olmo (1970) during his visit to North-Western Himalayan foothills recorded two of the native species like *V. lanata* and *V. parviflora.* He also noted that an indigenous cultivar Rangspey and many intermediary types derived from natural crosses between *V. lanata* and *V. parviflora*, grown in Kinnaur (Kalpa), Himachal Pradesh along the upper basin of river Sutlej. These intermediate types though resembled *V. vinifera* but differed distinctly with the *viniferas* of Middle-Asian origin. The following 6 native species are briefly described :

***V. lanata* Roxb.**

A woody deciduous species found throughout the Himalayas at altitudes ranging 1500-2100 m amsl. The berries and roots are edible, besides the berries are resistant to rain damage. The leaves are large, greenish and shining upper side and heavily felted (copper to silvery hairs) underneath. Intra specific variation is abundant in this species.

***V. parviflora* Roxb.**

It has small, globose black or white coloured berries. Berries are sweet and delicately flavoured and are sold in the markets of North-Western Himalayan villages and townships from Kashmir to Nepal. The species is grown at altitudes between 600 and 1800 m amsl. The white berried strain has reflexed stamen (female type/male sterile) hence useful for mass breeding purposes. This species is moderately resistant to downy and powdery mildew diseases.

***V. barbata* Wall: (*Ampelocissus barbata* Wall.)**

This species is found in North- Bengal, parts of Assam and Khasi hills of Meghalaya upto an elevation of 1200 m amsl. It has edible berries and a possible source of new grape hybrids.

***V. tomentosa* Hayne. (*Ampelocissus tomentosa* Planch.)**

This species is found in peninsular India along the Western ghats. The berries are sub-globose and 2-3 seeded.

***V. himalayana* Br. (*Cayratia trilobata*)**

The leaves are trifoliate/palmate and serrated with long petiole. It has edible purplish berries with 2-4 seeds and usually ripens late October. The species is found throughout sub-Himalayan region.

***V. vinifera* Linn.**

The indigenous cultivars are not affected by fungal diseases (especially powdery mildew) unlike other introduced viniferas from Middle-Asia as reported by Olmo (1970). These are very promising for imparting disease resistance. They need to be evaluated under peninsular Indian conditions. These varieties are grown in Kinnaur (HP), named as Rangspey White and Black and other intermediate types, suspected natural hybrids and later segregants of *V.lanata* and *V. parviflora*. The varieties found near Malnai (Kashmir valley) though resembled Kinnaur varieties differed in certain aspects. Comparisons are hense required between varieties of Kinnaur and Malnai (Olmo,1970).

Most of the indigenous species and varieties being adapted to mild humid subtropical conditions, offer potential values for breeding for disease/insect resistance under peninsular Indian conditions. They prove useful under North Indian conditions also by providing resistance to berry cracking/rain damage and also for their earliness or late ripening .

Status of *Vitis* Under *Ex situ* Conditions in India

Most of the grape varieties grown in India belong to a species *Vitis vinifera* Linn. and a few of them are of *V. labrusca.* The former has its origin in the region between Black sea and Caspian sea of Middle Asia (Negrul, 1940; Vavilov, 1950), whereas the latter species is of North American.

Jindal and Singh (1993) stated that in India about 50 species of *Vitis* were introduced at various research stations which included 1217 cultivars/varieties and 40 rootstocks. A report of NBPGR, New Delhi indicated that 1850 accessions were introduced through the bureau during 1947-1991.

After the establishment of National Research Centre for Grapes, Pune in 1997 an inventory of authentic and active germplasm sites in India was made.

Table 9.1: Collection of Grape Germplasm at NRCG from Different Sources (1997–2003)

Sl.No	*Name of the Germplasm Centre*	*Project/Scheme/Jurisdiction*	*Peak Number of Accessions Made in the Past (Number Presently available)**	*Accessions Obtained by NRC Grapes, Pune*
1.	IIHR, Bangalore	ICAR Institute for tropical conditions	600 (135)	35
2.	Division of Fruit Crops, IARI, New Delhi	ICAR Institute for subtropical conditions	400(30)	30
3.	Fruit Research Station, Ganesh Khind, Pune/MPKV, Rahuri	Maharashtra	150 (nil)	nil
4.	Agharkar Research Institute, Pune	AICRP (STF) on Grapes, Centre for cytological Investigations on Indigenous genera and species	150 (60)	45
5.	PAU, Ludhiana and its Regional Fruit Research Stations at Abohar and Bhatinda, Punjab	AICRP (STF) on grapes as a coordinating centre for North Indian Conditions	600 (70)	30
6.	Regional Fruit Research station, Sharbo, Rekong Peo, Kinnaur District, Himachal Pradesh	Indigenous species and varieties and also introduced exotic varieties under humid temperate conditions	65 (20)	10
7.	Growers' vineyards, Private wineries, Individual growers etc. (Maharashtra, Karnataka, Andhra Pradesh, Tamil Nadu and Himachal Pradesh)	Commercial varieties introduced directly by them	45 (40)	40
8.	Introductions at NRCG through NBPGR, New Delhi (1997-2006)	Through bilateral agreements, requests/Purchases etc. Russia, Bulgaria, Oman, Ukrain, Uzbekistan, USA, Nigeria, Korea, Brazil, Italy	70 (68)	70

Ampelographic Studies

Genetic variability was studied with 149 accessions for 7 fruit characters, mean bunch weight, berries/bunch, juice percolate(per cent), TSS, acidity and pH. Variability was satisfactory for bunch and berry characters among the accessions maintained at NRCG. Further, the data were subjected to multivariate (cluster analysis) and based on squared Euclidean distances, ten well defined clusters were obtained (Table 9.2).

Molecular Characterization of Grape Germplasm

Inter Simple Sequence Repeat (ISSR) markers were used to characterize 34 grape varieties. Out of 93 ISSR primers screened initially, 11 showed good polymorphism.A total of 174 bands were obtained, out of which 145 were polymorphic. Cluster analysis of the varieties resulted in formation of two main clusters, one belonging to *Vitis vinifera* and other to *V. labrusca*. Varieties belonging to *V. vinifera* appeared more diverse and formed distinct sub-clusters based on colour, flavour and seed contents. Out of 34 varieties screened, 10 varieties with green/yellow berries were grouped together in one subcluster and 15 with red/black berries in the other. Three varieties with green/yellow berries, Italia, Queen of Vineyard and Thompson Seedless were grouped with varieties with red/black berries. The cluster of labrusca varieties showed homogeneity and had five varieties except Dakh, which belonged to *vinifera*. Concord separates initially from all other varieties. Incidentally, Concord is a pure selection from *V. labrusca*, while other varieties like Bangalore Blue, Black Muscat, Catawba and Muzzafar Nagar in labrusca group, may be the hybrids of *V.labrusca* x *V. vinifera*. The current study thus revealed that genetic relationships among grape cultivars could be assessed using ISSR markers. In another study (Argade *et al.*, 2006) evaluated 43 seedless grape varieties using ISSR markers. The amplification profiles were generated with 14 ISSR primers showing clear, polymorphic band patterns. Out of total 121 bands scored, 95 were polymorphic. The UPGMA cluster analysis revealed high degree of similarity among the varieties. Distinct subgroups based on berry colour could be distinguished. The varieties with yellow/green berries were separate from those with red/black/purple berries with a few exceptions. Among bud sport selections, Tas-e-Ganesh shared 100 per cent similarity with Thompson while other bud sports and selections of Thompson showed more than 95 per cent similarity with Thompson. Based on the fingerprinting key prepared from the band data,only three primers UBC 857, 890 and 888, in 37 out of a total 43 varieties could be identified. Overall groupings were in accordance with morphological characters and/or pedigree, proving usefulness of ISSR markers to assess genetic relationships among seedless grape varieties.

Table 9.2: Variability Among Bunch and Berry Characters in 149 Genotypes

Traits	*Minimum*	*Maximum*	*Mean*	*SEm*
Bunch weight (g)	5.00	600.00	138.91	7.8544
Berries/bunch	6.00	300.00	75.61	3.5447
Berry dia.(mm)	7.40	21.60	14.19	1.1707
TSS °Brix	7.50	27.80	18.64	0.2591
Juice per cent	20.00	92.00	59.42	1.0625
Acidity per cent	0.12	2.19	0.846	0.0350
pH	2.90	4.92	3.62	0.0233

In a study at NRCG (Anuradha Upadhyay *et al.*,2007) 21 rootstock accessions were analyzed with seven grape microsatellite (SSR) primers and seven AFLP primer combinations. SSR primers detected 56 alleles across 21 genotypes and primer heterozygosity varied from 0.617 to 0.856. Similarly 252 AFLP bands were obtained with seven primer pairs. The average similarity index for different rootstocks was relatively low and ranged from 0.068 to 0.36 for AFLP data and from 0.13 to 0.36 for SSR data. A combination of three SSR primers was sufficient to distinguish 21 rootstocks. In cluster analysis majority of rootstocks belonging to same species grouped together.

Screening of Germplasm for Disease Tolerance

Details of morpho-cytological and disease reaction studies are reported by Patil (1993) at MACS, Pune, for 12 cvs. of *V.vinifera*, 6 cvs. of *V.labrusca*, 3 cvs of *V.champini*, 1 cv. of *V. rotundifolia* and also included 13 other species like *V.lanata, V. parviflora, V. aestivalis, V.arizonica, V.berlandieri, V.californica, V.candicans, V.lincecumi, V.longii, V.palmata, V.riparia, V.rupestris and V.tiliefolia.* The disease reaction studies indicated that several varieties and species of *Vitis* were resistant to downy mildew, powdery mildew and anthracnose diseases (Patil *et al.*,1990). Cultivars; Bangalore Blue, Catawba, Concord and Champion of *V. labrusca*, James of *V. rotundifolia* found tolerant for both downy mildew and powdery mildew. Rootstocks like deGrasset and Dog Ridge of *V. champini* and all the above mentioned 13 species are tolerant to both downy and powdery mildews.

Screening for anthracnose resistance indicated that cvs Gulabi, Phakdi, Bangalore Purple of *V. vinifera*, Catawba, Concord and Champion of *V.labrusca*, James of *V. rotundifolia*, Dog Ridge of *V. champini* and species; *V.aestivalis, V.lanata, V.arizonica, V. candicans, V. longii, V. palmata, V. riparia, V. rupestris* and *V. tilifolia* are tolerant.

Screening under unsprayed field conditions at NRCG, Pune over a period of 3 years cv. Carolina Black Rose, 6 Seyve-Villard hybrids, Seibel 9836 and H533 were immune to downy mildew and also moderately resistant to powdery mildew.

Table 9.3: Evaluation of Grape Varieties for Downy Mildew Tolerance at NRCG, Pune

Sl.No	*Name of Variety/Accession*	*In Vitro Rating**	*Field Rating*
1.	TS (Cal.)	5.28	7.00
2.	Kali Sahebi	5.14	9.00
3.	Anab-E-Shahi	3.42	9.00
4.	Delight	5.00	9.00
5.	Gulabi(IIHR)	3.00	9.00
6.	Gulabi (TN)	1.48	7.00
7.	E–8/5	2.71	7.00
8.	E-2/1	2.90	9.00
9.	Alamwick	4.40	9.00
10.	Carolina Black Rose	1.38	1.00
11.	*V.parviflora*	3.57	5.00
12.	H-533 (*V.tilifolia* x Gulabi)	5.50	3.00
13.	SV-23501 (Poly-Vitis hybrid)	1.00	1.00
14.	SV-12309 (Poly-Vitis hybrid)	1.86	1.00
15.	SV-18315 (Poly-Vitis hybrid)	1.43	1.00
16.	SV-12364 (Poly-Vitis hybrid)	1.19	3.00
17.	SV-12375 (Poly-Vitis hybrid)	1.00	1.00
18.	Seibel 9813 (Poly-Vitis hybrid)	2.43	1.00
19.	Seibel 9308 (Poly-Vitis hybrid)	1.00	1.00
20.	Kanai Local (*V.lanata*)	4.14	5.00
21.	Lake Emerald (*V.labrusca*)	2.16	5.00
22.	Fruhroter Veltliner (*V.vinifera*)	3.00	5.00
23.	SV-18402 (Poly-Vitis hybrid)	3.14	1.00
24.	St.George (*V.rupestris*)	1.09	1.00
25.	Bharat Ruba Black (*V.lanata*)	5.14	9.00
26.	Flame seedless	5.70	9.00
27.	Concord (*V.labrusca*)	1.19	5.00
28.	Catawba (*V.labrusca*)	3.00	5.00

* Ratings: 1: Immune; 3: Resistant; 5: Moderately resistant; 7: Moderately susceptible; 9: Highly susceptible.

Drought and Salt Tolerance

At present 25 rootstock varieties are available at NRC for Grapes, Pune. Dog Ridge rootstock is widely recommended for table varieties in drought proned areas and saline soils. Other rootstocks like 110R, 140 RU, 1103 P, Teleki 5A and 99R are under field evaluation. All these rootstocks are under field evaluation for tolerance to moisture and salt stress conditions and especially suitability to wine varieties. Table grape varieties Marroo Seedless, Christmas Rose, Red Globe and Beauty seedless and wine varieties, Merlot and Carignane are drought hardy. Salt Creek, Dog Ridge, deGrasset, SO- 4 and Freedom are identified for salinity tolerance.

Utilization of Germplasm in Cross Breeding Programmes

There is an immediate need for developing seedless cultivars having tolerance to downy mildew under peninsular Indian conditions. Introgression of seedless character with downy mildew resistance, is taken up through cross breeding and embryo rescue techniques at NRCG in collaboration with NCL Pune (Bharaty *et al.*, 2003, 2005). The following crosses were effected and the plantlets of F1 progeny were raised by embryo rescue following Emershad and Ramming(1985). The F1 progeny is planted for further evaluation in the field.

Table 9.4: Number of Plantlets Obtained Through Embryo Rescue in F1 Progeny of Various Interspecific Crosses

Sl.No.	*Male Parents*	*Female Parents*	
		Thompson Seedless (V.vinifera)	*Flame Seedless (V.vinifera)*
1.	Lake Emerald (*V.labrusca*)	45	47
2.	S.V.18402(*V.spp/polyvitis hybrid*)	71	30
3.	*Vitis tiliefolia*	5	27
4.	*V.candicans*	33	1
5.	Concord (*V.labrusca*)	76	60
6.	Catawba (*V.labrusca*)	65	135
7.	St.George(*V.rupestris* var.du Lot)	24	0
8.	Fruhroter Veltliner (*V.vinifera)*	15	11
9.	Open Pollinated seedlings	5	3
	Total	339	314

Utilization of Male Sterile Varieties in Cross Breeding Programme

Six varieties *viz.* Spin Sahebi, Katta Kurgan, Neagra Vertis, Madelien Angevine, Arka Trishna and *V.parviflora* were identified as male sterile. Spin Sahebi and *V.parviflora* are crossed with seedless male parents like A 17-3 (clonal selection from Tokay), Flame Seedless, Arkavati, Sharad Seedless (A clone of Kishmish Chernyi), Beauty Seedless, Pusa Seedless, Sultanine-II, Perlette, Kishmish Rozavis and Superior Seedless and with seeded cv Red Globe. The F1 progeny is under field evaluation at NRCG, Pune.

Table 9.5: Active Field Germplasm of Indigenous Collections at NRC Grapes, Pune

Sl.No.	Name of Variety	Source of Augmentation/Collection
1.	V Bharat Prince	Pandarpur
2.	Bharat Ruba Black	--"--
3.	Country Bangalore	--"--
4.	Dakh	--"--
5.	Hussain Kadu	--"--
6.	Anab-e-Shahi	GRS,Hyd
7.	Arka Kanchan	--"--
8.	Pandari Sahebi	Borgaon
9.	Katta	Pandarpur
10.	Katta Kurgan	--"--
11.	Khalili	Borgaon
12.	Bangalore Blue	Borgaon
13.	Sahebi	IARI, New Delhi
14.	Champach	RFRS, Abohar
15.	Isabella	Sharbo, Kinnaur
16.	Thompson Seedless	Madurai, TN
17.	Gulabi	–do–-
18.	Motia	Pandarpur
19.	Fakri	IIHR, B'lore
20.	Khalili	IIHR, B'lore
21.	Bangalore Purple	–do–-
22.	Kali Sahebi	–do–-
23.	Bhokri	IARI, New Delhi
	Landrace	
1.	Choultu Black	Kinnaur
2.	Choultu Red (*V.labrusca*)	Kinnaur
3.	Choultu White (*V.labrusca*)	Kinnaur
4.	Rangspey White (*V.vinifera*)	Kinnaur
5.	Rangspey Black (*V.vinifera*)	Kinnaur
6.	Kanai Local (*V.lanata*)	Kinnaur
7.	Kinnauri–1 (*V.vinifera*)	Kinnaur
8.	Khwangi	IIHR, B'.lore
9.	Skibba Black	Kinnaur
	Hybrid (Variety/Rootstock)	
1.	Gulabi × Bangalore Purple	Borgaon
2.	E-4-30 (AES x BC)	--"--
3.	Arkavati (Black Champa x Thompson)	IIHR, B'lore

Contd...

Table 9.5–Contd...

Sl.No.	Name of Variety	Source of Augmentation/Collection
4.	E 19/4 (B'lore Blue x Cheema Sahebi)	GRS,Hyd
5.	E 5/20 (AES x BC)	--"--
6.	Arka Chitra (Angur Kalan x CLB)	IIHR,B'lore
7.	E 5/12 (AES x BC)	GRS,Hyd
8.	E 7/22 (AES x CLB)	--"--
9.	Gulabi x Fakri	--"--
10.	Arka Neelmani (BC x TS)	IIHR, B'lore
11.	E 12/2 (AES x TS)	GRS, Hyd
12.	E 29/6 (BC x TS)	GRS,Hyd
13.	E 32/8 (BC x TS)	--"--
14.	E 11/29(Arka Shweta) (AES x TS)	IIHR B'lore
15.	Pusa Urvashi (Hur x BS–76-64)	IARI
16.	E 30/14 (BC x TS)	GRS,Hyd
17.	E 31/4 (BC x TS)	GRS,Hyd
18.	E 31/5 (BC x TS)	--"--
19.	E 29/7 (BC x TS)	--"--
20.	E 9/3 (Arka Soma) (AES x QV)	--"--
21.	E 7/24 (AES x CLB)	--"--
22.	Arka Krishna (BC x TS)	IIHR, B'lore
23.	BA x BS–71-50	IARI
24.	E 8/24 (AES x CLB)	GRS, Hyd
25.	E 34/29 (CS x CLB)	GRS, Hyd
26.	E 21/28 (Arka Trishna) (BB x CLB)	--"--
27.	Arka Majestic (AK x BC)	IIHR, B'lore
28.	Arka Shyam (BB x BC)	Indage Narayangaon
29.	AES x Black Muscat	GRS,Hyd
30.	Pusa Navrang (MA x RR-76-2)	IARI
31.	MA x RR-76-3	--"--
32.	BA x Per-75-32	--"--
33.	HY-17-54-4-17	MPKV, Rahuri
34.	HY-23-14-23	–do-–
35.	E 26/5 (BC x AES)	GRS,Hyd
36.	E 6/2 (AES x BC)	
37.	E 5/4 (AES x BC)	—do-–
38.	Arka Hans (BB x AES)	—do-–
39.	E 12/3 (AES x TS)	GRS, Hyderabad
40.	F 26-8 (AK x BC)	GRS, Hyderabad

Contd...

Table 9.5–Contd...

Sl.No.	Name of Variety	Source of Augmentation/Collection
41.	V.P. x POC-7	GRS, Hyderabad
42.	Delight (Regina de vignette x SM)	RFRS,Abohar
43.	H-533 (*V.tilifolia* x Gulabi)	ARI, Pune
44.	E 8/5 (AES x CLB)	GRS,Hyderabad
45.	E 2/1 (AK x BC)	–do–-
46.	E 2/7 (AK x BC)	–do–-
47.	H-144 (*V.lanata* x Catawba)	ARI, Pune

Table 9.6: Improved Line/Advanced Line/Clonal Selection

Sl.No.	Name of Variety	Source of Augmentation/Collection
1.	Madhu Angoor	GRS, Hyd
2.	A17-3	NRCG,
3.	A18-3	NRCG,
4.	Sharad Seedless	–-"–-
5.	Tas-A-Ganesh	Arve,Borgaon
6.	Sonaka	Baramati
7.	Pusa Seedless	IARI
8.	Thompson Seedless(France)	Narayangaon
9.	PS-III-13-2	MPKV, Rahuri
10.	PS-II-2-28	–do–-
11.	PS-III-11-1	–do–-
12.	PS-II-11-4	–do–-
13.	PS-III-12-14	–do–-
14.	PS-III-8-6	MPKV,Rahuri
15.	PS-III-12-3	–do–-
16.	Manik Chaman	–do–-
17.	Sonaka (Mutant)	Ekunde,Sangli
18.	A3-3(25/26/36)	NRCG
19.	A 18-2(19/20)	NRCG
20.	A 39-2(14/15)	NRCG
22.	A 25-3(25/26/28)	–do–-
23.	Tas-A-Ganesh Mutant	Miraz,Sangli
24.	Kishmish Rozavis white Mutant	NRCG
25.	Nath Seedless	Kalamb, Narayangaon
26.	Sonaka Mutant	Negde, Miraj, Sangli dist.
27.	Krishna Seedless	Maisal, Miraj

Table 9.7: Status of Germplasm Exploration and Collections by NRCG, Pune

Crop	*Botanical Name*	*Area of Diversity Concentration*	*Areas explored*	*Germplasm Collected and held by N.R.C. for Grapes*	*Areas Yet to be Surveyed or Revisited*
Grapes	*Vitis* spp. and related genera of Vitaceae family, such as Ampelocissus, Cissus, Leea, Cayratia etc.	North-Western Himalayan region, Western Ghats, and North-East region and cultivated areas of peninsular and North-Western plains	Parts of cultivated areas of peninsular India and North-Western India, Kinnaur area of H.P. and Upper Konkan region of Maharashtra	415 accessions including the exotic collections	North-Western Himalayas, (Himachal Pradesh, parts of Kashmir) Western Ghat areas, North Eastern states.

Table 9.8: Status of Germplasm Holding

Name of the organisation: National Research Centre for Grapes, Pune

Name of the crop: Grape

Botanical name: *Vitis* spp. and related genera in Vitaceae family.

Accessioning Status			*Evaluation Status*		*Conservation Status*		*Promising Lines*		*Remarks, if any*
Total Number of Germplasm Available with Cooperating Center	*Number of Germplasm Assigned with IC/EC No. by NBPGR*	*Number of Germplasm Without IC/EC Number*	*Accessions Evaluated*	*Accessions to be Evaluated*	*Accessions sent to National Gene Bank*	*Accessions to be Sent to National Gene Bank*	*Promising Lines Identified*	*Promising Lines to be Registered with NBPGR*	
415	70	345 Approx.	215	200	All	215	9	22	The Centre was established in 1997, all the indigenous varieties registered subsequently.

Table 9.9: Other Wild Relatives/Related Species and Genera

Sl.No.	Name of Variety	Source of Augmentation/Collection
1.	*Cissus quadrangularis*	IIHR,B'lore
2.	*Cayratia trilobata* (*V. himalayana*)	Kinnaur,H.P.
3.	*V. parviflora*	MACS,Pune
4.	*V. champini*	MACS,Pune
5.	*V. palmata*	MACS,Pune
6.	*V. candicans*	MACS,Pune
7.	*V. tiliefoliae*	MACS,Pune
8.	*V. longii*	MACS,Pune
9.	*Leea angulata*	IIHR,B'lore
10.	*Ampelocissus sp.*	Northern Konkan Region
11.	*Leea sp.*	Northern Konkan Region

Popularization of Exotic Cultivars for Commercial Cultivation

National Research Centre in collaboration with Agricultural Produce Export Development Authority(APEDA) initiated a multilocation project under pilot scale in 8 different places; 3 in Maharashtra (Nashik, Phaltan in Satara district and Pune), 2 in Andhra Pradesh (Rangareddy and Anantpur districts), 2 in Tamil Nadu (Theni in Cumbam valley and Coimbatore) and one in Karnataka (Bangalore). Under this project, cultivars; Crimson Seedless, Red Globe and Italia having demand in international markets are included. Recently bred varieties like Autumn Royal, Autumn Seedless, Merquis and Blush Seedless with international market in table grapes are being introduced with financial help of National Horticulture Mission(NHM).

Conclusion

Grape being the most important fruit crop of the world includes ten thousand named cultivars spread around 100 countries. It has 70 related species. India has to develop its own varieties suited to tropical climate. Efforts are being made at NRCG, Pune in this direction to utilize indigenous species and varieties in the breeding and crop improvement program through biotechnological interventions.

References

Anuradha Upadhyay, Mamtha D.Saboji, Suvarna Reddy, Kiran Deokar and Karibasappa, G.S.2007. AFLP and SSR marker analysis of grape rootstocks in Indian Grape Germplasm. *Scientia Horticulturae* 112(2): 176-183.

Bharathy, P.V., Karibasappa, G.S., Biradar, A.B., Kulkarni, D.D., Solanke, A.U., Patil, S.G. and Agrawal, D.C. 2003. Influence of pre-bloom sprays of benzyladenine on *in-vitro* recovery of hybrid embryos from crosses of Thompson Seedless and 8 seeded varieties of grape (*Vitis spp*. L.). *Vitis*, 42(4): 199-202.

Bharathy, P.V., Karibasappa, G.S., Patil, S.G. and Agrawal, D.C. 2005. *In ovulo* rescue of hybrid embryos in Flame seedless grapes: Influence of pre-bloom sprays of benzyladenine. *Scientia Horticulturae*, 106(3): 353-359.

CMIE. 2001. Centre for Monitoring Indian Economy. *CMIE Monthly Bul. Nov.* pp. 259-265.

Cooke, T. 1903. *The Flora of the Presidency of Bombay.* Vol. 1. p. 263-279.

Emershad, R.L. and Ramming, D.W. 1984. In ovulo embryo culture of Vitis Vinifera L. cv. Thompson Seedless. *Am. J. Bot.* 71: 837-877.

Gamble, J.S. 1918. *Flora of the Presidency of Madras.* Vol. 1. p. 226-240.

Jindal, P.C. and Singh, Room 1993. Genetic Resources of Grape. *Ibid.* pp. 171-187.

Kanjilal, U.N., Kanjilal, P.C., Das, A. and Purkayastha, C. 1936. *Flora of Assam.* Part II. p. 287-308.

King, G. 1897. Notes on the Indian species of *Vitis. J. Agric. Soc. Bengal,* 65: 108-114.

Lawson, M.A. 1875. *The Flora of British India* (Ed. J.D.Hooker) Vol.1. pp.644. L. Reeve and Co., London.

Manjusha Dhane, Tamhankar, S.A., Patil, S.G., Karibasappa, G.S. and Rao, V.S. 2006. Assessment of genetic diversity and relationships among some grape varieties using ISSR markers. *Journal of Applied Horticulture,* 8(1): 50-52, January-June.

Nandini, C. Argade, Tamhankar, S.A., Karibasappa, G.S., Patil, S.G. and Rao, V.S. 2006. Assessment of Genetic Variation and Profiling of Important Seedless Grape Varieties in India using ISSR Markers. A Paper presented in the International Symposium on Grape production and processing, held at Baramati, India on 6-11 Feb., 2006. A*cta Horticulturae, ISHS*) Negrul, 1940. The evolution of cultivated grapes. *Priroda (Nature),* 4: 37-46.

Olmo, H.P. 1970. *Grape Culture: A Report to the Government of India*, UNDP, FAO. No. TA, 2825, FAO, Rome, pp.93

Patil, S.G., Honrao, B.K, Rao, V.G.and Patil, V.P. 1990. Screening of Grape (*Vitis species*) germplasm for resistance to three major fungal diseases. *Indian J. Agric. Sci.,* 60(12): 836-838.

Patil, V.P. 1993. Indigenous Germplasm of Vitaceae. In: *Advances in Horticulture* Vol. 1. (Eds.) K.L.Chadha and O.P. Pareek. Malhotra Publishing House, New Delhi, pp. 149-169.

Vavilov, 1950. The origin, variation, immunity and breeding of cultivated plants. *Chronica Botanica,* 13: 20-46.

Biodiversity in Horticultural Crops Vol. 2 *Pages* **202–225**
Editor: **Professor K.V. Peter**
Published by: **DAYA PUBLISHING HOUSE**

Chapter 10

Cocoa (*Theobroma cacao* L.)

S. Prasannakumari Amma, V.K. Mallika and E.K. Lalitha Bai
Cadbury- KAU Co-operative Cocoa Research Project,
College of Horticulture, Kerala Agricultural University, Vellanikkara, KAU P.O – 680 656,
Thrissur, Kerala, India

The bean of the tree *Theobroma cacao L.*, the original source of chocolate, is one of the greatest treasures ever discovered by man. It is an ancient crop cultivated, harvested and used by the indigenous people of Central and South America for thousands of years. According to legend, the cocoa tree was discovered by the Mayan king Quetzalcoatl three thousand years ago. There are evidences to show that cocoa beans played a very vital role in the economy in Central and South America. The beans were used as currency (1 slave = 100 beans) during that period. Indians of Northern South America, Central America, and Mexico consumed a strong bitter, semisolid aromatic beverage made from the seed.

Cocoa was introduced to Europe during 16th century. In 1828 technology to separate the cocoa butter from dried beans was developed by Coenraad Van Houten, a Dutch chemist using hydraulic press, which led to manufacture of cocoa powder. The inventor of chocolate as we know it today is unknown, but the first eating chocolate was sold in England in 1847 by Joseph Fry and Son. Two years later the Cadbury family began selling a similar product. In 1876 milk chocolate was invented by William Peter in Switzerland. This led to worldwide multi-billion-dollar chocolate processing industry. Since then, cultivation of cocoa spread to many tropical countries of the world. It is now grown in 58 countries on more than 6.9 million ha worldwide (Figure 10.1) and is worth an excess of $4 billion to the world's economy annually. Cocoa is the commodity produced in the developing countries of the tropics and consumed by the affluent class in developed countries of temperate zones. The major share (70 per cent) of global production of cocoa comes from West African countries like Côte d'Ivoire, Ghana, Nigeria and Cameroon. Of these, some countries rely heavily on cocoa exports for their economic development.

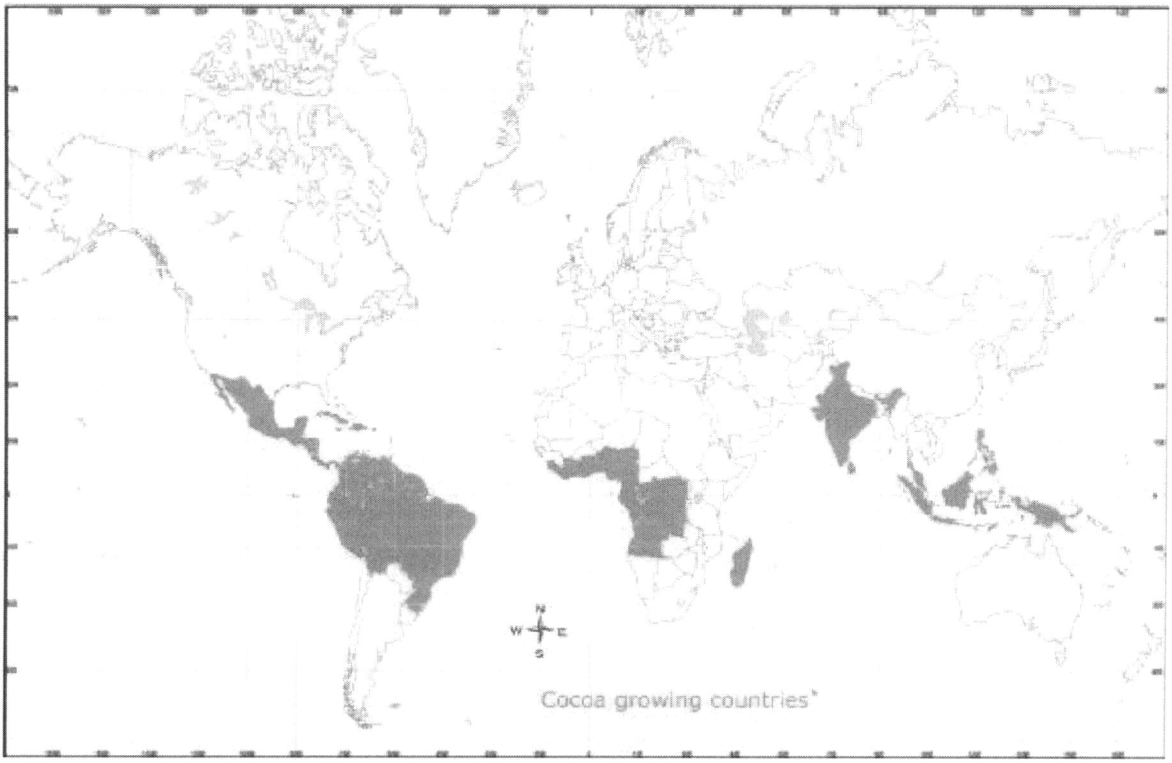

Figure: 10.1

From the level of 1.5 million t. in 1983-84, the world production of cocoa steadily increased and touched a peak of 3.5 million t. in 2003-04. This significant increase in production arose mainly due to expansion in area. Over 90 per cent of the world cocoa is in small holder sector and most of the farms are low yielding. The yield improvement, in spite of genetic improvement programmes, is nominal. Apart from improvement in yield, varieties with resistance to pests and diseases are essential.

Most of the genetic diversity is found in the Americas, centre of origin of the crop, whereas most of the production comes from West African countries. The genetic base in major producing countries is quite narrow and if the targeted goals are to be achieved, diverse genotypes are to be introduced from Amazon and neighbouring areas and exploited. A detailed assessment of biodiversity of cocoa in different areas of its cultivation, collection of valuable types, their conservation and utilization are relevant to bring about significant accomplishment through breeding.

Early History, Origin and Spread of Cultivated Cocoa

According to Cheesman (1944), cocoa is a native of tropical humid forests on the lower eastern equatorial slopes of the Andes in South America, occurring in primary and secondary regions of distribution in the Americas within a range between latitudes 20^{0}N and 20^0 S. The area drained by the river Amazon (70,00,000 km^2) includes the north-western tributaries of river Amazon and western tributaries of river Orinoco. Over 1000 watercourses make up the hydrologic systems of river Amazon

(Bartley, 2005). It is indicated that each component stream would represent a distinct cocoa population. The cocoa populations on the tributary rivers contain varieties that are unique to each river in terms of phenotypic characters. Prior to the arrival of Europeans, the evolution of cocoa diversity took place over the entire primary distribution area. Only part of the Amazon river basin and the adjacent areas on its periphery were occupied by cocoa at the start of European colonization. From 17th century, expansion in area took place. The greater part of primary cocoa diversity is located in Western and South- Western quadrant of the region.

The first civilization to cultivate the tree was the Olmec, living in the heart of equatorial Central America, three thousand years ago. After the Olmec, the Maya who took up cultivation of this crop in fourth century AD believed that the tree belonged to Gods and that the pods growing from its trunk were an offering from the Gods to man. The Toltecs and Aztecs, who later settled in the Mayan empire, also took interest to spread cultivation of cocoa. Cocoa travelled along the trade routes used by the Mayas, Aztecs and Pipil- Nicaraos (Young, 1994; Coe and Coe, 1996). Well–to–do Indians made a thick beverage by pounding roasted cocoa beans with maize and *Capsicum*. The first Europeans to drink cocoa were the Spanish who invaded and conquered the empire in Mexico in 16th century. The Spanish learnt from 'Aztecs' the technique of making `xocoatl', a drink made from cocoa beans after roasting and grinding. The word `chocolate' originated from `xocoatl'. The word `cacao' also was used by the Spanish and it probably originated from `cacahuatl', a word that Aztecs used for cocoa beans. The present distribution of diversity involves spontaneous and domesticated populations. In the course of time, as the inhabitants migrated to other areas, the species was extended to regions of the Americas beyond the range of spontaneous occurrence. Further expansion occurred some time after the conquest both within the Americas and the tropical regions of the Eastern Hemisphere.

Spanish explorers took cocoa to the Philippines, from where it spread throughout Southeast Asia, India and Sri Lanka. Amelonado cocoa was taken to West Africa. Criollo types spread to Central America and to large number of Caribbean Islands, including Trinidad in 1525 and thereafter to Jamaica. The dissemination to Venezuela and Costa Rica was made by the Spanish (Pittier, 1933). Introduction to Martinique and Haiti was by the French. Planting in Belem and Bahia in 1750, was attempted by Portuguese.

During 1822, cocoa seeds were taken from Portuguese colonies of South America to the island, Sao Tome of the West Coast of West Africa. It also spread to neighbouring island, Principe. Cocoa cultivation was started in Fernando Po in 1840. The most successful introduction into African mainland was made by the Ghanaian, Tetteh Quashie in 1879. He brought one pod from Fernando Po and the early population of cocoa in Ghana originated from this pod. From Ghana, it spread to other African

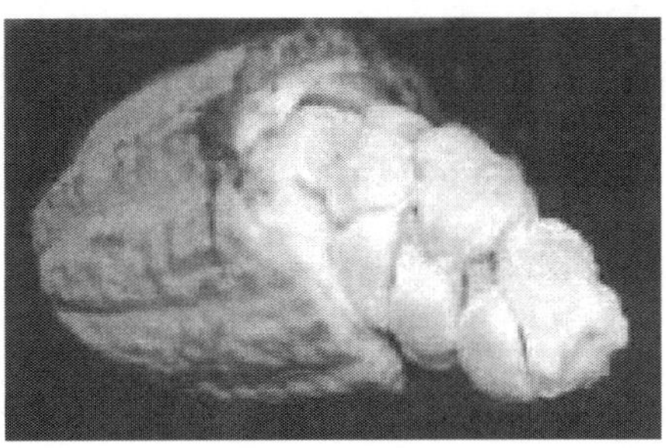

Figure 10.2

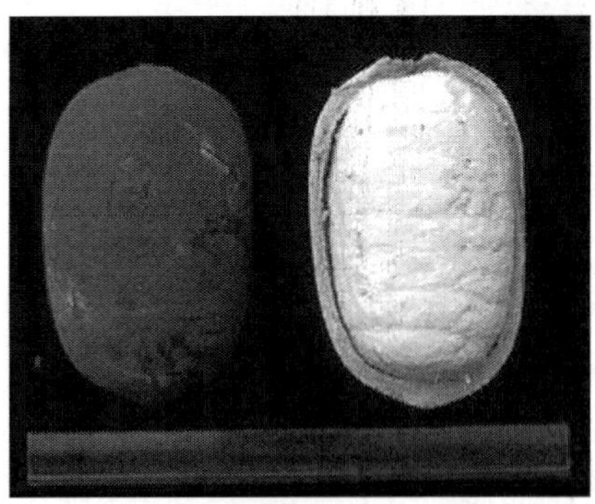

Figure 10.3

countries, the most important of which are Ivory Coast, Nigeria and Cameroon.

Cocoa was introduced in the 16th century to Asia and the Pacific. Venezuelan Criollo was introduced in Celebes by the Dutch in 1560. They also introduced the crop into Java. The Spanish took Criollo types from Mexico to Philippines in 1614. It was introduced to Sri Lanka also from Trinidad at about 1798. From Sri Lanka, cocoa was taken to Singapore and Fiji in 1880, Samoa in 1883, Queensland in 1886 and Bombay and Zanzibar in 1887. Cocoa was introduced to Malaysia in 1778 and in Hawaii in 1831. Cocoa was introduced to India in1791; but its cultivation started in a big way only in 1960's.

The Genus *Theobroma* and its Relatives

The scientific name of cacao tree is *Theobroma cacao* L. (2n=2x=20) as quoted by Linnaeus in 1753 in the first edition of *Species Plantarum*. The Mayas and Aztecs of Central America considered the tree and fruits to be of divine origin. The generic name *Theobroma* is derived from two Greek words *theos* (God) and *broma* (beverage) and thus the literal meaning is "drink of gods". Linnaeus felt that the tree deserved a name that reflected the Mayan belief that the tree belonged to the Gods.

The genus *Theobroma*, a member of family Sterculiaceae contains 22 species grouped into 6 sections (Cuatrecasas,1964). The family contains 50 genera and 750 species of trees and shrubs, mostly tropical. Of these *T. cacao* is the most important species producing cocoa. *Cola* spp. are sometimes used for beverage purposes, but more usually as a masticatory in West Africa. Except *T. cacao* and *T. bicolour*, most of the spp. have restricted geographical distribution. About 15 spp. are utilized for their edible pulp or seeds. *Theobroma grandiflorum* (cupuassu), *Theobroma gileri* (mountain cocoa), *T. bicolor* (macambo) and *T. subincanum* (wild cocoa) are other species utilized for their sweet, edible pulp and edible seeds. Beans of *T. angustifolium* Moc. and Sesse. are mixed with cocoa in Mexico and Costa Rica and sweet pulp around the beans of *T. grandiflorum* (Willd. ex Spreng.) K.Schum. is used for making a drink in parts of Brazil and is also eaten. A few other species are *T. mammosum, T. obovatum, T. subincanum, T. simiarum* and *T. sylvestre*. Figures 10.2 and 10.3 provide a comparison of the fruits of *T.cacao* and *T. grandiflorum* respectively.

The different *Theobroma* species are crossable (Posnette, 1945, Martison, 1966, Jacob and Opeke, 1971 and Williams,1977). The genus *Herrania* having more than 10 species is closely related to *Theobroma*. Williams (1977) made about 10,000 reciprocal crosses among *Theobroma cacao, T.grandiflorum, T.microcarpum, T. speciosa, T. bicolor* and *Herrania*. Success of hybridization was unidirectional, being more marked when cacao cultivar was used as female parent. The butterfat content of *Herrania* x *T. cacao* was 80.4 per cent, whereas *Herrania* sp x *Herrania* sp gave only 67.7 per cent butter fat. Cuatrecasas (1964) discussed results of his studies on interspecific crosses involving *T. cacao, T. grandiflorum, T.angustifolium, T. microcarpum, T. speciosa, T. bicolor, T.mammosum, T. obovatum, T. subincanum, T.simiarum* and *T.sylvestre*, where there were indications of heterosis in several characters in some of the crosses.

Ecological Adaptations

In wild state, cocoa grows to a height of about 12-15m. Seedling derived trees come to full bearing at 5 or 6 years, though production starts at 2-3 years. A well maintained plantation continues to be profitable for at least 25-30 years.

These are under-storey trees in the tropical rainforest and seasonally flooded sites from sea level to 1080 m. It usually grows in groups along river banks, where it may stand for 6 months in slow running water. As the cultivation started, cocoa was planted in environments similar to its natural habitat on the conviction that cocoa must be planted under such conditions. Experience over the last several years proved that heavily shaded cocoa produces low yield.

Purseglove (1968), Wood (1975) and Wood and Lass (1985) illustrated ecological needs of the crop. Limits of cultivation are 20°N and S, with bulk of the crop lying between 10°N and S of equator. It is grown mainly at low elevations, usually below 300 m. But experience suggests realization of good yield in still higher elevations. Areas of tropical evergreen and semi- evergreen rain forests are the most suitable ecological zones for cocoa. The range in mean monthly temperature of majority of cocoa growing regions is 15 to 32° C. The absolute minimum for any reasonable period is about 10° C below which frost injury is likely. Thus, temperature sets the latitude limit for the best growth of cocoa to within 8° North and South of equator.

The total annual rainfall ranges from 1500 to 3000 mm. Rainfall, below 1500 mm would necessitate regular irrigation to support unrestricted growth of crop. Rainfall above 3000 mm with torrential and continuous rain for long spells favour incidence of diseases like black pod (*Phytophthora palmivora*) and vascular streak die back. For ensuring the best growth of cocoa, even distribution of rainfall is more important than the total quantity. In most of the major South American, African and South-East Asian cocoa-producing countries, distribution is more or less even with minor peaks. It is so well distributed that around 10 cm of rain is received almost every month.

Cocoa survives in very heavy shade, which would kill many other species, but yield will be low. Photosynthesis is reduced by shading, but is partly compensated by large leaf area. The seedlings grow the best under 75 per cent shade and grown up plants must receive 50 per cent shade. Higher illumination rates produce higher yields, but during initial years of establishment, plants must be protected from scorching sun.

Cocoa is grown in a wide range of soils. The soil in high rainfall areas is relatively coarse-textured and acidic to neutral in reaction. Very coarse sandy soils with very low water holding capacity and low organic matter are not suitable. Usually virgin, freshly cleared forest soils are used for cultivation of this crop in most of the cocoa growing regions of the world, which are rich in organic matter and nitrogen, well-drained and acidic to neutral in reaction. Another basic soil requirement of this crop is a depth of up to 1.5 m., free from any hard rock.

Root System

The root system shows dimorphic growth characterized by the orthotropic tap root which goes deep into the soil and lateral plagiotropic ramification, which constitutes the feeding roots. The tap root of cocoa grows predominantly downwards with only a few branches. Under suitable growing conditions when the soil is deep, they grow to a depth of about 150–200 cm. The primary function of these roots is to provide anchorage. The main feeding roots arise from tap root at a depth of 15-20 cm and grow out in humus layer to a length of 5-6 m, giving rise to a dense mat of surface feeding roots. Middle of tap root is usually devoid of secondary roots, but these arise lower down and either grow upwards to surface mat or downwards into lower layers of soil. The rooting pattern is modified to an extent by environment.

Stem

Cocoa exhibits dimorphic branching and grows in tiers. Seedlings grow vertically with straight stem called "chupon" until it reaches 12-14 months or attains a height of 150-200 cm. Growth of chupon ceases and terminal bud breaks up into 3-5 giving rise to three to five plagiotropic branches. These lateral branches are called 'fans' or 'fan branches'. The point at which fans arise is called 'jorquette' and the process of formation of fans from jorquette is called 'jorquetting'. The height at which jorquetting occurs varies with genotype and is also influenced to an extent by environment, with plants growing under dense shade recording high jorquettes. A layer of fans is called a 'tier'. If the plants are allowed to grow under natural conditions, axillary buds just below the jorquette grow and continue to grow vertically and jorquettes again. In this way, several successive chupons arise sympodially, each producing a tier of branches. Eventually, the lowermost tier of plagiotropic branches tend to dry and fall. Chupons can be distinguished from fans from their nature of growth and leaf arrangement. Fan growth will be predominantly to the sides and chupons grow vertically up.

Leaves

Cocoa leaves show dimorphic characters depending upon type of stem on which they are produced. Leaves of fans are arranged in one plane and are alternate. Leaf arrangement of chupons will be spiral with a phyllotaxy of 3/8. Leaves are large, simple, pubescent with pronounced pulvinus and dark green when mature. Petioles are 1-4 cm long and are longer on chupons. Stipules are lanceolate 5-20 x 1-2 mm, pubescent dropping off early. Leaf lamina is entire, simple, elliptic or obovate- oblong, usually glabrous, 12-60 x 4-20 cm, base rounded and obtuse, apex acuminate, main vein prominent, lateral veins pinnate, 9-12 pairs. The leaves exposed to full sun light are stronger and thicker than those under shade.

Leaves in fan branches are produced in flushes, period between flushes depending upon temperature, tree health and soil moisture status. Under ideal conditions, new flushes appear only 2-4 times a year. During flushing, terminal bud grows out rapidly, producing 3 to 6 pairs of leaves. The young leaves are tender and delicate varying in shades from green to red. The leaves usually harden turning to green as they mature. Once the leaves are hardened, the terminal bud becomes dormant. Next flushing occurs when the environmental conditions become favourable. Flushing demands higher supply of nutrients, a part of which is met from older leaves. Flushing in cocoa is referred to as 'change of leaf'. The extent of leaf fall during flushing is an indication of health of tree. Leaves usually persists through two more flushes and drop on third. The period of active photosynthesis is the most marked during first 4-5 months of its existence.

Normally, buds arising on a chupon give rise to chupons, except that at the time of jorquetting. Similarly, fans generally produce fan branches. At times, fans produce chupons also.

Inflorescence

Cocoa produces flowers throughout the year, but flower production is not uniform with peaks during a certain months of year. These periods of peak flowering are often different for different regions, indicating strong association with climatic factors. For example, in Ghanaian cocoa, normal flowering reaches its peak during May-June, starting from March and extending through April. In addition to this normal flowering, 'crazy' flowering also occurs during any period of year. Flowering starts after one or two years of planting.

Cocoa flowers are cauliflorous, borne on older leafless wood of main stem and fan branches and never on new flushes. Inflorescence is a much compressed cincinnial cyme with greatly reduced branch and produced on thickened leaf axils on stem called "cushion" (Figure 10.4). The number of flowers/cushion/season is up to 50. A succession of flowering occurs, depending on environmental conditions and health of tree. Some cocoa trees may have pronounced two flowering peaks separated by small but continuous flowering periods as in Upper Amazonian selections. In West African Amelonados, the flowering peaks are separated by periods of no flowering.

Flowers

Flowers are regular, pentamerous and hermaphrodite. Size of flower varies with genotype (Figure 10.5). Each flower has a pedicel of 1-3 cm length, greenish, whitish or reddish. Sepals are five in number, pink or whitish, 7-10x 1.5 x2.5 mm, triangular, valvate and shortly united at base. There are five petals, base of which is ovate, 3-4 mm long, expanding into a concave cup-shaped pouch, white with 2 prominent purple guidelines. Tip of the petal is spatulate, 2-3 mm long and attached to the pouch by a narrow connective. Androecium consists of 5 outer staminodes, which are erect, pointed with dark purple centres and white ciliate margins and form a ring around style. The five fertile stamens have 4 pollen sacks which dehisce longitudinally. The filaments bend outwards so that anthers are concealed in pouch of corresponding petal. Ovary is superior, five carpelled with numerous ovules. Ovule is anatropous with axile placentation at base and parietal above. Ovules/flower ranges 20 to 60. Style is single, 2-3 mm long, hollow and shorter than surrounding ring of staminodes. The style has five stigmatic lobes. The floral formula is $K_5C_5A_{5+5}G(5)$.

Fruits

The fruit of cocoa is botanically a drupe, but commonly referred a pod. The pods are indehiscent and show wide variability in size, colour, shape and texture (Figure 10.6). Length of pods varies 10-35 cm. Pod shape is determined by the ratio between pod length and width and also by shape of two ends of pod. Pods may be spherical (Calabacillo), elongated and pointed (Cundeamor), oval (Angoletta) and regular melon shaped (Amelonado). Pod surface may be smooth or warty, with or without 5 or 10 furrows. Young pods are green or red and on ripening turn to yellow. Pericarp consists of three distinct layers; hairy and thick epicarp, mesocarp, thin and hard and more or less woody; and the hardy endocarp of varying thickness. Pods attain full size in 4-5 months after pollination and take one more month for ripening. Fruit weight is variable and ranges 200 to more than 1000 g, but the average pod weight is 300-500 g.

Self pollination of cocoa flowers is made practically impossible due to peculiar flower structure (the fertile stamen is concealed in the pouched portion of petal and stigma is surrounded by a ring of staminodes). Flowers are devoid of scent or nectar and pollen grains are sticky. Natural pollination occurs only with help of small crawling insects. The most important pollinating insect is the flying female midge (*Ceratopogonid* midges of genus *Forcipomyia*). The insects are small and barely visible to naked eye. The midges are attracted by purple pigmented tissues of staminodes and guidelines of petals. The midges moving on guidelines near the anther pick up the pollen grains and when they crawl to the staminodes, some of the pollen get transferred from their body to the stigma. The midges fly to about 75 m only. Though the midges are the most important pollinating agents, other insects like ants, aphids, fruit flies and thrips are also implicated as probable pollinating agents. There is a possobility of wind pollination also. Flowers open late in the afternoon and are fully open by forenoon, next day. As such, most of the pollination occurs in early hours of the day. Anthesis commences between 14.00 and 16.00 hrs and is complete 14.00 and 16.00 hrs in the next day. Anther dehiscence starts between 04.00 and 05.00 hrs and is completed between 08.00 and 10.00 hrs.

Taking into account the complicated flower structure and problems of self and cross incompatibility, extent of successful pollination varies considerably. Cocoa flowers are produced in large numbers but only a few of them (1-5 per cent) develop into fruits due to above problems. Unfertilized flowers fall off within 24 hours. Instances of delayed self and cross incompatibility are also noticed.

Figure 10.4

Figure 10.5

Figure 10.6

Seeds

Seeds are called beans. Each pod contains 20-60 beans. The beans/pod are higher in Forastero. Beans are arranged in 5 rows, variable in size (2-4 x 1.2-2 cm), ovoid or elliptic. A mature bean has no endosperm and from outside to the centre, consists of 1) a thin, resistant, pink- veined husk or a leathery testa, originating from tissues comprising seed coats of ovule 2) a fine translucent and shiny silver skin, which is the remains of endosperm and 3) two large convoluted cotyledons called 'nib' and a small embryo (germ). The cotyledons vary in colour based on type of cultivar and it varies from white to dark purple (Figure 10.7). Fresh seeds are surrounded by mucilaginous, whitish, sugary, acid pulp which develops from outer integument of ovule. Fresh bean weight/pod varies from 60-250 g. The fresh weight varies from 1.3 to 4.0 g and dry weight of bean, from 0.5-2.5 g. The international cocoa standards specify the bean count (Number of cured beans/100g.) to be 95. Beans constitute 25 per cent by weight of mature pod. Shell content of cured beans is around 10-12 per cent.

Self Incompatibility

A special feature of cocoa is self-incompatibility shown by a few cocoa types. This was first reported by Harland in 1925. Upper Amazon and Ecuador types introduced in Trinidad were self incompatible; but most of the self incompatible plants were cross-incompatible also. Many of the homozygous types like West African Amelonados are self-compatible. Even though the self compatible types may have the advantage of better fruit set under varied situations, self incompatibility is important

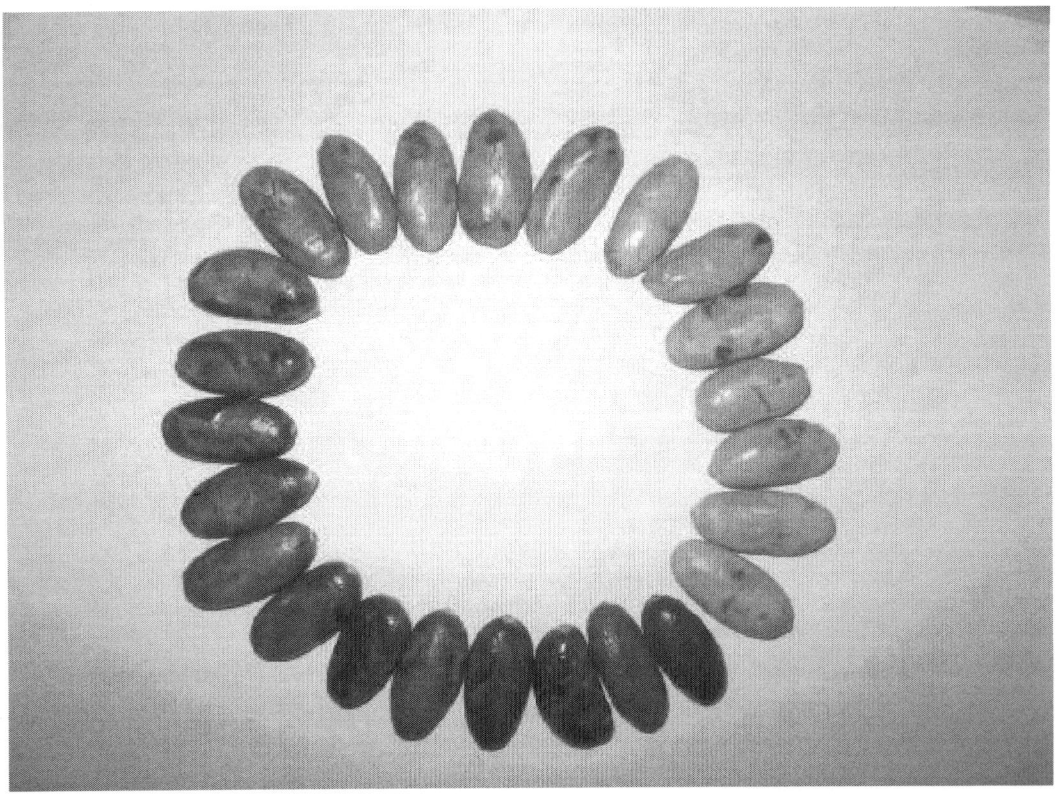

Figure 10.7

in commercial hybrid seed production. Incompatibility in cocoa is unique in that, site of incompatibility is the embryo sac (Cope, 1962). After incompatible pollination, the pollen tube grows faster and delivers gametes into embryo sac in a normal fashion. The embryo sac is in no way abnormal and the rejection is due to failure of male nuclei to unite with egg. This incompatibility is referred to as "prefertilization inhibition in the ovule" and it is genetically controlled. Fusion or non fusion is controlled by a series of alleles operating at a single locus (S), showing dominance or independence relationships (Purseglove, 1968). In incompatible matings, flowers drop two to four days after pollination. If a population of cocoa is examined for self incompatibility reactions, it could be observed that majority of the plants belong to self incompatible group. Cross incompatible types frequently occur between two individuals with different genotypes and it occurs only in diploid gametophytic systems when individuals share the same S genotype (Richards, 1986). Mallika *et al.*, 2002) while working out compatibility relations among sixteen selected parents, observed 23 crosses to be cross incompatible among 128 crosses attempted. Cross incompatibility is an indirect measure of degree of closeness between the genotypes. When the parents used in crossing happen to be genetically similar, incompatibility mechanism operates.

Classification

Classification of cocoa is difficult as it has undergone a substantial genetic mixing over past several years. Like many species in Amazon river basin, cocoa appears in a great variety of forms. Because of this, for a long time, there was lot of confusion on classification of species.

Pittier (1933) designated each group as a different species: *T. leiocarpum* for Forastero and *T. cacao* for Criollo. Since many genotypes are interfertile, inclusion of all wild and cultivated cocoa into a single species is well justified. The overall classification into two morpho-geographic groups, Criollo and Forastero, is in widespread use.

Cuatrecasas (1964) classified cocoa into two subspecies of *Theobroma cacao:*

T.cacao subsp. *cacao*

The criollo like forms are included under this category and it has four forms.

1. *T.cacao* subsp. *cacao* f. cacao): This is the original Central American Criollo, found in Mexico and British Honduras. The fruits are oblong, surface warty, 5 deep narrow furrows with five shallower furrows between, immature pods green or dark red, mesocarp thin, seeds round, cotyledons white and of the highest quality.
2. *T.cacao* subsp. *cacao* f. *pentagonum* (Bern.) Cuatr.: This is the original South Mexican and Central American Criollo, known as Alligator cocoa, characterized by highly rough and warty fruits. Fruits are oblong- oval, with prominent ribs, immature pods reddish yellow, mesocarp not woody, seeds large and round, cotyledons white and of the best quality.
3. *T.cacao* subsp. *cacao* f. *leiocarpum* (Bern.) Ducke.: This form is cultivated mainly on Atlantic coast of Guatemala. Fruits are ovoid, shallowly furrowed, almost smooth, thin smooth shell, nibs plump, cotyledons white or pale violet and of high quality. Porcelain cocoa cultivated in Surinam belongs to this form.
4. *T.cacao* subsp. *cacao* f. *lacandonense* Cuatr. This is a wild half–wine known only in Chiapas in Mexico.

***T.cacao* subsp. *sphaerocarpum* (Chev.) Cuatr. (*T.sphaerocarpum* Chev.)**

This subspecies includes all Amazonian and Amelonado forasteros. These are native to South America found in the Hylaea from the Guianas and middle Amazon north and west ward to the Andes. Because of its adaptability, it spread throughout the tropics. Fruits are ovoid, smooth, shallowly 10 furrowed, rounded at both ends, unripe pods green, pericarp very thick, mesocarp woody, seeds ovoid, compressed, cotyledons dark purple. Quality is variable but lower than other subspecies.

The two subspecies and forms intercross freely to give fertile F_1 hybrids and hence contributed to evolution of a large number of distinct local populations.

Cheesman (1944) classified the cultivated and wild cocoas into three groups based on morphological characteristics of the pods, flowers or seeds.

The Criollo Group

Domesticated for a very longtime, probably by the Mayas, Criollo cocoas are cultivated in Mexico, Guatemala, Colombia, Venezuela, Madagascar, Nicaragua, Sri Lanka, Indonesia and Samoa islands. Criollo pods are yellow or red when ripe, deeply 10 furrowed, markedly warty, conspicuously pointed, pod wall too thin, seeds large, plump and almost round and cotyledons white or pale violet (Figure 10.8). The beans ferment quickly, but yield is poor. It produces the highest quality cocoa and is very much sought for its strong aroma and low astringency. It is used in chocolate industry for luxury products. Criollo trees are found in isolated groups in very old plantations. The main types of Criollo include Pentagona or Lagarto cocoa, Real cocoa and Porcelana cocoa. It is susceptible to stress, major diseases and not adaptable to all situations. The trees are less vigorous and slow growing. This type constitutes only 5 per cent of global production.

It can be subdivided into two:

1. *Central American Criollo*: This is the original cocoa cultivated in Central America and Mexico. Selection over past several years resulted in fixing the recessive white cotyledons, with no astringency and requiring little fermentation. The unripe pod wall is green, turns yellow on ripening.
2. *Venezuelan Criollo*: This represents the population derived from Criollo introduced from Central America into Venezuela. This type exhibits wide variability in colour, size and shape of pods. Unripe pod wall is usually red.

The Forastero Group

This is a very variable group, found in an indigenous or semi indigenous state in High Amazonia (Peru, Ecuador and Colombia) in Amazon river basin in the Guyanas and along the Orinoco river in Venezuela. These are now widely planted throughout the cocoa producing countries. This group constitutes bulk of world's cocoa and almost all of the production currently coming from Brazil, West Africa and South East Asia.

The Amazonian Forasteros are characterized by green unripe pods, which turn to yellow on ripening. The pods are inconspicuously ridged and furrowed, surface smooth, ends rounded or bluntly pointed, pod wall thick, seeds flattened, fresh cotyledons deeply pigmented and dark violet giving an astringent product (Figure 10.8). The trees give high yields and are hardy. Quality is not comparable with Criollo. The beans take 5-6 days for fermentation.

Forastero **Criollo**

Figure 10.8

The West African type 'Amelonado', Ecuadorian type 'nacional' and Brazilian types 'Maranhao', 'Comum' and 'Para' types belong to this group. The Upper Amazonian cocoas include Forasteros collected during expeditions in twentieth century from upper part of Amazon basin, mainly to west of town of Iquitos. There is large variability among Upper Amazonians as compared to Lower Amazonian Forasteros.

The Trinitario Group

Both the above types were distributed throughout the Caribbean, where they hybridized in Trinidad, creating a distinct hybrid called Trinitario. These are very heterogeneous and exhibit a wide range of morphological and physiological characters. It is difficult to specify characters of Trinitarios as they may have pod and bean characters ranging from those of typical Criollos to those of Forasteros. Figure 10.9 depicts ICS 95, a trinitario tree with profuse yield. This type was introduced into Venezuela about 1825 and subsequently to most of the cocoa growing countries. There are clones combining the vigour of Forastero with the quality of criollo. Trinitarios are of great importance in breeding and in trade, it is regarded as 'fine' cocoa. The Trinitarios are grown mainly in all the countries where the Criollos were formerly grown (Mexico and Central America, Trinidad, Colombia and Venezuela) and in many African and South East Asian countries.

Theories Associated with Origin of Cocoa Types

Many theories are put forward about origin of Criollo group. According to Cheesman (1944) Upper Amazon is the centre of origin of Criollo and Forastero. The spread of Criollo throughout Central America began from a small population in Upper reaches of Amazon, which may have crossed

Figure 10.9

the Andes with help of man. Cuatrecasas (1964) on the other hand, suggested that the species was indigenous in the Amazon region to Mexico.

Alvim (1987) proposed that from the centre of origin (Amazon river basin), the species spread out naturally creating two main groups (1) the "Criollos" (*cacao dulce*) that developed north of Panama isthmus through the Andes towards lowlands of Venezuela, Colombia and Ecuador and towards North to Central America and Mexico and (2) the "Forasteros" (*'cacao amargo'*) which resulted from dissemination towards the Amazon valley in Northern Brazil and the Guayanas. Criollo types were cultivated by indigenous people of Central and South America and Europeans were first exposed to these types. Commercial production commenced in Brazil using Forastero types, mainly a uniform type called Amelonado. Both types were distributed throughout the Caribbean, where they hybridized in Trinidad, creating a distinct hybrid called Trinitario. It is recorded that Criollo population from Venezuela and the Amelonado type Forastero from Guayana could have been involved in hybridization leading to production of Trinitario. The general names used are based on Venezuelan terminology, namely Criollo = native, Forastero = foreign and Trinitario = native of Trinidad.

A map showing origin and evolution of cultivated types of cocoa is provided in Figure 10.10.

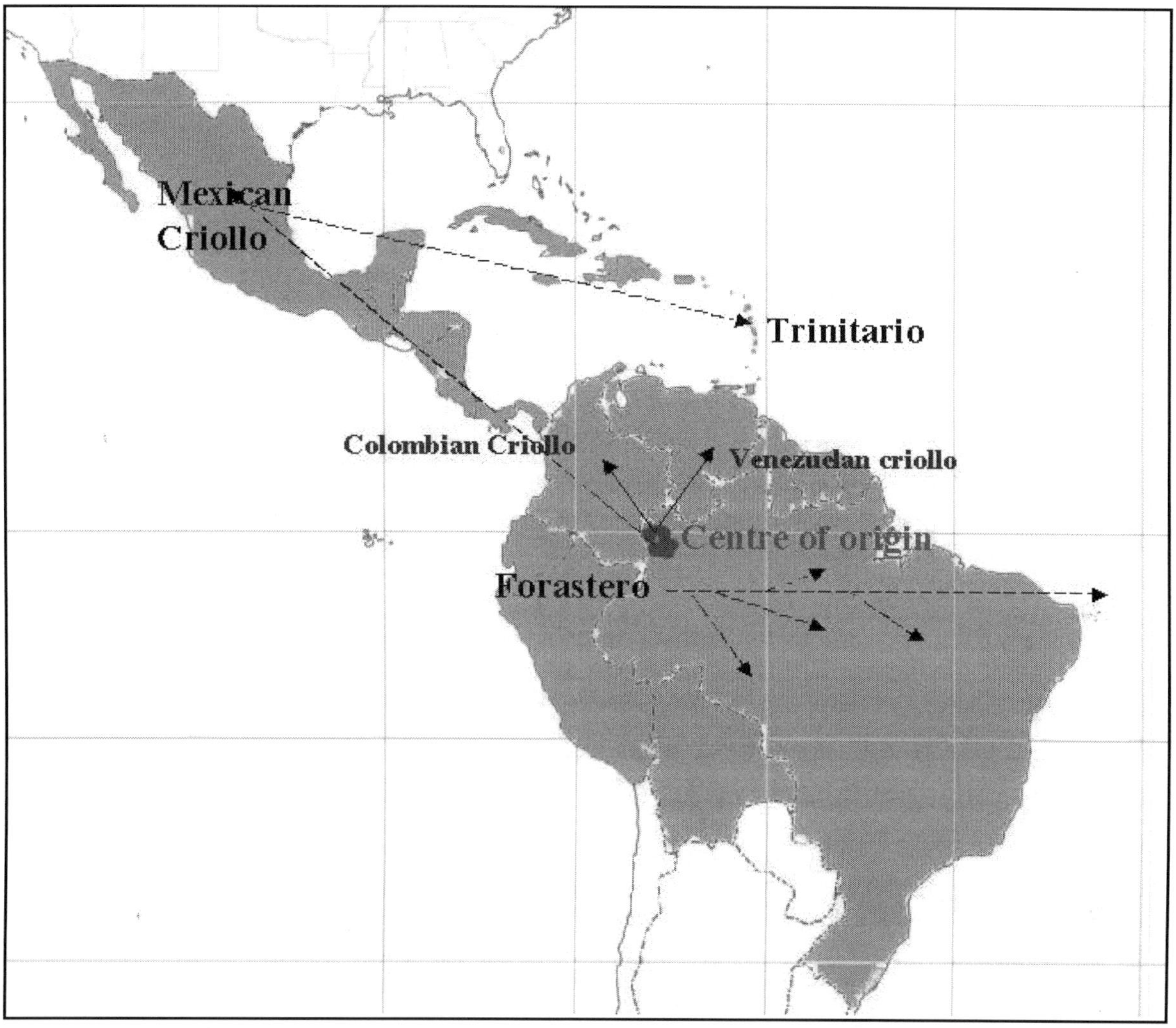

Figure: 10.10

Varieties/Clones

Each country has its own unique cultivar. The cultivars in different countries vary with the region from where they are introduced and the extent of hybridization. There are thousands of clones of cacao in field gene banks in different research institutions of world. Selection and hybridization work in the cocoa growing countries of the world yielded a lot of clones under each category with distinct and unique features. Apart from this, a lot of expeditions are made mainly in centre of origin and neighbouring countries and a large number of useful clones are released. Wadsworth *et al*. (1997) provided the database of about 29,000 clones.

The clones bear name of place or of river in the region in which they have traditionally been harvested or the research station where it was evolved: Iquitos, Nanay, Parinari, Scavina, ICS etc.

Guidelines for providing clone names to avoid duplication are provided by Wadsworth *et al.* (1997). Some of the important clones coming under each type are

1. Criollo
   (*a*) RIM 189 (MEX)- Rosario Izapa Mexico
   (*b*) CAT 201 (ARA)–Cata Aragua State
   (*c*) ICS 39, 40, 45, 47, 60, 84, 89 and 91- Imperial College Selections
2. Forastero
   (*a*) AMAZ 2/1, 3/2 (Amelonado)–Amazonas
   (*b*) APA–4 (Amazonica Palmira)
   (*c*) BAL 209 (Borneo Albacca Plantation)
   (*d*) BE 3 (Belem)
   (*e*) CAS-1 (Campo Agricola do Santarem)
   (*f*) CATIE 11, 1000 (Centro de Agronomico de Investigacion, Turrialba, Costa Rica)
   (*g*) CC 18, 34 (Cacao Centre, Costa Rica)
   (*h*) UF 4, 29, 36, 700, 701, 703, 705, 710 (United Fruit Company)
   (*i*) SCA 6, 8, 9, 12 (Scavina)
   (*j*) CCRP 1, 2, 3, 4, 5, 6, 7, 8, 9,10 (Cadbury-KAU Co-operative Cocoa Research Project, Kerala Agricultural University).

   The Scavina accessions, collected from the Upper Amazon in the 1930's, are the main source of resistance to witches' broom, cocoa swollen shoot virus, vascular streak die back and black pod diseases of cocoa (Soria, 1977; Kennedy *et al.*, 1987). These accessions, especially Scavina-6 and Scavina-12, transmit to their descendants not only resistance but also high hybrid vigor and good yielding performance. SCA 6 is the clone exploited very much in breeding programmes. It is indicated that this clone is currently used in 39 countries of the world for imparting resistance to major diseases of cocoa.
3. Trinitario
   (*a*) BL Z 9/R (MEX), BL Z 23/R (MEX), BL Z 34/R (MEX), BL Z 52/R (MEX), BL Z 56/R (MEX)- BeLize
   (*b*) C 26, 37, 38, 39, 40, 41, 42, 43, 44, 45, 46, 47, 48, 49, 52 (Cabruca, Brazil)
   (*c*) UF 1, 10, 168, 221, 296, 613, 650, 654, 666, 667, 668, 676, 677, 688, 707, 708, 709,711, 712 (United Fruit Company)
   (*d*) ICS 6, 8, 16, 43, 61, 95 (Imperial College Selections)

Biodiversity of Cocoa in Major Cocoa Growing Countries

Bartley (2005) elaborated population structure of cocoa in major cocoa producing countries and a brief outline of the same is furnished below:

South America: Population Derived from Amazonian Region Germplasm Base

Ecuador

Cocoa populations comprise the oldest population cultivated for commercial purposes, outside Circum Caribbean region. The population up to the end of 19th century is from a single variety with a narrow genetic base. Germplasm was introduced near the end of 19th century probably from foothills of Andes and the name "Nacional/Arriba" was introduced to identify original variety that existed previously. This population was uniform except in terms of compatibility. The general characteristic is thick husk and large seeds, intermediate purple pigment of cotyledons, reflexed sepals and staminodes that bent outwards. Ecuador's arriba has long been prized for its earthy but floral aroma and flavour.

Several farmers introduced red fruited criollo from Venezuela, some time later and are referred to as Venzolano. Appearance of diseases in the Americas led to introduction of diverse types with disease resistance. A search was made for trees that escaped infection by *Crinipellis perniciosa* and some trees called refractarios could be identified. Disease free seedlings were distributed. Thus a complex population was established. Some original plantings of the "Nacional" types are still maintained as these produce quality product. The present population consists of hybrids between Nacional types and the introduced ones and their progenies. The population created by private companies like United Fruit Company (UFCo) and Hacienda Clementia consist mainly of hybrid types.

One important feature of Ecuador derived genotypes is the short stature and small leaves. Estacion Experimental Tropicale, Pichilinge maintains a very diverse germplasm with Nacional and their hybrids.

Brazil

The area covered by the cacao population of Amazon extension zone is large and ill-defined. Knowledge about diversity is extremely scanty and only a small portion of genotypes could be collected for conservation and characterization. One important clone is BE2 with pale green flush leaves, pink staminodes, guidelines and cotyledons. Pound (1938) established that population of this region is composed of heterogenous nature of varieties. Le Cointe (1934) described four fruit types prevalent in the area before it got contaminated by introduced varieties.

1. Crioulo or Criollo: The variety produces large oblong fruits with well marked ridges with large seeds of 1.26g.
2. Jacare (Alligator): Fruits of this variety are elongate, sometimes with apex curved with a pronounced constriction below the base, thick husk, very rough with deep furrows and seeds weigh 0.95 g.
3. Amelonado: Fruits are ovoid with smooth surface and slightly prominent ridges. Seeds are well developed, more rounded with an average seed weight of 1.0g.
4. Calacinho (equivalent to Calabacillo): Fruits are almost rounded, smooth, ridges imperceptible, seeds small weighing 0.6 g. Trees are vigorous.

In the extra Amazon area, it is assumed that a single variety 'Comum' was involved during first century of establishment of crop in this region. The fruits are large and elliptic in shape, apex short, acute and straight and with a tendency for a constriction at the base. The fruit surface had fairly prominent ridges and varied from being rough to smooth. Subsequent introductions were made using seeds from trees of unknown parentage on Amazon river basin. This population is referred to as 'paras'. One of the varieties named ' Maranhao' introduced from Trinidad had different types *viz.* Maranhao rugos, Maranhao Liso and Maranhao Melao. Maranhao and Paras types can be distinguished from Comum types by fruit characters and habit of trees. In contrast to erect habit of Comum, the other types had wide canopies with long branches that tended to curve towards the ground.

Several mutations are encountered in the population- mainly with undesirable traits. 'Folha de Louro' is characterized by slow growing plants of small stature and small leaves. Types with anthocyanin inhibitor alleles are called 'Catongo' and 'Almeida'. 'Maracuja' is another remarkable mutant with flowers resembling those of the genus *Passiflora* with sterile stamens replaced by normal ones. Significance of this mutant is that it would represent the wild ancestor of *Theobroma* with ten fertile stamens. 'Jaca' is another mutant genotype with only three whorls of plagiotropic branches as against five in *Theobroma cacao*. Another mutation which is entirely sterile but produces a few small red parthenocarpic fruits resembling *Capsicum* is called Pimenta.

Attempts to introduce Venezuelan and Nicaraguan cocoa in 1907 did not make any impact on genetic composition of Bahia population.

Surinam

The cocoa industry in Surinam is based on a single variety ' Porcelain'/'Suriname' resembling forasteros of Trinidad. This variety is uniform distinguished by a thin smooth shell and fullness of its beans. The unripe fruit colour is light green with short smooth pods tending to be spherical with a rounded apex. The variety is esteemed for its good qualities. Subsequently 'Caracas' cacao belonging to Venezuelan Criollo group and the Alligator Cacao were introduced and these entered into cultivation. At present, a number of variable types are available in the country.

Circum Caribbean Region and Neighbouring Territories: Population Evolving from a Criollo Germplasm Base

Trinidad and Tobago

The first plantations were raised from materials collected from Venezuela. The original Criollo populations had white cotyledons, smooth pod surface and thin husk. Some trees with rough fruit surfaces were also described. A phenomenon described as 'blast' occurred in the area which destroyed many of the cocoa plantations. Another variety was subsequently introduced from Amazon river valley. Hybridization between the two populations occurred. Later introductions were 'Ocumare'- a variety belonging to the Venezuelan Criollo group. Selection was practiced for better quality of Criollo and adaptability and high yield of Amazonian Forastero. Some varieties namely, 'Ochroe', Verdilico, Sangre Toro and Five grooved Cundeamor were reported from this area in early years.

Introductions were made from Nicaragua which comprised of Alligator, Pentagonum and *Cacao del Pais*. Other types introduced from Nicaragua included types with greater vigour and purple cotyledons segregating for Pentagonum fruit type and single plagiotropic fruit character.

In Tobago, the earlier plantations belonged to the Criollo group with distinct mamillate apex. Extensive cultivation started by beginning of 20th century and composition of population would have been similar to that of Trinidad.

Venezuela

Since 19th century, the criollo population of Venezuela underwent a lot of changes with introduction of germplasm mainly of Calabacillo from Trinidad. The introduced variety would have hybridized with criollo and in time, supplanted the criollo and also spread to lands to which criollo was not adapted. Trees with Amelonado and Calabacillo fruits are also encountered. Trinitario forms with different degrees of intermixing and breeding with local populations are observed in the country and the gene base would be the same as in Trinidad.

Colombia

No evidence is found for existence of cocoa in the pre-Colombian times and all the cocoa in Colombia are introduced. The first plantations were derived from Criollo population originating in Venezuelan Andean foothills. Some genotypes with green fruits and single orthotropic branch character resembling Venezuelan types were observed. The genetic composition of the criollo underwent many changes with the introduction of varieties like 'Pajarito' from Trinidad and Calabacillo types from West Indian islands. The present population consists of mainly hybrids with prominent criollo characters.

Guyana

It is presumed that the varieties are introduced from Venezuela and therefore belonged to the Criollo group. Green fruited criollo types with white cotyledons are encountered in this region and differ from normal concept of Venezuelan Criollo. During the last two decades of 19th century, cocoa varieties were introduced from Trinidad, Surinam 'Porcelaine' and Grenada.

The Lesser Antilles

The major islands that make up the chain on windward side of Caribbean sea have cultivated cocoa during the past 200 or 300 years. Varieties like 'Caracas" were introduced to the islands and the population developed from the introduced material have a large component of hybrid types. The principal varieties in Martinique and Guadeloupe were Criollo types with high variability. There are indications to show that Criollo, Calabacillo and Amelonado existed on this island.

The population structure in Grenada is based on presence of Amazonian variety or Calabacillo and Criollo from Venezuela. Trinidad was probably the source of seed for these plantings but some natural hybrids between Calabacillo and Criollo would have been produced in Grenada. Some genotypes with elongate green fruits classifiable as Criollo are found in a few locations and these may be the progenies of varieties introduced to Trinidad from Nicaragua.

In St. Lucia, Calabacillo types are found and there are indications to show that these are introduced from Martinique. Criollo also occurs in other locations. Variable hybrids between Calabacillo and Criollo types are also recorded.

In the island of Dominica, the oldest trees were of Calabacillo, similar to those of 'Creole' of Martinique. Planting material was imported from Trinidad comprising of a sample of the Criollo and Forastero hybrids. The cocoa population on some farms comprise hybrid types similar to those of Trinidad.

The Greater Antilles

The largest of the islands is Hispaniola and cocoa cultivation commenced using planting material introduced from Atlantic coast cacao areas of Mesoamerica. Criollo types were introduced from Martinique and Venezuela during early days. Subsequently, hybrid types were introduced from Trinidad, Ecuador and Venezuela which resulted in the present population being highly variable.

In Jamaica, seeds of 'Caracas' variety were introduced from Venezuela, followed by Creole or Santo Domingo, Forastero from Amazon river basin and Calabacillo. Hybrids between these types are encountered at present.

Cuba has a long history of cocoa cultivation and the first plants were derived from trees on the main land, either on coast of Mexico or that of Gulf of Honduras.

In Puerto Rico, seeds for first plantations are from Venezuela. Plantations composed of introduced hybrids and Calabacillo types with red and green fruits are recorded. The Criollo plants are also encountered.

Mesoamerica

The original plantings along entire length of coast of Panama northwards and the interior of a few countries were using Amazonian types same as the Martinique 'Creole'. The first plants along the Atlantic coast were Criollo types. The self-compatible type identified as 'Matina' was also recorded in its original form and has some characteristics as Creole. This was followed by introductions from Trinidad. The United Fruit Company took up large scale cultivation of cocoa using 'Matina' variety followed by hybrids with Criollo types. Introductions were also made from Ecuador, especially, 'Nacional' variety.

The 'Creole' type got established along the coast of Honduras. This genotype possesses anthocyanin inhibitor gene.

Along the Atlantic coast of Mexico, Criollo, Forastero and Calabacillo existed. The traditional Criollo varieties were replaced to a large extent by an introduced type or hybrid between it and Criollo. The population was described as being a mixture of native Criollo with an Amelonado of high productivity. The fruits were smaller in size and light green in colour.

Old World Countries

Various attempts were made to establish the crop in other regions. The first attempt to plant cocoa outside the American continent is attributed to introduction of plants from Mexico into the Philippines in 1670. Growth of plantings in the old world was gradual until end of 19th century. The planting materials for cultivation were derived from a few sources and with limited genetic content. Variability developed as a result of mutation and during the later periods from hybridization following the addition of other germplasm.

South Asian and Pacific Region

Cocoa cultivated in the first two centuries following the first introduction was the Mesoamerican Criollo group. Further introductions were made from Mexico and there is evidence to show presence of Amelonado genotypes.

Criollo from the Philippines and Indonesia probably is responsible for first cocoa varieties in Sri Lanka in the beginning of 19th century. Cocoa arrived the island of Mauritius fairly early, from where it spread to Madagascar, Thailand and India.

The most important producer in the South East Asia in the beginning of the 20th century was Java and the major genotype was 'Java porcelaine'. A mutant with anthocyanin inhibitor gene is reported with green fruits, unpigmented flowers and flush leaves. A natural hybrid between Venezuelan Criollo and Trinidad varieties was introduced. Later Forastero cocoa was also introduced.

In Sri Lanka, cocoa trees existed in the Botanic Garden in 1819, which probably derived from Philippines or Java or Trinidad. Hybrid populations imported from Trinidad and varieties like Cundeamor, Forastero, Criollo, Cayenne, Verdilico, Sangre Toro and Pentagonum were available in this country.

Introduction of cocoa to South India was made as seeds from Ambon island in Moluccas and was received in Madras in 1791. The progenies of these trees were distributed to other localities in South India. The best known of these plantations is one near Coimbatore, mainly of a pure Criollo type. Subsequent introduction was made in the early 20th century but its cultivation was limited to a few Government farms. Both Criollos and Forasteros were introduced into the country. In 1930s, it was decided to remove all Forastero plants in the country to maintain genetic purity of Criollos. The Criollos which were maintained in the farms, failed to come up well and many plants were damaged by pests and diseases. A few continued to survive, but yields were low. Cocoa cultivation resumed in a big way in 1960s with pods of Forastero type from Malaysia, followed by types from African countries.

Cocoa was planted in Peninsular Malaysia in 1777, but it did not gain importance until recent times. The first introductions belonged to Criollo group resembling Guatemalan and Colombian Criollo. Cocoa varieties were later brought from Sri Lanka which consisted varieties from Trinidad.

Cocoa cultivation in Papua New Guinea commenced in the beginning of 20th century. The first plantings would have comprised of Criollo, followed by Forastero and West African Amelonado.

In Fiji islands, the first introduction was made from varieties sent from Trinidad. After 1948, clones were introduced from Trinidad, Papua New Guinea, Western Samoa and Grenada. It was decided to grow Amelonado and Cundeamor types.

West African Countries

A single fruit was concerned in first introduction since it was stated that the original planting consisted of 30 plants. This provided the basic material for spread of cocoa cultivation in the African continent. In 1832, cultivation spread to Sao Tome. The variety existed in the region was self compatible and due to several generations of inbreeding, the variety became homozygous. This variety was related to 'Comum' of Brazil and was called 'Sao Tome Creoulo'. This variety was taken to the island of Fernando Po and it became the basis for the cocoa produced in this continent and it came to be known as ' West African Amelonado'. This introduction and the first encouraging crops from it coincided with the global rise in demand for cocoa in this era of first industrialization. At the same time, political instability and crop diseases threatened cocoa production in Latin America and so traders started looking for new territories where the cultivation of cocoa could be developed.

Varieties were later introduced into Sao Tome from Ecuador, Trinidad and Venezuela. Natural hybridization of all these genotypes with original variety developed a wide range of phenotypic variability. The Sao Tome population contains several unique genotypes that are probably unknown in other countries. Some mutants are also located in this country. One such is 'Laranja'. Another mutant is with thick tree trunks, short broad green fruits with short pointed apex and short peduncle with erect fruits. Another mutant carries a recessive gene in the homozygous state segregates for progenies with yellow leaves.

The islands of Sao Tome and Principe, on account of its central position played a significant role in spreading cocoa along coastal region of West Africa.

The first attempt to grow cocoa in Cameroon was in 1876 using 13 plants from the Royal Botanic Gardens. There are no records of these plants and subsequent introduction was from Sao Tome or Fernando Po by Baptist Missionaries and formed the source of planting material for beginning cocoa cultivation in Ghana in 1889. There are evidences to show that Forastero, Criollo, 'Puerto Cabello', 'Soconusco', 'Venezuela', 'La Guaira', 'Maracaibo', 'Guayaquil', 'Nueva Grenada' and 'Suriname' and other varieties from Trinidad existed in Cameroon. These varieties were planted mixed with the type known in Cameroon as 'Victoria-kakao'.

In Gabon, the variety cultivated in early years was 'Sao Tome Creoulo'. Preferred types like Soconusco were imported from Cameroon in 1898. Later studies indicated a large variability in the population.

In Angola, the first experiment with cocoa was initiated in 1860 utilizing seeds of 'Sao Tome Creoulo' and this variety continues to be the major one in areas where the crop was extended. In Belgium Congo, planting was taken up in 1884 with seeds obtained from Sao Tome, Venezuela and Colombia. In Ghana, Ivory Coast and Nigeria, origins of cultivation were based on import of seed of Creoulo from Sao Tome and the variety continued to occupy the prominent position for about six decades. Later introductions were made from Trinidad like Ocumare, Trinidad Criollo and Nicaraguan Criollo. Criollo types, Jamaican white variety and hybrid types were also introduced. Some mutants with albino leaf and anthocyanin inhibitor gene are also reported in Ghana. Hybrid varieties were later produced in these countries to overcome problem of low yield.

In Ivory Coast, cocoa became a cash crop only in 1912, when colonial authorities forced Africans to cultivate it. Cocoa cultivation initially occurred only in Southern forest zone. Then, cocoa was grown mainly on small family-owned farms with labor supplied principally by immigrants from other African countries. It produces 40 per cent of the world's cocoa and that previous success boosted the country's economic development.

Spread of Biodiversity: Germplasm Collection and Exchange

Distribution of cocoa germplasm is through internationally recognized agencies. The International Cacao Gene bank, Trinidad and the collection at CATIE Regional Germplasm Centre, Turrialba, Costa Rica are designated as "universal collection depositories". The core of Trinidad collection is Pound's Ecuadorian and Peruvian collection which forms 70 per cent of it, the 1952 Anglo Colombian collection, representatives of Chalmers' and Allen's material and selections from cultivated cocoa in Trinidad and other Caribbean Islands. The core of Turrialba collection is selections from cultivated cocoa, especially the United Fruit Company clones and their derivatives from Costa Rica, similar material from other American countries and Criollo. Large collections of primary material are also maintained in Colombia, Ecuador, French Guiana, Venezuela and Brazil. Field collections are maintained in Puerto Rico, Cote d' Ivoire, Jamaica, Malaysia, Grenada, Nigeria, Papua New Guinea, Ghana and India.

There are thousands of clones of cacao in field gene banks in different areas of the world. Some of the largest collections are at the Cocoa Research Institute in Tafo, Ghana (6,000 accessions), the International Cocoa Gene bank in Trinidad (1,872 accessions), and CEPLAC in Brazil (1,749 accessions). Long distance distribution is done using intermediate quarantine facilities at the Royal Botanical Gardens, Kew (University of Reading, from 1983) and the United States Department of Agriculture (USDA) in Miami, Florida. These transfers are carried out by authorised organizations like IBPGR, now Bioversity International, Rome.

Important cocoa germplasm repositories of the world as provided in ICGD (2003) by Wadsworth *et. al.*(2003) are furnished below:

1. CEPEC, Centrode Pesquisas do Cacau, Superindendencia Regional da Bahia do Espirito Santo, Rodovia Ileus- Itabuna Km 22, CEP 45 600-000 Ilheus, Bahia, Brazil
2. CEPLAC, Dep. Especial da Amazonia, Rodovia Augusto Montenegro, Km 7, Caixa Postal 1801, CEP 66, 823-110, Belem, Para, Brazil
3. IRAD, BP. 2067, Yaounde, Cameroon

4. N'koemvone Research Station, par Ebolowa, Cameroon
5. Instituto Colombiano Agropecuaria, Apartado Aero 233, Centro Nacional de Investigacion (CNI), Palmira, Valle, Colombia
6. CATIE Regional Germplasm Centre, 7170, Turrialba, Costa Rica
7. Centre National de Recherche Agricole, 01 BP 1827, Bingerville, Abidjan, 01, Cote d'Ivoire
8. Instituto Nacional de Reforma Agrarian, Cuba
9. Dominican Republic- Department of Coffee and Cacao, Center De Los Heroes, Santo Domingo, Dominican Republic
10. Instituto Nacional de Investigaciones Agropecuarias, Estation Experimental Tropical Pichilingue, Casilla Postale 24, Quevedo, Ecuador
11. Aburi Botanic Garden, P.O.Box 23, Aburi, Akwiapim, Ghana
12. Cocoa Research Institute of Ghana, Private Mail Bag, International Airport, Accra, Ghana
13. Ministry of Agriculture, Lands and Fisheries, Technical Department, Mt. Horne, St. Andrews, Greneda
14. Cadbury- KAU Co-operative Cocoa Research Project, Kerala Agricultural University, KAU P.O. Thrissur, 680 656, Kerala, India
15. Central Plantation Crops Research Institute, Kasaragod, 671 124, Kerala, India
16. Estate Crops Research Institute, Jember Station, East Java, Indonesia
17. London Sumatra Plantations- PT PP Lon- Sum Plantations, Jin Jend. A. Yani No.2, P.O.Box 1154, Medan 20111, Sumatra, Indonesia
18. Agricultural Research Center Tuaran P.O, Box 3, Tuaran 89207, Sabah, Malaysia
19. Golden Hope Plantations, P.O, Box 11043, 50734 Kuala Lumpur, Malaysia
20. Malaysian Cocoa Board, Centre for Research and Production Hilir Perak, P.O. Box 25, 36307, Sungai Sumun, Perak, Malaysia
21. Campo Experimental Rosario Lzapa, Apdo Postal 96, Tapachula, Chiapas 30700, Mexico
22. Cocoa Research Institute of Nigeria, Idi- Ayunre, PMB 5244, Ibadan, Nigeria
23. Lowlands Agricultural Research Station, P.O Keravat, East New Britain, Papua New Guinea
24. Instituto Nacional de Investigacion y Promocion Agropecuaria, San Rogue, Iquitos, Peru
25. CIRAD- CP/E.E Agronomica de Poto, CP 375, Sao Tome and Principe
26. Agricultural Experiment Station- Ministry of Agriculture, Animal Husbandry and Fishery, Agricultural Experiment Station, P.O Box 1807, Paramaribo, Suriname
27. Cocoa Research Unit, University of West Indies, St. Augustine, Trinidad and Tobago
28. Imperial College of Tropical Agriculture, St. Augustine, Trinidad and Tobago
29. Trinidad Cocoa Board, Trinidad and Tobago
30. Royal Botanic Gardens, Kew, Richmond, Surrey, TW 9 3 AB, United Kingdom
31. The BCCCA/University of Reading Intermediate Quarantine Station, Department of Horticulture and Landscape, Plant Science Laboratories, Whiteknights, P.O Box 221, Reading RG6 6 AS, United Kingdom
32. Institut National pour l'Etude et la Recherche Agronomics, B.P.1513, Kisingani, Zaire

Conclusion

The future of world cocoa mainly depends on the ready availability of improved planting materials. A lot of improvement is warranted in resistance to major diseases like witches' broom, cocoa swollen shoot virus, vascular streak die back and *Phytophthora* pod rot, bean quality traits, bean size and adaptability to adverse weather conditions. To obtain any significant breakthrough in breeding, the cocoa breeders should have a very strong assembly of diverse germplasm. One of the serious limitations of cocoa breeding programmes in major producing countries is the narrow genetic base. Clones with unique characters are being collected through well organized explorations to the center of origin and neighbouring areas. Apart from this, many useful materials are being produced in different cocoa research institutes. All these improved and diverse genotypes should be accessible to the breeder to make effective genetic manipulations and evolve more desirable genotypes. Wonderful results through breeding can be achieved only through systematic collection, conservation and utilization of biodiversity.

References

Alvim, P. de. T. 1977. Cacao. In: *Ecophysiology of Tropical Crops* (Ed). Alvim, P.de.T. and Kozlowski, T.T.) Academic Press, New York, pp.502

Bartley, B.G.D. 2005. *The genetic diversity of cacao and its utilization.* CABI Publishing, CAB International, Wallingford, Oxfordshire OX 10 8DE, UK. pp 1-341

Cheesman, E.E., 1944. Notes on the nomenclature, classification and possible relationship of cacao populations. *Trop. Agric. Trin.* 21(8): 144-159

Coe, S.D. and. Coe, M.D. 1996. *The true history of chocolate*. Thames and Hudson Ltd, London, United Kingdom

Cope, F.W. 1962. The mechanism of pollen incompatibility in *Theobroma cacao*. L. *J.Hered.* 17: 157-182

Cuatrecasas, J. 1964. Cacao and its allies: a taxonomic revision of the genus *Theobroma*. *Contributions from the U.S National Herbarium* 35(6): 379-614

Jacob, V.J. and Opeke, L.K. 1971. Interspecific hybridization in *Theobroma*. *Proc. 3rd Internat. Cacao Res. Conf.* Accra, Ghana, 1969: 552-555

Kennedy, A.J.; Lockwood, G.; Mossu, G.; Simmonds, N.W. and Tan, G.Y. 1987. Cocoa breeding: past, present and future. *Cocoa Growers' Bulletin*, Birmingham, 38:5-22, 1987.

Le Coinnte, P. 1934. *A Cultura do Cacau na Amazonia Ministerio de Agricultura*, Rio de Janeiro, Brazil pp.35

Mallika, V.K., Amma, S.P.K., Nair, RV and Namboothiri, R. 2002. Cross compatibility relationship within selected clones of cocoa. *National Seminar on Technologies for enhancing productivity in cocoa (Ext. Sum.)* Central Plantation Crops Research Institute, Regional Station, Vittal, Karnataka, India 29-30 November 2002. pp 44-46

Martison, V.A. 1966. Hybridization of cacao and *Theobroma grandiflora*. *J.Hered.* 57: 134-136

Pittier H. 1933. Degeneration of cacao through natural hybridization. *J.Hered.*, 36: 385-390.

Posnette, A.F. 1945. Interspecific pollination in *Theobroma*. *Tropic. Agric.* Trinidad, 22:188-190

Pound, F.J. 1938. Cacao and witches' broom disease of South America (*Marasmius perniciosus*). Report on a visit to Ecuador, the Amazon valley and Colombia. April 1937- April 1938. Yuilles' Printerie, Port-of- Spain, Trinidad and Tobago pp.58.

Purseglove, J.W. 1968. *Theobroma cacao L.* (In) *Tropical Crops- Dicotyledons* (Ed) Purseglove, J.W. John Wiley and Sons Inc. London, United Kingdom pp. 571-599.

Richards, A.J. 1986. *Plant Breeding Systems*. London. George Allen and Unwin, Boston, Sydney pp.189-232

Soria, J. 1977. The genetics and breeding of cacao. *Proc. V Internat. Cocoa Res. Conf.*, Cocoa Research Institute of Nigeria, Ibadan, Nigeria 1975, pp.18-24.

Wadsworth, R.M., Ford, C.S., End, M.J and Hadley, P. 1997. *International Cocoa Germplasm Database Vol 1,2& 3*. London International Financial Futures and Options Exchange (LIFFE), Cannon Bridge, London, EC 4R 3XX, U.K

Wadsworth, R.M., Ford, C.S., Turnbull, C.J. and Hadley, P. 2003. *International Cocoa Germplasm Database [ICGD]* v 5.2 Euronext. Liffe/University of Reading U.K

Williams, J.A. 1977. Hybridization between *Theobroma* and *Herrania* species. *Proc. V Internat. Cocoa Res. Conf.*, Ibadan, Nigeria, 1975: 171-175

Wood, G.A.R. 1975. *Cocoa*. Longman Group Ltd., London. 3rd Edn.

Wood, G.A.R and Lass, R.A. 1985. *Cocoa.* 4th edn. Tropical Agriculture Series, Longman Publications, New York.

Young A.M. 1994. *The chocolate tree: a natural history of cacao*. Smithsonian Institution Press, Washington, United States, pp. 200.

Part III

Biodiversity of Spices

Biodiversity in Horticultural Crops Vol. 2
Editor: **Professor K.V. Peter**
Published by: **DAYA PUBLISHING HOUSE**

Pages **229–251**

Chapter 11

Tamarind (*Tamarindus indica* L.)

D. Singh[1], L. Wangchu[2], V.C. Prahalad[1], P.S. Chauhan[1] and A. Kavita[1]
[1]*College of Horticulture and Forestry, Jhalawar*
Maharana Pratap University of Agriculture and Technology, Udaipur
[2]*College of Horticulture and Forestry, Pasighat*
Central Agricultural University, Imphal

Tamarind (*Tamarindus indica*) is the only species of the genus *Tamarindus* in the family: Fabaceae, division: Magnoliophyta; class: Magnoliopsida; order: Fabales; subfamily: Caesalpinioidae; Tribe: Detarieae. It is a tropical tree, native to eastern Africa where it grows wild throughout Sudan but has now been introduced into most of tropical Asia, Latin America and the Caribbean. Of all the fruit trees of the tropics tamarind isthe most widely distributed and appreciated as an ornamental plant also. Most of its colloquial names are variations of the common English term. In Spanish and Portuguese, it is *tamarindo;* in French, *tamarin, tamarinier, tamarinier des Indes,* or *tamarindier;* in Dutch and German, *tamarinde;* in Italian, *tamarandizio;* in Papiamiento of the Lesser Antilles, *tamarijn.* In the Virgin Islands, it is sometimes called *taman;* in the Philippines, *sampalok* or various other dialectal names, in India, it is tamarind or *ambli, imli, chinch.*; in Cambodia, it is *ampil* or *khoua me;* in Laos, *mak kham;* The name "tamarind" with a qualifying adjective is often applied to other members of the family Fabaceae having somewhat similar foliage. Alternative names include Indian date, translation of Arabic *tamr hindî.* In Malaysia it is called *asam* in Malay and *swee boey* in Hokkien. In Indonesia it is called *asem* (or *asam*) *Jawa* (means *Javanese asam*,in the Philippines it is called sampaloc. In Oriya it is called Tentuli. In Hindi it is called *imli.* In Marathi it is *chinch.* In Bangla, the term is *tetul.* In Sinhala the name is *siyambala,* in Telugu it is Chintachettu (tree) and Chintapandu (fruit extract) and in Tamil and Malayalam it is *puli.* In Kannada it is *hunase.* In Malagasy it is *voamadilo.* The Vietnamese term is *me.* The tamarind is the provincial tree of the Phetchabun province of Thailand (in Thailand it is called *Ma-kham*).

Taxonomy

Tamarindus indica L. is a member of the dicotyledonous family Fabaceae (Leguminosae), which is the third largest family of flowering plants with 700 genera and approximately 17,000 species (Chant, 1993).It belongs to subfamily Caesalpinioideae, divided into five to nine tribes or groups of genera based on the morphological characters, but all attempts so far at division of subfamily are unsatisfactory. The genus *Tamarindus* contains only one species (monotypic genus). It is a diploid species with a chromosome number x = 12 and $2n$ = 24 (Purseglove, 1987). In the past, a distinction was made between tamarinds of the West and East Indies. The tamarinds from the West Indies belonged to *Tamarindus occidentalis*, in which the pods are up to three times longer than their width and contain 1–4 seeds/pod. (Grisebach, 1864). The East Indian tamarinds were recognised as *Tamarindus officinalis*, in which the pods are 6 times longer than the width and contain 6–12 seeds/pod (Coronel, 1991). These differences were insufficient for separation into two species and *Index Kewensis* (Jackson, 1895) considers the names *Tamarindus occidentalis* and *Tamarindus officinalis* as synonyms of the Linnaean name *Tamarindus indica*.

Origin and Distribution

Tamarindus indica originated in tropical East Africa probably in Madagascar from where it spread to South and East Asia (recorded in the Bhuddist sources in 1200–200 BC and in 650 AD in Indian Brahmasamhita scriptures). It is native to Burkina Faso, Cameroon, Central African Republic, Chad, Ethiopia, Gambia, Guinea, Guinea–Bissau, Kenya, Madagascar, Mali, Niger, Nigeria, Senegal, Sudan, Tanzania and Uganda. *T.indica* is exotic to Afghanistan, Australia, Bangladesh, Brazil, Brunei, Cambodia, China, Colombia, Cote d' Voire, Cuba, Dominican Republic, Egypt, Ghana, Greater and Lesser Antilles, Haiti, Hawaii, Honduras, India, Indonesia, Iran, Jamaica, Laos, Liberia, Malaysia, Mauritania, Mexico, Mexico, Myanmar, Nicaragua, Pakistan, Papua New Guinea, Philippines, Puerto Rico, Sri Lanka, Thailand, Trinidad, Togo, United States of America, Vietnam and Zambia (Salim *et al.*, 1998). *T.indica* is the most appreciated ornamental and it is even found in the royal palace in Thailand as shade tree, grows wild throughout Sudan, was introduced and adopted in India from where it spread to the Persians and Arabs (NAS, 1979; Storrs, 1995). It existed in ancient Egypt and Greek in 4th Century B.C, is naturalized in East Indies and the island of pacific and was planted in Hawaii in 1797. It was introduced to tropical America, Bermuda, the Bahamas and South Florida where it is grown as shade and fruit trees along roadsides, dooryards and parks. In Mexico, 4,440 ha.of tamarind are grown in Chiapas, Colina, Guerrero, Jalisco, Oaxala and Vera Cruz, In Motagua Valley of Guatemala, and Northern Malaya. In Kenya *T. indica* is found on farms in the coastal districts, valued for shade and used as a meeting venue in villages (Vogt, 1995) and one time a protected tree by local laws and customs. Its canopies provide suitable microenvironment for crops like beans in drought conditions and the wood is used for making rafters, walking sticks, oxen yokes and tool handles. Tree parts are used in medicines. *T.indica* occurs in Zanzibar in homesteads. It is cultivated in home gardens, farmlands, roadsides, common lands, and in commercial plantation in India, Thailand, Brazil, Puertorico, Latin America, Belize and Mexico, Malaysia, Penang, Myanmar, the Caribbean island, Jamaica, Cuba, the great and lesser Antilles and the Dominican Republic (Bueso, 1980; Jayaweera, 1981; Sozolnoki, 1985 and Coates-Palgrave, 1988).

Botanical Description

Vegetative Structures

Tamarind is a slow-growing, long-lived evergreen or semi-evergreen massive tree which under favorable conditions reaches a height of 24-30 m and attains a spread of 12 m with a trunk circumference of 7.5 m. It is highly wind-resistant, with strong, supple branches, gracefully drooping at the ends and has dark-gray, rough and fissured bark. The trunk grows up to 8 m in circumference, with a spreading, rounded crown, up to 12 m in diameter. The tree has poor form, the bole forking at about 1 m from the ground and often multi-stemmed, though it has fairly good coppicing ability. The trunk is short fissured with bark light grey to brown, very rough and scaly; the inner bark is 1-2 cm thick and green-white in colour. The branches are often crooked, thick, wide-spreading and drooping at the ends to form a dense crown (Storrs, 1995). A blood-red-coloured gum exudes from the bole and branches when they are damaged.

Tamarind is a long-lived tree and individuals with ages of 150 years in Hawaii and 200 years in Sri Lanka are reported (Anon, 1991). Tamarind trees are very common in South India, particularly in Tamil Nadu and Andhra Pradesh. They are used as ornamental trees and to provide shade on the country roads and highways.

Wood

Tamarind timber consists of hard, dark red heartwood and softer, yellowish sapwood. It is dense, very hard and close grained, and takes a fine polish. It is resistant and durable but susceptible to termite attack. Due to its density and durability, tamarind heartwood can be used in making furniture and wood flooring. A tamarind switch is sometimes used as an implement for corporal punishment.

Root

Tamarind produces a deep tap root and has an extensive lateral root system under favourable conditions. The main root is wiry and flexuose, while the lateral roots are numerous, moderately long and distributed down the main root. In the seedling, the *hypocotyl* is distinct from and thicker than the root, 3.8–8.9 cm long, slightly compressed and finely tomentose. The *cotyledons* are sessile 1.3-1.8 x 0.95-1.3 cm in size, plano-convex, thick, fleshy and unequally ovate, orbicular or obovate with the apex rounded and the base projecting about 0.3 cm behind the point of insertion.

Leaves

The mass of bright-green, fine, feathery foliage is composed of pinnate leaves, 7.5-15 cm in length, each having 10 to 20 pairs of oblong leaflets 1.25-2.5 cm long and 5-6 mm wide, which fold at night. The leaflets are, narrowly oblong, unequal and entire, rounded or slightly emarginate at the apex and rounded at the base, they are glabrous or slightly puberlous, glaucous beneath and darker green above with reticulate venation and are sessile to sub-sessile.The leaves are normally evergreen but may be shed briefly in very dry areas during the hot season.

Inflorescences and Flowers

The flowers are produced in racemes which are small, 5-10 cm long, terminal and lateral drooping. Inconspicuous, flowers 3cm wide, are 5-petalled (2 reduced to bristles), yellow with orange or red streaks. The flower buds are distinctly pink due to outer colour of 4 sepals which are shed when the flower opens. Flowers are bisexual and 2-2.5 cm in diameter. Pedicels are about 5 mm long, nodose and jointed at the apex. The bracts are caducous, fragrant, zygomorphic and deciduous. There are two boat-shaped bractioles and four sepals which are unequal, ovate and pink, cream or pale yellow. Petals are five in total; the posterior and lateral ones are large and showy, slightly exceeding the calyx while the anterior ones are much reduced. There are 3 fertile stamens. Anthers are transverse, reddish brown and dehisce longitudinally. The ovary is superior and unicarpellary on a sheath coming from the posterior part of the calyx tube (Purseglove, 1987).

Figure 11.1: Flower and Inflorescence of Tamarind

Pollination and Fruit-set

Tamarind is protogynous, entomophilous and highly cross-pollinated. Honeybees are the main pollinating agents. Sometimes self-pollination also results in seed set.

Fruit

The fruits, flattish, beanlike, irregularly curved and bulged pods, are borne in great abundance along the new branches and usually vary from 2 to 7 in long and from 2-3.2 cm in diameter. Exceptionally large tamarinds are found on individual trees. The pods may be cinnamon-brown or grayish-brown externally and, at first, are tender-skinned with green, highly acid flesh and soft, whitish, under-developed seeds. As they mature, the pods fill out somewhat and the juicy, acidulous pulp turns brown or reddish-brown. Thereafter, the skin becomes a brittle, easily-cracked shell and the pulp dehydrates naturally to a sticky paste enclosed by a few coarse strands of fiber extending lengthwise from the stalk. The fruit is pendulous, pods are 5-10 (-14) cm x 2 cm, oblong or sausage-shaped, curved or straight with rounded ends. The shell is light greyish or scurfy brown and minutely scaly, indehiscent, often irregularly constricted between seeds, brittle and easily broken if pressed. It is filled with firm soft pulp surrounding the seed cavities, the outer surface of the pulp has 3 tough, woody, branching fibrous cords from the base to the apex and the pulp is thick and blackish brown in colour (Coronel, 1991). The indehiscent pods ripen about 10 months from flowering and remain on the tree until the next flowering period. In Mexico the fruits begin to dehydrate 203 days after fruit set and continue to 245 days when they reach full ripeness with up to half the fruit original water content lost. The fully ripened fruits remain on the tree for six months. One tree produces 150 to over 800 kg of fruits annually with 30-50 per cent being pulp, 11–30 per cent shells and fibre, 33–40 per cent seeds.

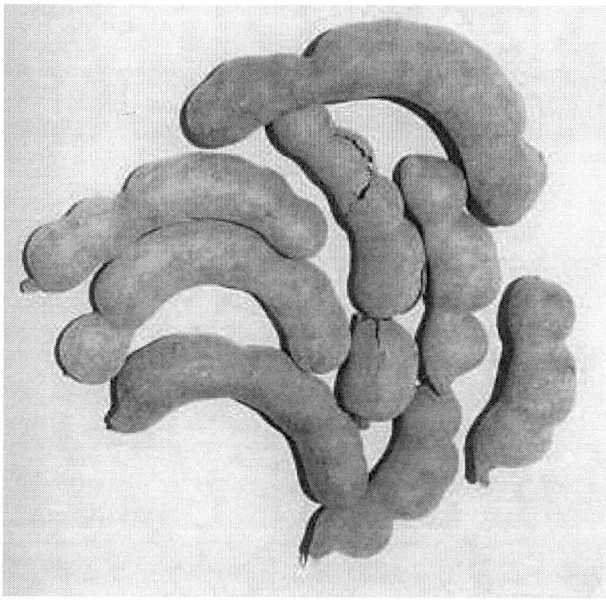

Figure 11.2: Tamarind Pods

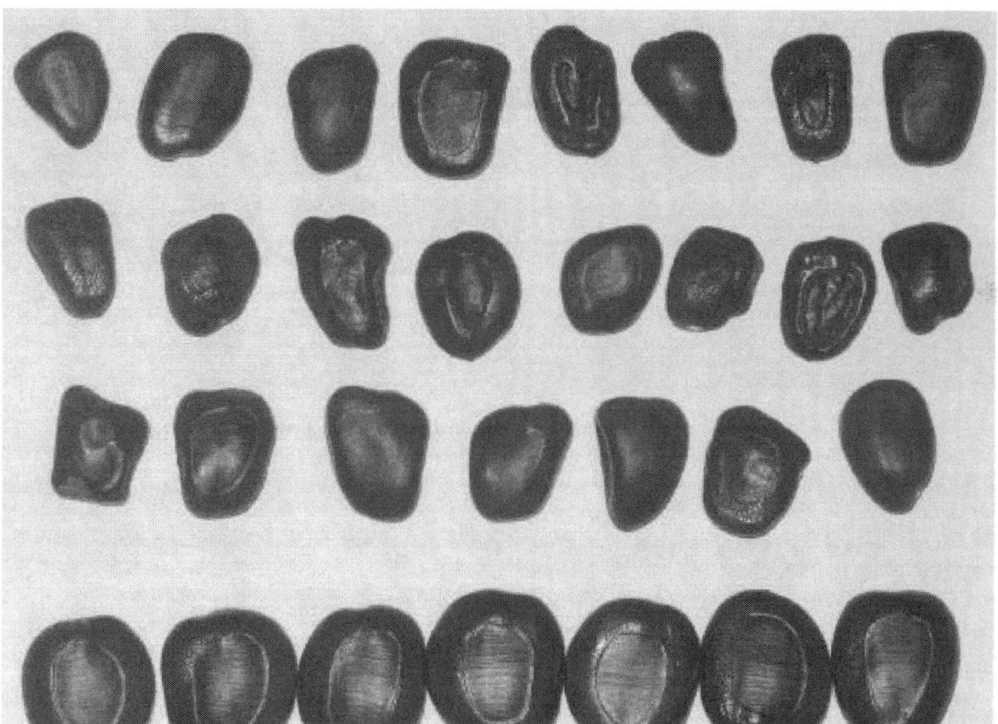

Figure 11.3: Variation in Tamarind Seeds

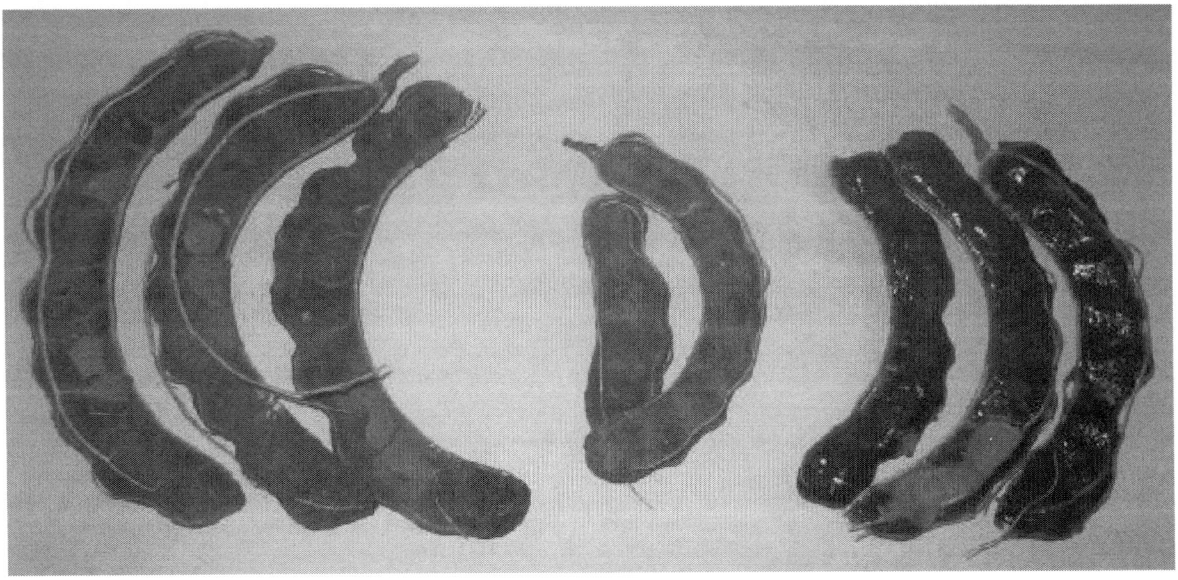

Figure 11.4: Variation in Colour of Tamarind Pulp

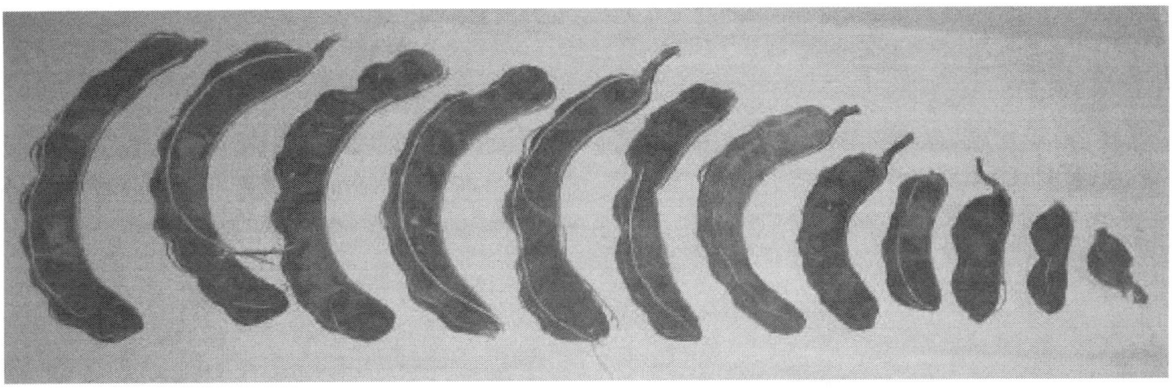

Figure 11.5: Variation in Tamarind Pods

Seed

The fruit is a brown pod-like legume, which contains pulp and 1-12 hard-coated seeds. The seeds are very hard, shiny, reddish or purplish brown, non-arillate exalbuminous, 1.1-1.25 cm in diameter, and each is enclosed in a parchment like membrane. The seeds can be scarified to enhance germination. Each seed is about 3-10 x 1.3 cm, irregularly shaped, flattened, rhomboid, with the centre of each flat side of the seed marked with a large central depression. There are about 720 seeds/kg of fruit (Hong *et al.*, 1996). However, this number may vary from 700-1000 seeds per kilogram depending on the conditions under which the trees are grown (Chaturvedi, 1985). One kg. of fruits has upto 1000 seeds in India and Sri Lanka (FAO 1988, Coronel 1991, Hanamashetti and Madiwalar, 1992).

Tamarind Varieties

Since tamarind is propagated through seeds no standard cultivars are available, however some selections are reported from India, Thailand and the Philippines and two major types of tamarind are recognised based on the sweetness of fruit pulp. These are "sweet fruit" and "sour flavoured" types. Sometimes the cultivars are also differentiated by pulp colour, which may be red or brown. The brown pulp type tamarind is harvested after full maturity and is mostly used as an acidulent in curries, gravy and many dishes whereas reddish pulp type is generally harvested when the fruits are immature and is generally used for making preserve. In most parts the type with reddish flesh is regarded as superior in quality from the ordinary brown-fleshed type due to its sweet taste. The largest producers and exporters of tamarind however continue to grow the sour types of improved cultivars. Mostly unselected tamarind cultivars are grown and these vary in fruit size and quality. Major tamarind growing countries have selected improved cultivars from among the natural populations as shown below.

Table 11.1: Selected Tamarind Cultivars Grown in Some Countries

Country	*Cultivars*
Thailand	'Makham waan', Muen Chong, Sri Tong, Nam Pleung, Jae Hom, Kun Sun, Kru Sen, Nazi Zad, Sri Chompoo
Philippines	Cavite, Batangas, Bulacan, Laguna
India	Prathisthan, Periyakulam (PKM 1), Urigam,Yogeshwari

In Thailand the tartaric acid content of sweet tamarind ranges from 2.0 to 3.2 per cent with sugar content of 39.1 to 47.7 per cent whereas the sugar content of sour tamarind is only 4.80 to 38.94 per cent (Feungchan *et al.*, 1996 b).

Table 11.2: Chemical Composition of Sweet Tamarind Pulp from Thailand

Cultivar	*Tartaric Acid (%)*	*Sugar (%)*	*Acid/Sugar Ratio*	*Fibre (%)*
Sithong	3.18	41.07	1 : 13	1.44
Piyai	2.01	47.19	1 : 23	1.97
Praroj	2.70	43.09	1 : 16	0.88
Srichompoo	2.39	42.52	1 : 18	1.96
Kru-in	2.70	39.06	1 : 14	1.02
Jaehom	44.68	47.71	1 : 16	1.33
Pannanikom	2.34	47.71	1 : 20	1.10

Source: Feungchan *et al.* (1996 a).

Genetic Diversity of Tamarind

Tamarindus indica is a domesticated species of high economic value but despite this importance, very a few data are available on its diversity as well as its structure leading to controversial discussions on its origin. Study on Tamarind population using markers RAPDs with the seeds collected from Asia (India and Thailand), Africa (Burkina Faso,

Senegal, Kenya and Tanzania) and from three islands (Madagascar, Reunion and Guadeloupe) revealed that *T. indica* has a high intra population genetic variability with a higher value obtained in the population from Cameroon.

Genetic diversity is basic plant material for improvement of any crop. Detailed information on genetic variation of the species are a pre-requisite for its effective utilisation. However, very little is known about the genetic variation of tamarind. Although there is a wealth of tamarind germplasm across the regions where it is naturally grown, no systematic germplasm collection and evaluation is attempted to date in spite of its future potential. Several countries in Africa prioritised this species for conservation. Wide variation is fundamental for the development of new varieties with good quality and higher yields (Holden and Williams, 1984). This diversity could be in land races, local selections, elite cultivars and wild relatives of crop plants (Simmonds, 1984).

The genetic variation of tamarind is primarily based on fruit characteristics, like length of pod, pod weight, seed number, pod colour and sweetness of pulp. The variations are also reported for tolerance to drought, wind, poor soils, water logging, high and low pH and grazing. Phenological diversity also exists in tamarind and tree to tree variations are common in flowering and maturing of fruits (Mahadevan, 1991), which may reflect either genetic variation or genotype x environment interactions and/or both.Pathak *et al.* (1992) and Feungchan *et al.* (1996 a) reported considerable variation in fruit yield (from 150-200 kg/tree/year to 800 kg/tree/year) of tamarind in different countries showing the potential for improvement of fruit production. Chundawat (1990) also reported wide variation in fruit size and pulp content in tamarind. Wide phenotypic variation in tamarind germplasm is attributed to geographic isolation and gene mutation (Feungchan *et al.,* 1996 a).Origin of sweet tamarind is attributed to a point mutation. It is also assumed that the sweetness in tamarind is a rare trait, which may be governed by recessive genes (Fuengchan *et al.,* 1996 a). Budspots, which are isolated branches bearing sweet fruits from normally sour tamarind trees provide ample opportunities for genetic improvement of tamarind. In Sudan, prioritisation for genetic conservation was studied by collecting information through on-market surveys of non-wood products in relation to export and home consumption (FNC/FAO, 1995). Under the sponsorship of ICRAF, Nairobi, Kenya, an African Tamarind Network is formed to study genetic variation in the African tamarind.

It is essential to initiate studies on genetic variation, genetic structure and effective size of breeding populations, both in natural and naturalised areas. Molecular techniques (DNA finger printing, isozyme studies etc) will be ideal tools to assess the genetic variation among and within populations and could help to formulate management strategies for genetic resources on-farm. From such diverse germplasm, plants with sweeter, more juicy pulp and other desirable characteristics could be selected in tree improvement programmes. In recent years less acidic, sweet tamarind types are selected in the Philippines and Thailand and are becoming popular. According to Challapilli et al. (1995), the fruit weight is positively and significantly associated with pulp, fibre, seed weight, fruit length and breadth. Fruit length is reported correlated with fruit weight, pulp, and number of seeds. The fruit thickness was negatively correlated with fibre weight, seed weight and seed number. Since fruit length, weight and thickness are measures of fruit size, larger the fruit, heavier the pulp weight (Shivanandam and Raju, 1988) It is suggested that fruit size is a good measure for evaluating tamarind trees and an indicator for selection of superior trees for improvement and breeding. The study undertaken to select high yielding varieties based on their flowering pattern also indicated that mid and late flowering trees should be selected for planting to obtain high yields (Usha and Singh, 1994).

Germplasm Evaluation in Thailand

In Thailand, wide genetic diversity in tamarind germplasm both sweet and sour types have been observed in all regions. In Thailand, there are more than 50 sweet tamarind cultivars under cultivation. They include Muen Chong, Sri Tong, Nam Pleung, Jae Hom, Kun Sun, Kru Sen, Nazi Zad and Sri

Chompoo. A cultivar commonly grown by the Thai farmers is known as Makham Waan (Morton, 1987). Pod specimens were analysed in Thailand to formulate a criterion for primary screening. Several accessions from Thailand are selected for improvement for industrial purposes. One accession (No: 86-2-23-017) gave the highest content of tartaric acid up to 11.2 per cent and has potential for commercially extraction of tartaric acid. Accession (No: 86-13-008) had a sugar content of 39.94 per cent and could be valuable for fructose production (Feungchan *et al.*, 1996 e). Another accession (No: 87-2-01-035) gave the highest concentrated tamarind flesh, suitable for making sauces, drinks and confectionery.

Table 11.3: General Characters Useful to Distinguish Tamarind Cultivars

Plant Part	*Character*
Flower	Colour of petals (dark pink to whitish cream), Flowering pattern–early-mid-late
Pod	Pod form, length, breadth, curvature, shape, pod size, pod weight, colour of shell
Pulp	Colour (red-whitish), pulp:shell ratio, real pulp value, pulp yield/pulp recovery per cent, fibre content
Seed	Colour, size, weight, seeds/pod, seed:pulp ratio
Yield	Total yield of pods(kg/tree/year), alternate/regular bearing habit Early, 4th or 5th year,
Biochemical	Sweetness of pulp (ratio of tartaric acid:sugar:protein and mineral constituents, amino acid composition)
Others	Resistance to: salinity, drought, degraded soils, water logging, high pH, low pH, grazing, diseases and pests

Germplasm Evaluation in the Philippines

In Philippines vast diversity in shape and sweetness of tamarind and traits like short oblong, long curved, sweet and sour types are reported. Sweet tamarind varieties, namely Cavite, Batangas, Bulacan, and Laguna are identified at Los Banos by the University of the Philippines having long plump fruits with thick sweet pulp. Some of the sweet types have pods 7-10 cm long, 2-3 cm wide, weighing 20-30 g with 6-7 seeds. Forty-six accessions of tamarind germplasm are presently available in the Plant Breeding Institute, Los Baños, Philippines. Several tamarind lines introduced from the Philippines were evaluated at the US Department of Agriculture, Subtropical Horticulture Research Station at Miami, Florida (Knight, 1980) and the station developed a variety called "Manila Sweet".

Germplasm Evaluation in India

BAIF's Central Research Station identified 15 superior tamarind provenances and a multiplication stand was established in Pune.Among them, selections from Maharashtra like Prathisthan and Periyakulam (PKM-1) and Urigam from Tamil Nadu show good results and are now covering large areas. Periyakulam 1 (PKM-1), a sour type clonal selection from a local variety of a village namely Endapalli near Periakulam is one of the most preferred cultivar of South India. The Horticultural Research Station of Tamil Nadu Agricultural University, Periyakulam, released this variety. It is an early bearer and suitable for high density plantation (160 plants/0.4 ha) as against 40 plants/0.4ha under conventional planting.) It has a pulp recovery of 39 per cent compared to local cultivars (28 per cent). with a yield of 263 kg/tree against 167 kg/tree in local cultivar.A high yielding red type tamarind Yogeshwari has also been released by Marthawada Agricultural University, Parbhani, Maharashtra, India. The tamarind cultivar Prstisthan released by Fruit Research Station, Aurangabad is a sweet type and constitutes 61 per cent pulp,12 per cent seed and 27 per cent shell. Seedling populations of tamarind are exploited and several selections are made in Karnataka, Maharashtra and Tamil Nadu. Selections T 9, T 10, T 11, T 12 and T 13 exhibited superior quality in Maharashtra.

Ex situ Conservation

Conservation of tamarind germplasm is an important pre-requisite to prevent the loss of genetic diversity of this species. *Ex-situ* management is one of the best and economically and socially acceptable method for conservation of tamarind germplasm (Singh, 1995). Seed germplasm banks can be maintained since seeds can be stored for longer periods. Recently ICRAF organised an African Tamarind Network to collect germplasm for characterisation and evaluation. Only a few collections are available in the Philippines, Thailand, India and Sri Lanka, but focus on these is on conservation rather than utilisation. The major limitation for domestication of this under-utilised species is the lack of a research agenda for its improvement at local, regional and international levels. Improved germplasm is not freely available, although efforts are made to develop improved cultivars in some of the producing countries. Conservation of tamarind germplasm is an important pre-requisite to prevent loss of genetic diversity of this species. Use of germplasm for tree improvement is limited in developing countries and in tamarind is not fully exploited (Gill, 1985). It is strongly recommended that countrywide collections are undertaken to determine the most appropriate *in situ* and *ex situ* conservation mechanisms. These approaches should be based upon farm/community-level management and use of genetic resources for local and regional economic development.

Participatory Improvement

Breeding of multipurpose trees like tamarind should also be participatory, involving farmers in all phases of the programme. As tamarind is presently a smallholder crop, this approach is even more important at this stage, before steps are taken for commercialisation. The local people with their indigenous knowledge could provide information on why they favour a particular selection, and characterisation based on users' perspectives will be an essential starting point for improvement of tamarind. The preferences will also vary from country to country. As such the use of Participatory Rapid Appraisal (PRA) or Diagnosis and Design (D&D) could provide technical information on reproductive biology, propagation techniques, existing conservation, management practices and processing technology. Scientists working on tamarind improvement should develop partnerships with local people and industry to identify the characters that should be incorporated into the breeding programmes. This approach may be the most important for countries which do not have strong national programmes for genetic improvement of under-utilised tree species. In Sri Lanka, the Plant Genetic Resources Centre undertook collection of tamarind germplasm for conservation purposes while propagation techniques are being studied by the Postgraduate Institute of Agriculture, University of Peradeniya (Gunasena, 1999).

Tamarind and Population Structure of Other Tree Species

Gunasena and Hughes,(2000) reported that information on basic biology of flowering, fruiting phenology, mode of pollination, breeding systems, seed dispersal mechanisms and regeneration potential are needed. Raising seedlings from nurseries are reported from India where seedlings are spaced 20 to 25 cm. The germination per cent of well stored seeds is between 65–75 and varies from 30 to 70 per cent (Muabamba and Sitaubi, 1995). Seeds cleaned and soaked overnight in water showed enhanced (80 per cent) germination in Malawi and slicing off the end of seed resulted in 92 per cent germination (Maseno, 1994 andMagehembe, 1995). The best medium for germination is sand or soil mixed with cow dung. Nursery potting requires three parts of topsoil, one part of sand and one part of compost. Germination is epigeal and under natural conditions, regeneration occurs and seedlings survive under parent trees when not eaten by grazing animals while grazing animals and monkeys eat the fruit and disperse the seeds. In commercial plantations, mature *T.indica* grows as orchard crop in pure stand in Central America, Mexico and Brazil, as agroforestry species in mixed cropping systems in Thailand and in home gardens or as hedgerow tree. Nursery produced juveniles are transplanted to the field at 12 to 14 months at 80 cm height and planted at 5 to 13 m or wider in fertile soils where trees grow bigger. They are also planted along land borders, canal banks, on field boundaries, in village groves or in home gardens as individual trees. In plantations, the juvenile phase generally lasts 4 to 5 years or longer, growth slows with age and annual growth is 50 to 80 cm (CATIE, 1986). According to these descriptions, population of tamarind in commercial plantations, home gardens and other managed land use systems are objectively controlled. It is known that species with large populations are less prone to extinction than those with small populations in habitats. Habitats supporting large populations of species also support more species than those supporting small populations (Preston, 1992). Presence of seedlings of species in habitats determines composition and abundance. Species have sufficient regeneration potential if their individuals in the different growth stages are in sufficient abundance *e.g.* there is continuous regeneration of seedlings that grow into juveniles and mature into trees (Scuppe, 1990; Seydack, 1995 and Ackerly, 1996).

Climate

Tamarind is adapted to Semi-arid tropical and sub-tropical region but can be grown in monsoon region with good drainage. It prefers warm climate. Tamarind is usually evergreen but sheds its leaves in very dry conditions during the hot/dry season (Morton, 1987). Essentially a tree of tropical climate, tamarind thrives under a maximum annual temperature ranging from 33-37° C and a minimum of 9.5-20 °C.

Soil Requirements

The tree tolerates a great diversity of soil types, from deep alluvial soil to rocky land and porous, oolitic limestone. It easily grows in poor soils available in arid region.In India, it tolerates sodic and saline soils where it grows in ravines and on degraded land (Anon, 1991). Tamarind, however, thrives the best in loamy, deep, well drained alluvial soil, which favours development of a long tap root. The optimum pH for tamarind is 5.5-6.8, which is slightly acidic (FAO, 1988), though it also grows well in alkaline soils (Singh *et al.*, 1997 quoted from Rao *et al.*, 1999).

Water

Tamarind grows well with an evenly distributed mean annual rainfall of 500-1500 mm.Minimum rainfall requirement is 250 mm. Tamarind can be grown in very dry areas with supplementary irrigation and withstands up to 6 months without rainfall; this is due in part to its deep and extensive root system (Coronel, 1991).

Propagation

Seed propagation is common. Budding is also successful. Cuttings also root easily during rainy season or in glass house. The most appropriate technique for arid region is *in situ* soft wood grafting on one year old seedlings when new growth takes place for which activated scion should be employed. Summer is the appropriate season when new growth takes place.

Planting

Tamarind tree attains a huge size; therefore it should be planted at 12 m in a field or 10 m apart as avenue or farm boundary. For this pits of 1x1x1 m size are prepared in summer, refilled with soil + 50 kg compost and planting should be done in the beginning of monsoon. It would be desirable to develop micro-catchment around trees to collect runoff water from monsoon rains.

Intercropping

This is a slow growing tree and would take about 15 years to occupy allotted space at 12 x 12 m. During rainy season, green manure crops or appropriate vegetables can be taken to economize early years of orcharding. Choice of the crop will be determined by the climate of the region.

Physiography

Due to diversity of habitats and soil types in which tamarind grows, it can be found growing in association with a number of different species. These include, *Terminalia arjuna* W. and A., *Anogeissus acuminata* Wall, *Pongamia glabra* Vent., *Barringtonia acutangula* Gaertn. and *Alangium landmarkii* Thw. (Champion, 1936). In the dry deciduous forests of Western Ghats, India, tamarind also grows near *Chloroxylon swietenia* DC., *Givotia rottleriformis* Griff., *Gyrocarpus jacquini* Gaertn., *Commiphora caudata, Moringa oleifera* Lam. and *Holoptelia integrifolia* Planch (Sharma *et al.*, 1986). Occurrence of tamarind with baobab (*Adansonia digitata* L.) is observed frequently (Jansen, 1981 and Mahony, 1990). Auberville (1950) noted that the seedlings prefer soils that are relatively rich in organic matter in the shadow of the baobab. During its growth and development, tamarind can push over the baobab tree and finally takes its place (Jansen, 1981).

Flowering and Fruiting

In tamarind foliage changes in March to April, flowering in May to June and the fruits available in February to March. A tree propagated by bud-grafting comes to bearing in 3 to 4 years, whereas trees propagated by seed take up to 12 years. Practical management and local conditions also affect the time for trees to bear.As tree grows in size and age, productivity increases and continues to be productive for more than 60 years. Mexican studies reveal that the fruits begin to dehydrate 203 days after fruit-set, losing approximately 1/2 moisture up to the stage of full ripeness, about 245 days from fruit-set. In Florida, Central America, and the West Indies, the flowers appear in summer, the green fruits are found in December and January and ripening takes place from April through June. In Hawaii the fruits ripen in late summer and fall.

Maturity Indices and Yield

Fruits generally mature in February to March in which exocarp gets hardened and separated from the pulp. Pod skin colour does not change rapidly with maturity and individual fruits mature at different times. Harvesting should be carried out selectively. Mature fruits should have a brown shell, while immature pods have a green skin. At maturity, fruits are filled with a sticky brown to reddish-brown pulp and the seeds become hard and glossy. The pod skin becomes brittle as the pulp shrinks and the shell can be broken easily by hand. The ripe pod produces a hollow sound when tapped with the finger. The yield of a tamarind tree varies considerably from country to country, and depends on genetic and environmental factors. Pod yield can also be cyclic, with bumper yields in every third year. A young tree may yield 20-30 kg of fruit/tree/year and a full-grown adult tree can yield about 150-200 kg of fruits/tree/year of which the pulp constitutes 30 to 55 per cent, the shells and fiber, 11 to 30 per cent, and the seeds, 33 to 40 per cent. An average tamarind tree yields 100 kg of fruit/tree/year. Pod yield stabilizes after 15 years. The tree has a pod bearing capacity of 50-60 years, but yields fruit for over 200 years

Harvesting Techniques

Fruits are plucked individually or shaken off and collected. The time for a tamarind tree to reach its first harvest varies, depending on method of propagation. Tamarind fruit can be harvested by hand picking, clipping with a hook mounted on a stick or by shaking the branches. Hand picking is used for sweet tamarind to avoid pod damage, as the fruit is sold whole. This requires climbing the trees and can be very time consuming. When the trees are shaken or clipped using a stick, a sheet is usually placed beneath the tree to collect the fallen fruits.This provides protection against hard earth and prevents damage of the shell. The sheet also allows easy collection of fruits. Fruit left on the tree will eventually fall off naturally.

Pulp

The most valuable and commonly used part of the tamarind tree is the fruit. The pulp constitutes 30-50 per cent of ripe fruit (Shankarachrya, 1998), the shell and fibre account for 11-30 per cent and seed about 25-40 per cent (Chapman, 1984 and Shankarachrya, 1998). Fruit composition is variable, depending on locality.Tamarind has low water content and a high level of protein, carbohydrates (60-72 per cent) and minerals. The soluble solids content varies from 54-69.9? Brix (Benero *et al.*, 1974 and Baragano de Mosqueda, 1980). The large range shown here may be associated with genetic variation in the cultivars grown in different countries (Benero *et al.*, 1974).

Table 11.4: Proximate Composition and Food Energy Value of Ripe Tamarind Pulp (%)

Constituents	*a*	*b*	*c*	*d*	*e*	*f*
Water	20.60	15.00-47.00	22.60	33.89	17.8-35.8	15.00-30.00
Protein	3.10	14.00-3.40	3.10	3.28	2-3	2.00-8.79
Lipids			0.40	0.50		
Fat	0.40	0.90-1.00	0.40	0.50	0.6	0.50-2.53
Carbohydrate	70.80	62.50	71.80	59.76	41.1-61.4	56.70-70.70
Fibre	3.00	5.10	3.00	1.79	2.9	2.20-18.30
Ash	2.10	1.50-4.20	2.10	2.57	2.6-3.9	2.10-2.90
Energy (kcal/100g)			272.0	230.0		

Source: (a) Purseglove (1987); (b) FAO (1988); (c) Leung and Flores (1961); (d) Wenkam and Miller (1965); (e) Coronel (1991); Feungchan *et al.* (1996 a).(f) Meillon (1974); Anon (1976); Duke (1981); Ishola *et al.* (1990).

The fruit contains a variety of pigments. Colour of red type is due to water-soluble red-rose anthocyanin pigment, while in the common type leuco-cyanidin is present (Lewis and Neelakantan, 1964 a and Bhattacharyya, 1974).The most outstanding characteristic of tamarind is its acidity due mostly to tartaric acid, ranging from 12.2-23.8 per cent, uncommon in other plant tissues. As the acidity does not disappear with ripening, but is more or less matched with increasing sugar levels, tamarind is simultaneously the most acidic and the sweetest fruit. In general, the dried tamarind pulp of commerce contains 8-18 per cent tartaric acid (as potassium bitartarate) and 25-45 per cent reducing sugars of which 70 per cent is glucose and 30 per cent fructose. The ascorbic acid in tamarind is very small and varies from 2-20 mg/100g (Lefevre,1971 and Ishola *et al.*, 1990).Tamarind pulp is also rich in minerals: high in potassium (62-570 mg/100g); phosphorus (86-190 mg/100 g); and calcium (81-

466 mg/100g), and a fair source of iron (1.3-10.9 mg/100g). It also excels in riboflavin and is a good source of thiamine and niacin, but is poor in vitamin A and vitamin C (Leung and Flores, 1961).

Seeds

The seed comprises the seed coat or testa (20-30 per cent) and the kernel or endosperm (70-75 per cent) (Shankaracharya, 1998). Tamarind seed is the raw material used in the manufacture of tamarind seed kernel powder (TKP), polysaccharide (jellose), adhesive and tannin. The seeds are also used for other purposes and are presently gaining importance as an alternative source of protein, rich in essential amino acids. Unlike the pulp the seed is a good source of protein and oil.

Table 11.5: Composition of Tamarind Seed, Kernel and Testa (per cent)

Constituent	*Whole Seed*	*Seed Kernel (Cotyledons)*	*Testa (Seed Coat)*
Moisture	9.4-11.3	11.4-22.7	11.0
Protein	13.3-26.9	15.0-20.9	
Fat/oil	4.5-16.2	3.9-16.2	
Crude fibre	7.4-8.8	2.5-8.2	21.6
Carbohydrates	50.0-57.0	65.1-72.2	
Total Ash	1.60-4.2	2.4-4.2	7.4
Yield of TKP	50.0-60.0		
Total sugar	11.3-25.3		
Reducing sugars	7.4		
Starch	33.1		
Tannin			20.2

Source: Anon. (1976), Morad *et al.* (1978); Ishola *et al.* (1990); Bhattacharyya *et al.* (1994).

Whole tamarind seed and kernels are rich in protein (13-20 per cent), and the seed coat is rich in fibre (20 per cent) and tannins (20 per cent).Panigrahi *et al.* (1989) reported that whole tamarind seed contains 131.3 g/kg crude protein, 67.1g/kg crude fibre, 48.2 g/kg crude fat and 56.2 g/kg tannins with most of carbohydrate in the form of sugars. Tamarind seeds are reported as a source of food or food ingredients due to presence of proteins.

Table 11.6: Mineral Content of Tamarind Pulp, Seed, Kernel and Testa

Mineral mg/100g	*Pulp*	*Seed*	*Kernel*	*Testa*
Calcium	81.0-466.0	9.3-786.0	120.0	100.0
Phosphorus	86.0-190.0	68.4-165.0		
Magnesium	72.0	17.5-118.3	180.0	120.0
Potassium	62.0-570.0	272.8-610.0	1020.0	240.0
Sodium	3.0-76.7	19.2-28.8	210.0	240.0
Copper	21.8	1.6-19.0		
Iron	1.3-10.9	6.5	80.0	80.0
Zinc	1.1	2.8	100.0	120.0

Source: Ishola *et al.* (1990); Bhattacharya *et al.* (1994).

Storage

After harvest the fruits are allowed to dry and their hard shell removed. Such fruits are marketed.In most rural households, the pods are dried in the sun for 5-7 days. Small-scale dehydrators can also be used to dry fresh fruit, if available. Once dried, the shells are hand-cracked and separated from the pulp. The fibres, seeds and shell pieces are removed from pulp by hand. The pulp is then dried for 3-4 days before being compressed and made ready for storage. The pulp can be mixed with salt or sugar according to preference, prior to storage. Seeds can be used in industrial processing and should be dried for about 2 days. The compressed pulp can be stored in plastic bags, jute bags or closed clay pots. The pulp is usually stored with the seeds, however, when produced commercially, the seeds are removed and fruits are made into tamarind balls and marketed. The freshly prepared, dried pulp is light brown in colour, but darkens with time in storage. Under cool, dry conditions the pulp remains good for about one year when stored with salt in a clay jar or in polythene, after which it becomes almost black, soft and sticky in texture. Products sold in local markets are usually in the form of fresh fruits, pulp, juice, sauce and paste. Products in international markets include pulp, juice, paste and TKP (Tamarind Kernel Powder). India is the world's largest exporter of Tamarind and in 1993, exported 11,145 t. Thailand exported 7,006 t. in 1999 and Sri Lanka exported 6903 t. in 1997. Other countries also export tamarind, including Indonesia, the Philippines, Nicaragua, Venezuela and other Asian and South/Central American countries.

Food Uses

The food uses of tamarind are many. The tender, immature and very sour pods are cooked as seasoning with rice, fish and meats in India. Immature fruits are used for chutney and ripe fruit pulp is an important ingredient in South Indian vegetable dishes to induce sourness. These fruits also make excellent chutney. In Thailand, there is a carefully cultivated sweet variety, consumed as a fresh fruit. Continuous use of tamarind in daily food reduces chances of stone formation in urinary system of human beings. Pulp is mixed with sugar and made into tamarind balls (sweetmeat). It is used for seasoning other food and in curries, preserves, chutneys and sauces. Seeds are eaten roasted or boiled. The fully-grown, but still unripe fruits, called "swells" in the Bahamas, are roasted in coals until they burst and the skin is then peeled back and the sizzling pulp dipped in wood ashes and eaten. The fully ripe, fresh fruit is relished out-of-hand by children and adults, alike. The dehydrated fruits are easily recognized when picking by their comparatively light weight. It is used as a spice and condiment in both Asian and Latin American cuisines, and is also an important ingredient in Worcestershire sauce and HP sauce. The hard green pulp of a young fruit is very sour and acidic and is most often used as a component of savory dishes. The ripened fruit is sweeter, yet still distictively sour, and can be used in desserts and sweetened drinks, or as a snack. The pulp, leaves, and bark also have medical applications. For example, in the Philippines, the leaves are traditionally used in herbal tea for reducing malaria fever. Due to its medicinal value, tamarind is used as an Ayurvedic medicine for gastric and/or digestion problems. Tamarind is a staple in South Indian diet, where it is used to prepare Sambhar, a soupy preparation and various types of chutneys. It is also sold as a candy in Mexico (see for example pulparindo), and in various snack forms in Southeast Asia (dried and salted, dried and candied, as a cold drink). Pad Thai, a Thai dish popular with Europeans and Americans, sometimes includes tamarind for its sour taste (though lime juice and/or white vinegar are more commonly used). A tamarind-based sweet-and-sour sauce served over deep-fried fish is also a common dish in Central Thailand. In Latin America, especially Mexico, and Latin American immigrant communities in the US, the fruit is wildly popular and is fashioned into a drink, "Tamarindo", and many kinds of treats. Tamarind is also a popular food in Mexico and is used in many Mexican candies. Likewise

Sino-Peruvian food uses tamarind-based juice for its distinctive sweet flavour. The wood is used in cart wheels, rice pounders, oil mills etc. The decoction of bark is useful in diarrhoea.

Table 11.7: Miscellaneous Uses of Tamarind in Food

Sl.No	*Miscellaneous Uses*	*References*
1.	Tamarind pulp blended with soursop (*Annona muricata*) to yield a refreshing drink	Bueso (1980)
2.	Tamarind pulp for making wine-like beverages	Benk (1987), Latno and Vega (1986) and Sanchez (1985)
3.	Pulp for preparing spiced sauces and beverages	Girdhari *et al.* (1958)
4.	Pulp is mixed with sugar and honey to make a sweet drink. "*Jugo*" and "*fresco de tamarindo*"	FAO, 1988
5.	Juice is used to preserve fish, for up to 6 months	Macmillan 1943
6.	Tamarind kernels as a stabiliser in ice cream, mayonnaise and cheese and as an ingredient in a number of pharmaceutical products	Morton, 1987
7.	Tamarind kernels as a substitute for pectin for making jellies, jams and marmalades	Bhattacharya *et al.,* 1994
8.	Tamarind xyloglucan, major component of tamarind kernel powder is used. for thickening, stabilizing and gelling in foods	Niemann*et al.,* 1997
9.	Roasted seeds as flour for bread, Indian chapattis and cake making	Purseglove, 1987 Shankaracharya, 1998
10.	Roasted seeds as betel nut in India	FAO, 1988
11.	Tamarind extract used as a replacement for phosphoric acid, citric acid and other acids added to soft drinks	Zablocki (1995)
12.	Leaves and flowers are used to make curries, salads, stews and soups	Benthall, 1933
13.	Young leaves of tamarind are used as a seasoning vegetable in some Thai food recipes	Coronel, 1991
14.	Acid leaves mixed with gum from fig trees to make a "chewing gum"	Sozolnoki, 1985
15.	Flowers as nectar resource for honeybees for preparation of honey of golden yellow colour and slightly acidic in flavour	NAS, 1979; Sozolnoki, 1985

Table 11.8: Miscellaneous Uses of Tamarind in Industry

Sl.No	*Miscellaneous Uses*	*References*
1.	Tamarind pulp as polishing agent for polishing brass, copper and silver . in Asian and African countries	Jayaweera, 1981, Mahony, 1990
2.	Seed testa for tanning leather and imparting colourfast shades to wool.	Coates-Palgrave, 1988;
3.	Tamarinds bark and leaves for tanning hides and dyeing.	Morton, 1987.
4.	Tamarind extract "*tamarindienal* as fungicidal agent and a strong bactericidal agent	Imbabi *et al.,* 1992 a; 1992 b
5.	Fruits as anti-fungal and anti-bacterial agents.	Guerin and Reveillere, 1984;
6.	Tamarind plant extracts as drinking water purifier.	Bleach *et al.,* 1991
7.	The seed husk as fish poison.	Roy *et al.* (1987). Jena (1991)

Contd...

Table 11.8–Contd...

Sl.No	Miscellaneous Uses	References
8.	Tamarind extract as anti-viral and anti-fungal agent	Singh *et al.,* 1989
9.	Tamarind seed as Antioxidant agent.	Osawa *et al.* (1994).
10.	Seed used as filler for adhesives in the plywood industry, a stabiliser for bricks, as a binder for sawdust briquettes and producing strong wood cement.	Rama Rao, 1975
11.	Humidification and compression of tamarind seed gum and the cellulose rich sisal plant fibre for construction materials such as false roofing and room partitioning.	Veluraja *et al.* (1997)
12.	Seeds as a cheaper source of protein to alleviate protein malnutrition	Sidduraju *et al.,* 1995
13.	Powdered, dehulled seeds are used as adhesive or binding agents in paper, textile sizing and weaving.	Niemann *et al.,* 1997
14.	Seeds oil, for making varnishes, paints and burning in oil lamps	Rama Rao, 1975; Anon, 1976
15.	Seed kernels for the extraction of polysaccharides	Gordon (1968).
16.	Seeds for the manufacture of Tamarind Kernel Powder (TKP) used as good paper adhesive.	Anon., 1976
17.	TKP as a source of carbohydrate for binding agent in paper and textile sizing, and weaving and jute products.	Anon., 1976; Shankaracharya, 1998.
18.	Tamarind pulp used as a raw material for the manufacture of several industrial products, such as Tamarind Juice Concentrate (TJC), Tamarind Pulp Powder (TPP), tartaric acid, pectin, tartarates and alcohol.	Anon., 1982 a; 1982 b.
19.	Bark tannins in the preparation of ink and for fixing dyes.	Storrs, 1995
20.	Purplish brown hardwood used for furniture, well construction, tent pegs, canoes, side planks, boats, cart shafts and axles, naves for wheels, toys, hubs, oil, sugar presses and tool handles.Fuel with high calorific value (4850 k cal/kg) used for brick making; excellent charcoal.	

Table 11.9: Medicinal Uses of Tamarind

Fruit and pulp	Fruit pulp improves appetite and is a gentle laxative.
Seed	Powdered seed is used to treat chronic diarrhoea, jaundice and dysentery.
Leaf	Used in cardiac and blood sugar reducing medicines.
	Ground with salt to prevent throat infections, coughs, fever, intestinal worms and liver ailments.
	Applied to boils/abscesses to reduce inflammation.
Flowers and immature pods	Used internally as a remedy to cure jaundice and externally to cure eye diseases and skin ulcers.
Twigs and bark	Bark is used as an astringent for the treatment of diarrhoea, as a cure for asthma and acts as a digestive aid.
Root	Component in medicines for leprosy and dysentery.

Pests and Diseases

Tamarind trees do not suffer badly from insect pests or diseases. There are however a number of pests (categorised as major and minor).

Pests

One of the major pests of the tamarind tree in India is the Oriental yellow scale. The mealybug, is a leading pest of tamarind in India, causing leaf-fall and sometimes shedding of young fruits. White grubs of *Holotrichia insularis* feed on the roots of young seedlings. Other predators attacking the leaves or flowers include the caterpillars, the black citrus aphid, the whitefly, thrips, and cow bugs. Fruit borers include larvae of the cigarette beetle. It also affects ripening pods on the tree and persists in the stored fruits, as do the tamarind beetle, and tamarind seed borer. The rice weevil, *Sitophilus oryzae,* the rice moth, *Corcyra cepholonica,* and the fig moth, *Ephestia cautella,* infest the fruits in storage. The lesser grain borer, *Rhyzopertha dominica* bores into stored seeds. Several insects attack the fruits and seeds of tamarind both on the tree and when in storage. These include: the beetles and larvae of *Lasioderma serricome, Pachymerus gonagra,* and *Culandra linearis. Aspidiotlls* spp. feeds on fruit sap and damage.

The most common pests are the leaf feeding caterpillars, bag worms, mealy bugs and scale insects. Sap suckers like white flies, thrips and coccid aphids, affect tender shoots and fresh foliage, and defoliators like caterpillars and chafer beetles also cause considerable damage. They can all be controlled with chemical sprays or physical removal of infected parts.

Table 11.10: Major Pests of Tamarind

Common Name	*Scientific Name*	*Nature of Attack*	*Chemical Control*
Scale insect	*Aonidiella orientalis, Aspidiotus destructor*	Sucks the sap of tender shoots.	For serious infestations, spray with diazinon or carbosulfan at 0.1 per cent solution.
Mealy bugs	*Planococcus lilacinus, Nipaecoccu viridis*	Sucks the sap of leaflets, mature and tender shoots and leaf petiole bases.	Spray with diazinon or carbosulfan at 0.1 per cent solution.
Thrips	*Scirtothrips dorsalis, Halothrips ceylonicus, Eublemma ungulifera*	Attacks flowers, bores into flower buds causing bud death and can web flowers together.	Spray with dimethioate at 20-40 ml/10 litres of water or Fenthion at 10-15 ml/10 litres of water.
Coccids Soft scale insects	*Aspidiotus tamarindi, Cardiococu castilleac, Hemiberlesia lataniae, Howasdia biclavis, Pinnaspia temporaria, Unaspis articolor, Saissetia oleae*	Sucks the sap of semi-hard and tender shoots.	Spray carbonsulphan at 0.1 per cent solution or dimethioate at 3 per cent solution (20-40 ml/10 liters of water).
Aphids	*Toxoptera aurantii, Acaudaleyrodes rachipora*	Sucks cell sap of tender leaves and stems. Leaflets are distorted and covered with moulds.	Spray endosulphan at 10 ml/10 litres of water or spray dimethioate at 20-40 ml per 10 litres of water.
Semilooper caterpillar	*Achaea janata*	Infests flowers and causes a lot of damage.	Spray permethrin at 0.5 per cent solution (5 ml per 10 liters of water).

Diseases

There are no major diseases of tamarind. The most common minor disease affects nursery seedlings and is downy mildew caused by *Oidium* spp. Other diseases, like powdery mildew, collar rot, stem rot, stem canker, trunk rot and root rot, are also caused by fungi. Spraying fungicides at the recommended rates can effectively control occurrence of fungal diseases. The tree is affected by a number of rots, like saprot (*Xylaria euglossa*), brownish saprot (*Polyportls calclitlensis*) and white rot (*Trametes floccosa*).In India, a bacterial leaf-spot occurs. Sooty mold is caused by *Meliola tamarindi*.In serious outbreaks, chemical methods may be the only feasible means of control.

Table 11.11: Major Diseases of Tamarind

Common Name	*Causal Agent*	*Chemical Control*
Leaf spot	*Bartalinia robillardoides, Exosporium tamarindi, Hendersonia tamarindi, Pestalotia poonensis, Phyllosticta tamarindicola, Prathigada tamarindi*	Spray maneb 80 wp at 200/100 litres of water.
Powdery mildew	*Erysiphe polygoni, Oidium* spp.	Spray benomyl 50 wp at 10 g/20 litres of water.
Collar rot	*Phytophthora* spp.	Spray benomyl 50 wp at 10 g/20 litres of water
Stem rot	*Pholiota gollani,*	
Stem canker	*Hypoxlon nectriodes*	Spray captan 80 per cent wp at 200 g/100 litres of water.
Trunk and root Rot	*Stereum nitidulum*	Spray captan 80 per cent wp at 200 g/100 litres of water.
Root and wood Rot	*Ganoderma lucidum*	Spray captan 80 per cent wp at 200 g/100 litres of water.
Stony fruit	*Pestalotia macrotricha*	Spray captan 80 per cent wp at 200 g/100 litres of

References

Ackerly, D.D. 1996. Tropical forest plant ecophysiology. (Ed). S.S. Mulkey, R.L. Chazdon and A.P. Smith. Chapman and Hall, New York, USA, 618–658.

Anonymous, 1976.*Tamarindus indica* L. (In) *The Wealth of India* (Raw Materials Series) Vol. X: 144-122. Council of Scientific and Industrial Research, New Delhi.

Anonymous, 1982 a. Some recent developments, Central Food Technological Research Institute, Mysore, India,.

Anonymous, 1982 b. Tamarind juice concentrate plant starts in Mysore. *Indian Food Industry*, 1: 43-44.

Anon., 1991.*Revised Handbook of the Flora of Ceylon*. Vol V11. Fabaceae. *Tamarindus indica* L. eds. M.D.Dassanayake and F.R. Fosberg. Oxford and IBH Publishing Co. Pvt. Ltd. New Delhi: 96-97.

Auberville, A. 1950. *Flore forestiere Soudano-Guineense*. 226-227.

Baragano de Mosqueda, M. (1980). Technology of clarified tamarind juice. (In) *Tropical and Subtropical Fruits*. (Ed.), Nagy, S. and Shaw, P.E., AVI Publishing, Westport, Conn., 375p.

Benero, J. R., Collazo de Rivera, A. L. and Georage, L. M. I. 1974. Studies on the preparation and shelf life of soursop, tamarind and blended soursop-tamarind soft drinks. *Journal of the Agricultural University, Puerto Rico*, 58-99.

Benk E, 1987. Tropical and subtropical wine-like beverages, *Alcoholic Industries*, 100: 87-129.

Benthall, A.P. 1933. *The Trees of Calcutta and its Neighbourhood*, Thacker Spink and Co, 513 p.

Bhattacharyya, P. K, Bal, S. and Mukherji, R. K. 1994. Studies on the characteristics of some products from tamarind (*Tamarindus indica* L.) kernels. *Journal of Food Science and Technology* (India), 31(5): 372-376.

Bhattacharyya, P. K. 1974. A note on the presence of anthocyanin pigment in the stem of red fruited variety of tamarind. *Indian Forester*, 100(4): 255-258.

Bleach M.F, Guillemin F, Baure L and Hartemann P, 1991. Preliminary study of antimicrobial activity of traditional plants against *E. coli*, *Zentralblatt hygiene und umweltmedizin*, 192(1): 45-56 (English summary).

Bueso, C. E. 1980. Soursop, Tamarind and Chironka. In: *Tropical and Subtropical Fruits*. Nagy, S. and Shaw, P.E. (Eds.), AVI Publishing, Westport, Conn., 375.

CATIE 1986. Growth and Yield of Firewood Species in Dry Land Humid Areas of Central America. Technical Report 79. Turrialba, Costa Rica: 691.

Challapilli, A. P. Chimmad,V. P. and Hulamini, N. C. (1995). Studies on correlation of some fruit characters in tamarind fruits. *Karnataka Journal of Agricultural Science*, 8(1): 114-115.

Champion, H. G. 1936. *A Preliminary Survey of the Forest Types in India and Burma*. Indian Forest Records (Siviculture), New Delhi. Government of India Press, 1(1): 286 p.

Chant, S. R. 1993. Fabales. (In) *Flowering Plants of the World*. (Ed) V. H. Heywood. B. T. Batsford Ltd. London.

Chapman, K. R. 1984. Tamarind. In *Tropical Tree Fruits for Australia*. Complied by P. E. Page. Queensland Department of Primary Industries. Information series Q 183018, Brisbane: 83-86.

Chaturvedi, A. N.1985. Firewood farming on the degraded lands of the Gangetic plain, U.P. *Forest Bulletin* No.50. Lucknow, India Government of India Press (1): 286 p.

Chundawat, D.J. 1990. *Arid Fruit Culture*. Oxford IBH Publishers, New Delhi: 158-163.

Coates-Palgrave, K. 1988. *Trees of Southern Africa. 10. Tamarindus indica* L. C. S. Striuk Publishers, Cape Town: 278-279.

Coronel, R. E, 1991. *Tamarindus indica* L. (In) *Plant Resources of South East Asia*, Wageningen: Pudoc. No.2. Edible fruits and nuts. (Ed). E. W. M Verheij and R. E Coronel, PROSEA Foundation, Bogor, Indonesia: 298-301.

Duke, J. A., 1981. *Hankbook of Legumes of World Economic Importance*. Plenum Press, New York: 228-230.

FAO. 1988. Fruit Bearing Trees. Technical notes. FAO-SIDA Forestry Paper 34: 165-167.

Feungchan, S., Yimsawat, T., Chindaprasert, S. and Kitpowsong, P. 1996 a Tamarind (*Tamarindus indica* L.) Plant genetic resources in Thailand. *Thai Journal of Agricultural Science*, Special Issue No.(1): 1-11.

Feungchan, S., Yimsawat, T., Chindaprasert, S. and Kitpowsong, P. 1996 b Evaluation of tamarind cultivars on the chemical composition of pulp. *Thai Journal of Agricultural Science*, Special Issue, No.(1): 28-33.

FNC/FAO. 1995. Forest Products Consumption Survey in Sudan. Summary Findings. Forests National Corporation. Forestry Development Project FAO/CGP/SUD047/NET, Khartoum 8p.

Gill, K. S. 1985. Exchange and use of genetic resources for crop improvement. In: *Regional Conference on Plant Quarantine Support for Agricultural Development*. ASPEAN, Serdang, Malaysia.

Girdhari, L., Das, D.P. and Jain, N.L. 1958. Tamarind beverage and sauce, *Indian Food Packer*, 12, 12-16.

Gordon, A.L. 1968. Extract. United States Patent Office. Patented on Aug 29, 1968.

Grisebach, A.H.R. 1864.*Flora of the British West Indian Islands*. Lovell Reeve and CO, Henrietta Street, Covent Garden. London : 212-213.

Guerin, J.C and Reveillere HP., 1984. Activite antifongique d'extraits vegetaux a usage therapeutique.1. Etude de 41 extracts sur 9 souches fonguques(Antifungal activity of plant extracts used in therapy). 1. Study of 41 plant extracts against 9 fungal species). *Allalles Pharmaceutiques Francaises*, 42(6): 553-559 (English summary).

Gunasena, H. P. M. 1999. Tamarind, an unexploited species in Sri Lanka. Paper presented at the National Workshop on Fruits for the Future. Gannoruwa, Sri Lanka. 4 November. Gannoruwa, Peradeniya, Sri Lanka.

Gunasena, H.P.M. and Hughes, A. 2000.Tamarind, *Tamarindus indica* L. Fruits for the Future 1. (Ed) Hughes, A., N. Hag and R.W. Smith. International Centre for Underutilized Crops. UK.

Hanamashettti, S. I. and Madiwalar, S. L. 1992. *Multipurpose Tree Species for Agroforestry Systems*. P. S. Pathak, R. Debroy and P. Singh (Ed.) 113-118.

Holden, J. H. W and Williams, J.T. 1984. *Crop Genetic Resources: Conservation and Evaluation*. George Allen and Unwin. London.

Hong, T. D., Linnington, S. and Ellis R. H. 1996. *Seed Storage Behaviour: A Compendium. Handbook for Genebanks* No.4 International Plant Genetic Resources Institute, Rome.

Imbabi, E.S and Abu-Al-Futuh, I.M, 1992. Investigation of the molluscicidal activity of *Tamarindus indica* L. *International Journal of Pharmacology*, 30(2): 157-160.

Imbabi, E.S., Ibrahim, K.E., Abulfutah, I.M. and Hulbert, P. 1992. Chemical characterization of tamarind bitter principle. *Fitoterapia*, 63(6): 537-538.

Ishola, M. M., Agbaji, E. B. and Agbaji, A. S. 1990. A chemical study of *Tamarindus indica* (Tsamiya) fruits grown in Nigeria. *Journal of Science, Food and Agriculture*, 51: 141-143.

Jackson, B. D. 1895.*Index Kewensis*. Plantarum Phanerogamarum. Clarendon Press, Oxford.

Jansen P. C. M, 1981.Spices, condiments and medicinal plants in Ethiopia, their taxonomy and agricultural significance. Centre for Agricultural Publishing and Documentation (PUDOC) Wageningen Agricultural Research Reports N0906.

Jayaweera, D. M.A. 1981. *Medicinal Plants (Indigenous and Exotic) Used in Ceylon*. Part 111. *Flacourtiaceae-Lytharaceae*. A publication of the National Science Council of Sri Lanka : 244-246.

Jena, S. 1991.Preliminary observations on the effect of tamarind seed husk on fish. *Journal of Inland Fish Society of India*, 18(1): 1-3.

Knight, T., Jr, 1980.Origin and world importance of tropical and subtropical fruit crops. *In Tropical and Subtropical Fruits*, Nagy, S. and Shaw, P. E., Eds., AVI Publishing, Westport, Conn., 1.

Latno, S. and Vega, D. 1986.Wines from tropical fruits. *Boletin Tecnico-Labal*, 7: 13-17.

Lefevre, J.C. 1971. Review of the literature on the tamarind. *Fruits* (English Summary), 26: 687.

Leung, W. T. and Flores, M. 1961. *Food Composition*. Tables for Use in Latin America, National Institute of Health, Bethesda, Md.

Lewis,Y. S. and Neelakantan, S. 1964. The chemistry, biochemistry and technology of tamarind. *Journal of Science and Industrial Research*, 23: 204.

Macmillan, H.F. 1943. *Handbook of Tropical Plants. 362. Medicinal Plants*. Anmol Publications, New Delhi, India,1943.

Magehembe, J. A. 1995. Achievements in the establishment of indigenous fruit trees of the Nimbo woodlands of South Africa. SADC-ICRAF Agroforestry Project, Zomba, Malawi. eds. J.A.Magehembe, Y. Nitupanyama, and P.W. Chirwa : 39- 49.

Mahadevan, N. P. 1991 Phenological observations of some forest trees as an aid to seed collection. *Journal of Tropical Forestry*, 7(3): 243-247.

Mahony, D. 1990.*Trees of Somalia Tamarindus indica*. Oxfam, Print Unit. 274, Banbury Road, Oxford OX2 7DZ.

Masano, M. 1994 Effect of pre-treatment on the germination process of asam (*Tamarindus indica* L.) seeds. *Bulletin of Penelitian-Hitan*, 563: 33-42.

Meillon, S. 1974. Process for making drinks, syrups, juice and liquor and solid extracts based on tamarind products thus obtained. French Patent No. 2231322 (English Summary).

Morad, M. M., El Magoli, S. B. and Sedky, K. A. 1978. Physico-chemical properties of Egyptian tamarind seed oil. *Fette Seifen Anstrichmittel*, 80: 357-359.

Morton, J. 1987. *Fruits of Warm Climates*. Miami FL: 115-121.

Muabamba, L. and Sitaubi, L. A. 1995. *Seed Pre-treatment, Growth and Phenology of Some of the Indigenous Fruit Trees of the Niomboecozone at Ndungu Forestry Reserve, Malawi*. eds. J. A. Mageheme, Y. Ntupanyama and P. W. Chirwa: 66-72.

NAS, 1979. *Tropical Legumes: Resources for the Future*. Washington DC: 117-121.

Niemann, C., Carpita, N.C. and Whistler, R.L, 1997. Arabinose-containing oligosaccharides from tamarind xyloglucan, *Starch-Starke*, 49(4) 154-159; 19.

Osawa, T., Tsuda, T., Watanabe, M., Ohshima, K. and Yamamoto, A. 1994. Antioxidative components isolated from the seeds of tamarind (*Tamarindus indica* L), *Journal of Agriculture and Food Chemistry*, 42(12): 2671-2674.

Panigrahi, S., Bland, B. and Carlaw, P. M. 1989. Nutritive value of tamarind for broiler chicks. *Journal of Animal Feed Science and Technology*, 22(4): 285-293.

Pathak, R. K., Ojha C. M. and Dwivedi, R. 1992. Adopt patch budding for quicker multiplication in tamarind. *Indian Horticulture*, 36(2): 17.

Preston, F.W. 1992. The canonical distribution of commonness and rarity, part 1. Ecology 43: 185–215.

Purseglove, J. W., 1987. *Tropical Crops. Dicotyledons*, Longman Science and Technology: 204-206.

Rama Rao M, 1975. *Flowering Plants of Travancore,* Dehra Dun India, Bishen Singh Mahendra Pal Singh, 484 p.

Rao, Y. S., Mathew, K. M. and Potty, S. N. 1999. Tamarind (*Tamarindus indica* L.) research-a review. *Indian Journal of Arecanut, Spices and Medicinal Plants*, 1(4): 127-145.

Roy R.N, Maiti S.S and Mondal C.R, 1987. Tamarind seed husk as an effective fish poison, *Environment and Ecology*, 5(3) 467-471.

Salim, A., Simons, A., Waruhin, A. and Orwa, C. 1998. Agroforestry Tree Database: A tree species reference and selection guide and tree seed suppliers directory. International Council for Research in Agroforestry, PO Box 30677, Nairobi, Kenya.

Sanchez PC, 1985.Tropical fruit wines, A lucrative business. Research at Los Banos, The University of the Philippines, 3, 10-13 (FSTS: 86-05H0063).

Scuppe. E.W. 1990. Annual variation in seed fall, post dispersal predation and recruitment of a neo tropical tree. Ecology, 71: 504–514.

Seydack, H.W.A. (1995). Unconventional approach to timber yield regulation for multi aged, multi species forest. I. Fundamental considerations. Forest Ecology and Management. 1-14.

Shankaracharya, N. B. 1998. Tamarind-Chemistry, Technology and Uses-a critical appraisal. *Journal of Food Technology*, 35(3): 193-208.

Sharma, S. K, George, M and Prasad, K. G. 1986. Cholakarai-A confluence of two vegetation types, *Indian Journal of Forestry*, 9(4): 314-320.

Shivanandam, V. N. and Raju, K. R. T. 1988. Correlation between some fruit characters of four tamarind (*Tamarindus indica* L.) types. *Mysore Journal of Agricultural Science*, 22(2): 229-231.

Siddaraju P, Vijayakumari K and Janardhanan K, 1995. Nutritional and antinutritional properties of the under-exploited legumes, *Journal of Food Composition Analysis*, 1995,8, 351-362.

Simmonds, N. W. 1984.*Evolution of Crop Plants*. Longmans, London.

Singh M, Singh AK and Singh AK, 1989. Screening of extracts of higher plants for virus inhibitory action, *Geobios* (India), 16(2-3): 74-77.

Singh, B. G., Vijayachandran, S. N., Mahadevan, N. P. and Santhi, K. 1997. Analysis of variation in fruit and seed characteristics in *Tamarindus indica*. Linn. Proc. Nat. Sym. on *Tamarindus indica* L., Tirupathi (A.P) Organised by Forest Department of A.P. 27-28 June, 1997.

Singha, R. K. 1995. Biodiversity conservation through faith and tradition in India. Some case studies. *International Journal of Sustainable Development and World Ecology* (UK), 2(4): 278-284.

Sozolnoki, T. W. 1985. *Food and Fruit Trees of Gambia*. Hamburg, Federal Republic of Germany: 132 p Tiftung Walderhatung in Afrika.

Storrs, A. E. G. 1995.*Know Your Trees. Some common trees found in Zambia*. Regional Soil Conservation Unit (RSCU). Zambia.

Usha, K. and Singh B. 1994. Selection of tamarind tree types for higher yield. *Crop Improvement*, 23(2): 199-202.

Veluraja K, Ayyalnarayansubburaj S and Paulraj A.J, 1997. Preparation of gum from tamarind seed and its application in the preparation of composite material with sisal fibre. *Carbohydrate Polymers*, 34(4)377-379.

Vogt, K. 1995. *A field workers, guide to identification propagation and uses of Common Trees and Shrubs of Dry Land Sudan*. 52p. SOS Sahel International, UK 1, Tolpuddle Street, London NI OXT.

Wenkam, N. S. and Miller, C. D. 1965. *Composition of Hawaii Fruits*. Hawaii Agriculture Experimental Station University of Hawaii, Bulletin, 135: 1965.

Zablocki, L. Pecore. 1995. Beverages using tamarind extract and method of making such beverages. US Patent No. 547791, Nutrasweet Co. Deerfield, IL, USA.

Biodiversity in Horticultural Crops Vol. 2 *Pages* **252–283**
Editor: **Professor K.V. Peter**
Published by: **DAYA PUBLISHING HOUSE**

Chapter 12

Indian Gooseberry

D. Pandey, G. Pandey, S.K. Shukla and R.K. Pathak
Central Institute for Subtropical Horticulture,
Rehmankhera, P.O. Kakori, Lucknow

Aonla or Indian gooseberry is one of the important indigenous fruits of Indian subcontinent. In different parts of India, it is known by different vernacular names like Amla or aonla in Hindi, Dhatri, Dhatriphal or Amalaki in Sanskrit, Amla or Amlaki in Bengali and Oriya, Nelli in Malayalam and Tamil, Amlakamu in Telgu, Amophal in Punjabi and aonla, Myrobalan and Indian gooseberry in English. This fruit is known for its medicinal and therapeutic properties since time immorial and it finds a prominent place in ancient Indian mythological literatures like Vedas, Askandpuran, Shiv-puran, Padmapuran, Ramayana, Kadambari, Charak Samhita etc. and known as Amritphal. Owing to its hardy nature, ability to survive in various kinds of wastelands, high productivity/unit area (15-20t/ha.), nutritive and therapeutic values and suitability to various kinds of cosmetic, medicinal, herbal and value added products, aonla is becoming an important fruit of 21st century (Pathak *et al.*, 2006).

According to Hindu mythology, it is auspicious to arrange one day meal beneath the aonla tree during October-November when trees are laden with mature fruits. Hindu literatures also suggest that if the ripe fruits of aonla are eaten for forty days in the morning after fast, it restores health and vitality known as Kaya Kalp (Benthal, 1946). It is a common practice in Indian families to prepare the aonla preserve with sugar and some times with saffron to give one or two pieces to children and old age people every morning. It is not only the source of nutrients and medicines, but cultivation of this fruit is highly remunerative in marginal lands. Initially, aonla grows in forests or as household tree but during last one and half decade, there is large expansion in area under aonla cultivation across the country, utilizing waste lands. This resulted in efficient utilization of resources leading to better income to farmers, nutritional security coupled with enhanced employment generation and rehabilitation of waste lands.

Origin and Distribution

The aonla originated in tropical Southeast Asia, particularly Central or Southern India, Pakistan, Bangladesh, Sri Lanka, Malayasia, Southern China and Mascarene islands. Seedling trees are of common occurrence in mixed dry deciduous forests of India, ascending from sea level (Western and Eastern ghats, Aravalli, Vindhyaan hills) to 1300 m amsl, from Northwest to Eastern Himalayas.

Aonla is well distributed in India, Sri Lanka (Mac Millan, 1943) Cuba, Porto Rico (William and Williams, 1951), Hawaii, Florida, (Barret, 1956; Sturrock, 1959) Iran, Iraq (Hooper and Field, 1937), Java, West Indies, Trinidad (Webster, 1957), Singapore, Southern Thailand, Pakistan, Bangladesh, Malayasia and China (Benthal, 1946) and Panama canal regions. In India, it is more common in Uttar Pradesh, Jharkahnd, Bihar, Orissa, West Bengal, Karnataka and Western Ghats (Rawat and Uniyal, 2003). Its cultivation is spreading rapidly in the semi arid regions of Gujarat, Maharashtra, Rajasthan, Andhra Pradesh, Karnataka, Tamil Nadu and Aravali ranges in Haryana and Kandi area in Punjab and Himachal Pradesh. As per cent estimates, the total area under aonla plantation has spread over 50,000 ha. with approximate production of 1, 75,000 t. along with average productivity of 4.0 t/ha (Pathak, 2003).

Table 12.1: Area, Production and Productivity of Aonla (1999-2000)

Name of State	*Area (ha.)*	*Production (t)*	*Productivity (t/ha)*	*Source*
Uttar Pradesh	15,750	63,000	4.0	Hort. Directorate, U. P.
Gujarat	10,050	30,150	3.0	Hort. Directorate, Gujarat
Rajasthan	5,000	6,000	1.2	Hort. Directorate, Rajasthan
Maharashtra	4,000	5,600	1.4	National Hort. Board
Haryana	600	3,100	5.2	-do-
Mizoram	70	200	2.9	-do-
Tamil Nadu	5,500	16,500	3.0	Expert Estimation
Andhra Pradesh	3,000	9,000	3.0	-do-
Karnataka	1,800	5, 400	3.0	-do-
Bihar	1,350	4,050	3.0	-do-
Others	2,500	7,500	3.0	-do-
Total	49,620	1,50,500		

Uses and Nutritional Value

The pulp of fruit is highly nutritious and is one of the richest sources of vitamin 'C' (Shankar, 1969). Aonla fruits are acidic, acrid, cooling, refrigerant, diuretic and laxative (Gopalan and Mohanram, 1996). Normally, one aonla fruit contains 20 times vitamin 'C' in terms of antiscorbutic value than two oranges. The main constituents of aonla are tannins, polyphenolic compound 1,3, 6-trigalloylglucose, terehebin, corilagin, phyllantine, beta siotosturol, linoleic acid, ellagic acid and lupeol. Aonla fruit contains a fair amount of polyphenols and tannins, which, retard the oxidation of vitamin 'C' and render its value as antiscorbutic in fresh and in dried condition. It is not only a rich but also the cheapest source of vitamin 'C' and fair source of carbohydrates, carotene, thiamine, riboflavin and minerals like calcium, iron and phosphorus.

Aonla is a rare example of an edible material, rich in tannins and ascorbic acid (Kalra, 1988). The chemical composition of aonla fruit is presented below.

Table 12.2: Chemical Composition of Aonla Fruits

Constituents	*Amount (per cent)*	*Constituents*	*Amount (mg/100 g pulp)*
Moisture	77.1–82.20	Iron	1.20
Protein	0.50	Nicotinic acid	0.20
Fat	0.10	Vitamin C	200–1814
Minerals	0.5–0.70	Carotene	0.01
Fibre	1.9–34	Thiamine	0.03
Carbohydrate	14.10–21.89	Riboflavin	0.05
		Niacin	0.18
Calcium	0.012–0.050	Tryptophan	3.00
Phosphorous	0.020–0.026	Methionine	2.00
		Lysine	17.00

Considerable variation in vitamin 'C' in aonla is reported from various parts of the world. Bajpai (1969) observed variation in vitamin 'C' content of aonla fruit from September to February under North Indian conditions. He observed continuous increase in total ascorbic acid in fruits until second week of February and thereafter slight decrease was noticed. He reported (563.3 mg/100 g pulp) as the maximum amount of vitamin 'C', while Amal and Raghwan (1957) reported 900 mg/100 g of fruits. Mustard (1952) reported 1561–1814 mg/100g pulp in Cuban fruits. A mature aonla fruit contains more vitamin 'C' compared to developing and immature fruit (Ojha and Pathak, 1993), the highest vitamin 'C' content being observed in February which decreased thereafter.

Botanical Description

Aonla belongs to genus *Emblica* of family Euphorbiaceae and order Euphorbiales characterized by its usually indehiscent fruits. It is medium sized much branched tree occupying height of 10-20 m. In the tropical regions, it is supposed to be an evergreen tree but behaves as deciduous tree due to complete defoliation of leaves. Stem is smooth, greenish grey to brown, exfoliating bark, which peels off in thin flakes like that of guava. It is characterized by phyllanthoid branching habit with two types of shoots. On the basis of growth characteristics, these are characterized as long and short shoots. (Bajpai, 1965; Ram, 1971). These shoots are referred to as branches or branchlets. The indeterminate shoots are longer and continue to put new growth in season but do not bear the flowers, irrespective of period of their emergence. Determinate shoots appear on nodes of indeterminate shoots and their number at each node varies from 3 to 5 in different cultivars (Pathak, *et.al.*1993).

The internodes are much shorter in determinate shoots. These nodes are barren, floriferous with imbricate leaves. First a few proximal nodes on the determinate shoots are barren (without leaves), which are reduced to dark brown scarious cataphylls. Succeeding nodes are with green but reduced leaves. Subtending cymules of male flowers are flowed by nodes each with cymules of one central female flower (rarely two) and several lateral males, distal half floriferous, determinate, shoots are normally sterile with typical leaves.

Fruits are depressed, round globose or oblate, indented at the base. A capsular (drupaceous) berry with fleshy exocarp, smooth to obscurely 6 lobed, initially light green becoming dull greenish yellow or rarely brick red as it matures, 2-5 cm in diameter, skin thin, tunas or semi-translucent, flesh crispy, juicy, stone containing 6 small seeds, the edible part of the fruit is mesocarp and the endocarp forms the hard stone, which encages the seeds.

Genetic Diversity

Aonla is naturally spread, due to seed propagation, creating a lot of variation in existing population in tree vigour, fruiting behaviour, yield and quality characters. Great diversity in *Emblica officinalis* Gaertn. is available in the forest areas of Vindhyan hills, lower hills of Uttarakhand and Himachal Pradesh, Chhattisgarh, Jharkhand, Madhya Pradesh, Rajasthan and Bihar. Madhya Pradesh Forests in the districts of Sheopur, Betul, Belaghat, Satna, Chitrakoot and Panna have rich diversity. Local tribes of the regions harvest the fruits before they reach full maturity. Besides this, they harvest the fruits at immature stage, boil them and dry the segments in sun. This dried product is locally known as 'Amrethi'. This prevents the natural germination of seeds in the forest area. The villagers/forest tribes cut the branches along with fruits and very often they chop the whole tree for fruit harvesting. This is a matter of great concern. Looking into this problem, Madhya Pradesh Government now declared seedling aonla as endangered species in the state (Wizmin, 2000). Considering great diversity in M.P., an exploration was planned to collect the superior types having bigger fruit size, good yield potential, high vitamin 'C' content along with high content of total phenols. A total of 46 germplasm accessions were collected and seven accessions were the most promising. The details of these accessions are given below (Pandey *et al.*, 2006).

Table 12.3: Salient Fruit Characteristics of Promising Selections of Aonla

Characters	*Accessions Numbers*						
	DP/SKS-14	*DP/SKS-15*	*DP/SKS-16*	*DP/SKS-33*	*DP/SKS-34*	*DP/SKS-35*	*DP/GP-11*
Fruit colour	Greenish white	Whitish green	Greenish pink	Whitish green	Whitish green green	Greenish white	Greenish white
Fruit weight (g)	45.0	49.0	39.0	42.0	45.0	32.0	38.0
Fruit length (cm)	3.54	3.82	3.52	3.48	3.74	3.24	3.73
Fruit breadth (cm)	4.48	4.52	4.16	4.34	4.36	3.92	4.0
No. of segments	6-8	6-8	6	6	6	6	6
TSS (° B)	9.6	10.0	9.8	10.46	10.76	10.93	11.60
Acidity (per cent)	2.03	1.38	1.48	1.90	1.50	2.02	2.46
Vitamin 'C' (mg/100g pulp)	429.85	594.48	466.44	418.78	388.87	397.42	446.67
Tanins (per cent)	2.88	2.70	3.14	2.82	2.79	5.79	3.28
Reducing sugars (per cent)	3.93	3.90	4.08	3.91	3.64	4.09	3.02
Total sugars (per cent)	4.41	4.51	5.34	6.54	6.26	6.27	5.22

DP/SKS/GP = D. Pandey/S.K. Shukla/G. Pandey.

In Western and Eastern ghats, three species *viz.*, *Phyllanthus emblica, Phyllanthus indofisheri* and *Phyllanthus acidus* are of common occurrence. A wild strain of aonla grows in Himalayas upto an attitude of 1600m amsl in places which experience mild snow fall during winter indicating cold tolerance. The fruits are smaller in size (Parmar, 2000).

Germplasm Conservation

The aonla germplasm is being maintained at many places in the country in field gene banks. The main centres having aonla germplasm collection are given below.

Table 12.4: Aonla Holdings in Field Gene Bank at Different Places

Institute/center	*No. of Accessions*	*Reference*
Dryland Agricultural Research Station, Anantpur, Andhra Pradesh	12	Pareek, 1998
Regional Research Station, Aruppukottai, Tamil Nadu	05	Pareek, 1998
Dryland Agricultural Research Station, Sardarkrushinagr, Gujarat	12	Pareek, 1998
Narendra Deva University of Agriculture and Technology, Faizabad, U.P.	21	Pareek,1998
G.B. Pant University of Agriculture and Technology, Pantnagar, Uttarakhand	10	Ram, 1993
National Research Centre for Arid Horticulture, Bikaner, Rajasthan	16	More *et al.*, 2000
National Bureau of Plant Genetic Resources, Pusa, New Delhi	05	Anonymous, 1999
Central Institute for Subtropical Horticulture, Rehmankhera, Lucknow	12	Pathak *et al.*, 2003
Mahatma Phule Krishi Vidyapeeth, Rahuri, Maharashtra	06	Karale, 2000

Apart from that, a total of 57 aonla accessions along with two accessions of *P. indfisheri* are maintained at the NBPGR, Regional Station, Thrissur, Kerala (Anon, 1999).

Climate

Aonla is a subtropical tree and prefers dry subtropical climate. Tree growth is influenced by hot wind or frost. Its cultivation is now quite successful and becoming popular in coastal regions and at higher altitude up to 1500 m amsl. Frost during winter is injurious to plant. In Rajasthan, Gujarat and Maharashtra, extremely high temperature coupled with desiccating winds during summer are very common and sometimes it causes excessive blossom and pin head drop. A mature aonla tree tolerates freezing and high temperature up to 44°C. Temperature above 45°C and low humidity even for a few days may show injurious influence and adversely affect the bearing. Low temperature around 7-10°C is conducive for floral bud initiation, and warm temperature and humid atmosphere are conducive for fruit growth and development. Dry spell in rainy season results in heavy dropping of fruitlets and prolonged fruit dormancy. In the dry spell of 2002-03 during July- August, aonla showed better tolerance when compared with other fruit plants. Among varieties, Chakaiya showed better tolerance to drought as compared to Banarsi and Francis.

Soil

Aonla is a hardy fruit crop and is now cultivated in various kinds of soil conditions (Bajpai and Shukla, 1985). Aonla withstands in marginal lands from slightly acidic to saline/sodic having pH 6.0-9.5, ECe 5 dSm^{-1} and ESP 30-40. Now-a-days, its cultivation is more common in waste lands like *rakkar*, calcareous, morum soils with rocky substratum in Bundelkhand (UP, MP). In red laterite soils of Maharashtra and Tamil Nadu, aonla shows even better performance as compared to tamarind (Mehta, 2000, 2001; Rajput *et al.*, 2000). Aonla can be adopted successfully for rehabilitation programme. Introduction of aonla in the natural rangelands of Bundelkhand showed its suitability for rehabilitation,

Figure 12.1

Figure 12.2

Figure 12.3

particularly through popularization of agri-silvi horticulture land use system. Presently, it is identified as an ideal crop for various kinds of wastelands, *viz.* moisture stress, eroded ravines, uplands, riverbeds and areas with undulating topography (Pathak and Pandey, 1985; Pathak, 1990; Pathak and Pathak, 2001). It can be successfully planted on roadsides and strip lands. Clay soil with stagnated water or areas with high water table should be avoided. Studies indicated that aonla has immense potential and possibility for commercial cultivation in salt affected soils which cover an area of 7.0 m. h. in India (Pathak and Pandey, 1985).

Varietal Improvement

The varietal improvement programme on aonla is limited only to selection of superior types from existing seedling population (Pareek, 2001; Shukla *et al.*, 2004). The limiting factors for aonla improvement are:

1. Highly heterozygous nature,
2. Perennial growth habit,
3. Long juvenile phase,
4. Complex taxonomy,
5. Cross and self-incompatibility,
6. Higher number of staminate flowers on the determinate shoot and
7. Tiny flower size (Pathak *et al.*, 1993).

Variety improvement programme in aonla should have the following major objectives.

1. To develop dwarf tree stature for medium and high-density orchards.
2. To develop precocious and prolific bearing cultivars.
3. To develop excellent fruit quality for processing and other value added products.
4. To develop cultivars having higher content of vitamin 'C', total phenols and high anti-oxidant properties.
5. To develop cultivars with resistance to biotic stress *i.e.*, pests and diseases.
6. To develop cultivars with resistance to abiotic stresses like frost, heat, drought and sodic soils.

For achieving these objectives, Shukla *et al.* (2004) suggested the following breeding techniques, which can be exploited for variety improvement in aonla.

1. Selection from diversity area in the forest and aonla orchards.
2. Selection from seedlings obtained from open pollinated population of known cultivars.
3. Induction of variability through combination of breeding techniques including mutation followed by selection.
4. Selection from clonal variation.
5. Hybridization with specific objectives.
6. Polyploid breeding.

Cultivars

Aonla cultivars are generally known on the basis of size, colour or after the name of the place. Banarasi, Bansi Red, Francis and Chakaiya were the known aonla cultivars. All these cultivars have their own merits and demerits. Banarasi is shy bearing and prone to heavy fruit drop; Francis though a prolific bearer, has serious incidence of fruit necrosis. Chakaiya, the most popular cultivar has comparatively small size fibrous fruits with a tendency to bear in alternate years (Pathak *et al.*, 2003). Pathak *et al.* (1993) worked on varietal improvement in aonla through selection at Narendra Dev University of Agriculture and Technology, Faizabad and as a result cvs. NA-6, NA-7, Krishna, Kanchan and NA-10 were developed. From Gujarat Agriculture University, Anand, Anand-1,Anand-2 and Anand-3 were selected from the seedling population grown in the area (Chundawat, 1990). The chief characteristics of the important cultivars are under:

Kanchan (NA-4)

It is a seedling selection from Chakaiya. The cultivar is heavy and regular bearer (7.7 female flowers/branchlet). Fruits are medium sized with higher fibre content, preferred by industry for pulp extraction and manufacture of various products. It has a mid season maturity (mid November to mid December) and has been adopted very well in the arid and semi-arid regions of Gujarat and Maharashtra.

Krishna (NA-5)

Krishna is a seedling selection from Banarasi. It is an early maturing genotype, which matures during mid October to mid November. The fruits are large, triangular, conical, skin is smooth, whitish green to apricot yellow with red spot on exposed portion. Flesh is pinkish green, less fibrous and highly astringent with moderate keeping quality. It is an ideal variety for preparing candy and juice.

Narendra Aonla-6

This is a selection from Chakaiya cultivar. It has a mid season maturity (mid November to mid December). Bearing is heavy. Fruits are most attractive and shining, medium to large size, flattened and very low in fibre content. This is an excellent variety for making preserves, candy and jam.

Narendra Aonla-7

It is a seedling selection from Francis. It is precocious and prolific regular bearer (9-10 female flowers/branchlet). Incidence of necrosis has never been observed in this cultivar. It has mid season maturity. Fruits are of medium to large size with conical apex. Fibre content is slightly higher than NA-6. This variety has adopted well in Rajasthan, Bihar, Madhya Pradesh, Andhra Pradesh and Tamil Nadu. The major constraint with NA-7 is its brittle branches, which often break due to heavy fruit load. Hence, staking to the fruited branches is desirable. This is a good variety for making chyavanprash, chutney, pickle, jam and squash.

Narendra Aonla-10

This is a chance seedling selection from cultivar Banarasi, which bears profusely. Fruits are attractive, medium to large in size and flattened round in shape. Skin is rough, yellowish green with pink tinge. Flesh is whitish green, fibre content is higher and phenolic content is lower. This is an ideal cultivar for commercial growing because of its higher productivity, early maturity, suitability for dehydration and pickle making. It has a tendency of alternate bearing.

Apart from these cultivars, Anand-1, 2 and 3 are selected as promising strains from Gujarat.

Anand-1

Trees are tall will upright growth habit. It is a moderate bearer having 1-2 female flowers/branchlet. Fruits are small and rough with fibrous and hard flesh.

Anand-2

Trees are tall with upright growth habit. It is a moderate bearer having 2-2.5 female flowers/ branchlet. Fruits are small to medium, skin rough with medium keeping quality.

BSR-1

This is also a seedling selection from Tamil Nadu. It is self fruitful, gives crop round the year, small fruit size (27.0g), high vitamin 'C' content (620.00mg/100g), fibrous and low moisture content, heavy bearing. This variety has high demand in Ayurvedic industries.

Pandey *et al.* (2005), identified a promising genotype in Pratapgarh district. This is cultivated in a private orchard as superior chance seedling of Francis and known as Laxmi-52. The tree is semi erect in growth habit and branches do not drop like its parent Francis. Fruit size is large (40-60g), diameter (4.0-4.5cm) with 6 ridges. During early period of fruit growth, fruit colour, particularly ridges are light pink, which disappears on full development and maturity. It is a mid season maturing (mid November to mid December) and is free from necrosis with yield potential of 200-250 kg/tree (after 10 years). Owing to larger fruit size, attractive colour and suitability for candy and preserve, it fetches higher price in the market. Due to heavy fruit load, cracking in branches is observed.

Propagation

Aonla is raised by seed. Though seed propagation is easy and cheaper, it does not ensure true to type plants due to cross pollination. Besides, such trees take long time to come into bearing. To limited extent, aonla was earlier propagated through inarching, but owing to upright growth habit, availability of scion shoots had been the main limiting factor for inarching. Moreover, there are heavy casualty during transportation and establishment of inarched plants in the field. In the recent years, techniques of budding, veneer grafting and soft wood grafting were standardized and commercially adopted in different aonla growing regions of the country (Pathak, 2003).

Rootsotck

Aonla is generally propagated through budding/grafting on seedling rootstocks. Kadam *et al.* (2003), attempted use of star gooseberry (*Phyllanthus acidus*) as rootstock for aonla propagation. Sincere efforts are required for identification of dwarfing and frost tolerant root stocks. Generally, seeds of desi aonla are used for rootstock purpose.

Seed Extraction

For seed extraction, mature fruits collected from seedling aonla trees during January-February are dried in the sun and seeds are extracted exerting light pressure. On an average, there are six seeds/ fruit. About one kg seed is obtained from 100 kg of fruits and average test weight of seed varies from 30,000-50,000/kg or 300-500/10g of fresh seed weight.

Seed Longevity

Systematic study on seed longevity of aonla is lacking. Singh *et al.* (2002) conducted a preliminary study in which aonla seed having 10 per cent moisture packed in cloth bags, plastic bottles, glass jar, paper bags and 300 gauge polythene bag was stored for 22 months under ambient conditions. Aonla seed stored in polythene bags maintained higher seed viability and better germination (per cent) as compared to other packing materials/containers.

Seed Germination

Sowing of seed in raised beds (after soaking in water for 12 hours) during March-April results germination within two weeks and seedlings of approximately 10 cm height are transplanted for better establishment. Maximum germination (77.8 per cent) was obtained in seeds sown in mid July. With the use of polyhouses, seed sowing can be adjusted as per need. Rajamanickam *et al.* (2002) reported earliness in seed germination and germination (per cent) in fresh seeds (6.30 days and 69.33) and one year old seeds (6.64 days and 46.00) soaked in 0.5 per cent KNO_3 for 8 h. and in GA_3 (200 ppm) for 8 h., respectively in cv. BSR-1. In another study, Dhankar *et al.* (1997) reported that soaking seeds in GA_3 250 ppm improved germination to the tune of 75.98 per cent and 64.14 per cent in laboratory and in pots, respectively and earliness in cv. Gujarat Aonla-1. In arid and semi arid parts of the country, earth ball formation is difficult, hence raising seedlings in seedbed is not advocated. Use of container is gaining popularity. Saroj *et al.* (2001) standardized perforated polybags for rootstock raising and 5-6 months old seedlings become ready for budding. Similar results with polybag raised aonla seedlings were obtained at Jhansi (Tewari and Bajpai, 2002).

Hardening of Seedlings

Study showed that seedlings (10cm in height) transplanted in beds/polybags after holding it three days under shade gave better establishment as compared to direct transplanting after removal from nursery bed. Delay in transplanting from nursery causes heavy mortality after transplanting in field (Singh *et al.*, 1993).

Method of Propagation

Earlier aonla was propagated through seeds or inarching. Recently, newer and cheaper techniques are standardized. These include softwood/cleft/wedge grafting, patch/modified ring budding. Use of poly and net house, container and rooting media have made aonla propagation possible round the year.

Of the various methods of propagation, budding is the most practical. Patch budding showed better (90 per cent) response followed by modified ring budding (80 per cent). Budding should be performed from May to August for obtaining better success, which continues upto September with appreciable success (Sharma *et al.*, 2000 and Wagh *et al.*, 2001). Saroj *et al.* (2001) reported that more than 90 per cent success can be achieved through patch budding when perforated polythene bags were used for raising rootstocks under irrigated hot arid eco-system.

Amin (1978) recorded 70 per cent success with soft wood grafting on terminal shoot of rootstock raised *in-situ*. Tewari and Bajpai (2002b) reported that seed sown in polythene bag measuring 10 x25cm in size, filled with red soil + black soil + FYM in 1:1:1 ratio in February attained graftable stem diameter (5 mm collar diameter) during July. Cleft grafting was performed at 3-4 cm height on rootstock from July-September. After grafting, plant was capped with polythene tube. The maximum grafting success was obtained in second fortnight of August (87 per cent) followed by first fortnight of August (83 per cent) in cv. NA-7. In another study, Tewari *et al.* (2004) reported that appreciable success (97.5 per cent) can be obtained through bench grafting using cleft method, on bare rooted rootstocks performed in February when the temperature ranges from 10-30°C. Tewari *et al.* (2005) reported that bench grafting using cleft method performed during February gave 85 per cent success in polythene bags and capped with polythene tube in nursery. Grafted plants obtained excellent growth within 5 months and were ready for field transplant in July. One-year-old stocks were suited for bench grafting and bench grafting can be done on rootstocks upto 2 years of age.

In another study, Singh and Pandey (2003) got maximum transplanting success (90.00 per cent) and bud sprouting (63.54 per cent) in leaf detached bare rooted budlings treated with IBA 200 ppm in February.

Attempts were also made to propagate aonla through cutting, stool layerings and air layering with use of plant growth substances. Rajan and Ram (1982) recorded 87.30 per cent rooting in cuttings using IBA (1500ppm) with the use of bottom heat treatment (33°C). Rajamanickam *et al.* (2002) reported that, the highest shoot length, root length, root weight, maximum root spread, total number of leaves and biomass production were recorded with ground inarching followed by air inarching. Similarly, establishment (per cent) was the highest (91.08) with ground inarching followed by air inarching (71.11).

Micro Propagation

In a preliminary study, Sehgal and Khurana (1985) reported plantlet regeneration from cultured endosperm of *Emblica officinalis*. The regeneration of shoots was more frequent when both BA (0.2mg/l) and IBA (0.1mg/l) were present in the media. The embryo like structure produced plantlets. Mishra *et al.* (1998) reported that *in-vitro* oxidative browning and contamination are the major constraint in establishment of culture. *In vitro*, browning was effectively controlled by sealing the cut ends of explant with paraffin wax and method is termed as explant waxing. With this method around 20 per cent success in culture establishment was achieved. The contamination of explant was also controlled by treating with Bavistin (0.1 per cent) for 240-260 minutes followed by treatment of explant with $HgCl_2$ for 2 minutes under aseptic conditions (Mishra *et al.*, 1999). There is urgent demand to standardize complete protocol with efficient acclimatization procedure. Kadam *et al.* (2003) attempted *in-vivo* micrografting, a new technique, by using seedlings of star gooseberry as rootstock.

Planting

Grafted or budded plants are planted 8-10 m. apart in square system of planting during July-August or February. Planting of an additional plant in the centre of each square, known as filler plant and the system is known as quincunx, not only provides the additional income to the growers but also offers maximum soil coverage. Among the fruits, ber, guava, lemon, karonda and drumstick are ideal filler plants. Hedge row planting of aonla is also being tried at some of the places. In this method, line to line distance is kept 8 meters, while plant to plant distance is reduced to 4 meters. In wastelands, pits of one cubic meter size are dug two months prior to planting. The pit soil is divided into two equal parts, *i.e.*, 50 per cent top soil and 50 per cent bottom soil. In the top 50 per cent soil, 3-4 basketful (25-40 kg) of well rotten FYM and one kg neem cake or 500g bone-meal is mixed. This mixture is filled up to the height of 6-10 cm above the ground level in the pit. In sodic soils, as per gypsum requirement (GR) value, 5-8 kg gypsum or pyrite or "Usar Tor Khad" along with 20 kg sand is also incorporated in the filling mixture. Filled pits are irrigated thoroughly, if there is no rain so that the soil is properly set. The budded/grafted plants with earth ball should be planted in the centre of the pits. The soil of the pit is properly pressed around the earth ball. Light irrigation should be given just after planting. Self-incompatibility in aonla is a common problem among various cultivars, hence, two varieties in the alternate rows need to be planted for higher productivity. The better combinations are NA-6, NA-7 and NA-10 or Kanchan and Krishna.

In situ Orchard Establishment

In degraded, arid, semi arid lands, ravines, kandi and ghar (foothill of Himalayas) areas *in-situ* orchard establishment gave encouraging results. In this technique, sowing of 2-3 seeds in well prepared pits as per lay out is done. Care should be taken to encourage vigorous seedling growth. In saline/ sodic soils, seed germination is adversely affected. Under such conditions, seedlings are raised in polybags/containers. After gaining appropriate height, these seedlings are transferred and following proper establishment, these seedlings are budded with aonla cultivars. Since, scion of aonla can safely be stored/transported in moss grass/moist paper for 5-7 days, this practice is helpful in establishment of aonla orchards in distant places (Pathak *et al.,* 2003). Shukla and Solanki (2000) compared traditional and *in-situ* orchard establishment in 'rakkar soil' of Jhansi and reported that expenditure on aonla orchard establishment through *in situ* was slightly higher than the traditional one. But when the cost of budded plants, transport charges, damage during transport and subsequent poor survival and growth of the plants are considered, *in situ* technique is far better than the traditional one.

Training and Pruning

Aonla plants should be encouraged to develop a medium height tree. The young plants are allowed to grow straight up to 0.75 to 1 m. from the ground level, to provide good framework of the plant and then allowed to branch. Plants are trained in modified central leader system. Two to four branches with wide crotch angles, appearing in the opposite directions should be encouraged during early years. The unwanted branches need to be pinched off regularly. In the subsequent years, 4 to 6 branches are allowed to develop. Regular pruning of a bearing aonla tree is not required. As per the growth habit, shedding of all the determinate shoots encourage new growth in coming season. Dead, infested, broken, diseased, weak, criss-cross branches and suckers developing from rootstock are to be removed regularly.

Top Working

Older bigger trees of inferior types (old traditional cultivars, *viz.* Banarasi, Francis and seedling trees) can be rejuvenated and easily changed into superior types by top working. These plants are headed back 3 to 4 m. above from the ground level during December-January, leaving 4-6 main cut limbs in all the directions. The cut ends of the trees are then pasted with the tree paste (cow dung + soil + water) or with simple cow dung paste to avoid fungal infection. Only four to six shoots, developing in outer directions on main limbs, are allowed to develop. During June-July, patch budding with superior scion cultivars is done on these shoots. After sprouting, the top portion of the budded shoots is removed. Care is taken to manage the insect pest problems during this period, since these plants are prone to insect pest infestation and some times also to wind damage (Mishra *et al.,* 2007).

Manure and Fertilizer Application

The systematic work has not been done on nutritional requirement of aonla trees. The fertilizer doses depend on soil fertility, age of the plant and production. Based on experience, recommendations of manure and fertilizers are given below.

Table 12.5: Manure Requirement of Aonla

Age of Tree (Year)	N (g)	P (g)	K (g)	FYM (kg)
1	100	50	100	5
2	200	100	200	10
3	300	150	300	15
4	400	200	400	20
5	500	250	500	25
6	600	300	600	30
7	700	350	700	35
8	800	400	800	40
9	900	450	900	45
10 and above	1000	500	1000	50

Time of Application

Full dose of well rotten Farm Yard Manure (FYM), half dose of nitrogen and full doses of phosphorus and potash should be applied during January-February before flowering. Remaining half dose of nitrogen is given during July–August. The manure and mixure of fertilizers are spread under the whole canopy of the tree leaving one meter radius from tree trunk and mixed well in the surface soil. Light irrigation is given immediately after the application of fertilizers. In sodic soils, 100g each of borax, zinc sulphate and copper sulphate should be applied along with fertilizers in accordance with tree age and vigour. Besides this, spray of micronutrients, *viz.* B (0.4 per cent), Zn (0.4 per cent), Cu (0.4 per cent), along with lime is also done. It is helpful in reducing fruit drop and improving fruit quality.

Irrigation

Aonla is a hardy tree. It requires less irrigation. Established aonla orchards in general do not require additional irrigation except normal rains. Irrigation is not required during rainy season. However, irrigation at an interval of 15-20 days is required during winter and summer seasons (October to January and April to June), particularly in the early years of orchard establishment. It protects the damage against frost and also controls fruit drop during winter. Brackish water should not be used for irrigation. In bearing plantations, first irrigation should be given just after fertilizer application (January-February). Irrigation should be avoided during flowering (mid March-mid April) period.

Among the various systems employed for irrigation, basin irrigation is the best. The results obtained with drip irrigation show promising response. Drip irrigation with 60 CPE is appropriate with water saving of 40-50 per cent (Pathak, 2003). In water deficient areas, pitcher irrigation is successful in initial establishment of orchard.

Mulching

Mulching with organic wastes is very effective for establishment of aonla orchards. Rao and Pathak (1996) reported that mulching with locally available organic materials (paddy straw, sugarcane conductivity and field capacity. Similarly, higher organic carbon content, nitrogen, potassium and lower ECe, ESP and cations were observed. Among the organic mulches, paddy straw showed better response followed by sugarcane trash and grasses. Rao and Pathak (1998) again emphasized that paddy straw when used as mulch encouraged tree growth, checked weed growth, reduced moisture loss from soil and also regulated soil temperature.

Flowering and Fruiting

Aonla tree behaves as an evergreen plant in the tropics, but under subtropical conditions of North India, it begins to shed its determinate shoots along with leaves from November onwards and the indeterminate shoots are devoid of foliage by February- March. New determinate shoots begin to appear at the nodes by the end of February and continue to emerge till the first week of April (Bajpai, 1965; Ram, 1971). The flower buds appear on these newly emerged shoots. Aonla is sensitive to day length. In North India, flowering takes place from March to May, while in South India, it blooms in June-July and again in February-March with second flowering only a small crop is obtained (Naik, 1949). The determinate shoots, which emerge after 15 April, do not flower in North India. In recent years, off season flowering during September-October was noticed in Pratapgarh. Removal of determinate shoots during July to October induced flowering under Lucknow conditions (Ram, 1982). Owing to low temperature in November-December, fruit development is not possible. Off season flowering is detrimental to the bearing in following year (Nath *et al.*, 2000).

Flowering period in aonla is more affected by day length than temperature (Ram, 1982). Male flowers appear in clusters in the axils of leaf on lower parts of the determinate shoots followed by female flowers on same shoots. Some of the shoots bear only male flowers and those appearing late in the season do not flower. Early formed indeterminate shoots of the previous season, which have attained sufficient growth and maturity, invariably, bear both kinds of flowers. The per cent of determinate shoots bearing both male and female flowers may be 14 to 17 and the male flowers are many times more than the female ones (Ram, 1971; Pathak *et.al* 1993). Anthesis of male flowers occurs between 16.00 to 17.00 hrs in Banarsi, while in Chakaiya both in morning (8.30–11.00 h) and in the afternoon (15.30-18.30 h) as reported by Alleemullah and Ram (1990). Dehiscence of pollen grains coincides with the anthesis of male flowers in different cultivars and majority of pollen are fertile (Bajpai, 1957; Alleemullah and Ram, 1990). Stigma is receptive between second to fifth day after anthesis.

After pollination, pollen grains germinate within 8 hours of pollination and grow into styler tissues. Fertilization takes place in about 24 h of pollination (Bajpai, 1968, Ram, 1971). The styles of pollinated flower turn brown along with the perianth segments within 3-4 days after pollination. The dried stigma, however, persists for 2-3 months. The egg cell enlarges to some extent giving the appearance of fertilized egg. The ovary gets encased in a papery brown covering called 'capule'. The formation of capule signifies the fruit set (Ram, 1993).

Pollination

Aonla is a cross pollinated fruit plant and wind, honey bees and gravity play important role in effective pollination. A systematic research on (per cent) relative abundance (PRA) of different pollinators of aonla shows that *Melipona trigona* (47.5 per cent), *Apis florea* (21 per cent) and coccinelid beetles (14.50 per cent) cause pollination respectively. Similarly, survey of aonla orchards at Faizabad revealed that the PRA of different pollinators viz; *Melipona* spp., *Apis florae, Apis cerana indica* were 60.32, 17.41, 3.81 per cent respectively along with two species of calliphorid flies and one species of beetles was 18.46 per cent (Anon, 2002). Common visiting hours of honey bees are late evening and morning (Bajpai, 1957). The pollinators visitation order was *Melipona trigona* > *A. florae* > coccinelid beetles > *A. cerena indica*.

Earlier studies showed that there is no self incompatibility in aonla and poor fruit set might be due to high (per cent) of male flowers (Bajpai, 1957; Ram, 1971). Pollination studies show considerable variation with respect to fruit set and retention (Pathak and Srivastava, 1994). Open pollination showed maximum fruit set (42.5 per cent) followed by geitonogamy (29.4 per cent) and bagging (4.6 per cent) along with corresponding fruit retention to the tune of 8.9, 3.0 and 0.5 per cent, respectively. Self pollination under bagging resulted in poor fruit set (2.35 per cent) with no fruit retention at maturity as reported by Mohammad and Ram (1990) and Singh, *et al*. (1998). Significantly high fruit set (61.43 per cent) and fruit retention (18.54 per cent) were recorded under open pollination as reported by Balasubramaniam and Arulmozhiyan (2003) and confirmed self incompatibility problem in aonla. Singh *et.al* (1998) also reported that most of the aonla cultivars were self incompatible and needed provision of proper pollinizer to improve fruit set and retention. Higher fruit set and retention under open and cross pollination may be because of compatible mating and higher pollination stimulus provided by the pollen taking part in pollination and fertilization (Singh *et.al* 2001). To obtain higher fruit set and optimum economic yield along with quality fruits, NA-6 and Krishna were suitable pollinizer for NA-7.

Fruit Growth and Development

In aonla, after fruit set, fruitlets remain dormant for about 3-5 months after pollination in North India (Bajpai, 1963; Ram, 1971). During dormancy period, no division in zygote or endosperm nuclei occurs and the ovaries remain encased in the capule. Secondary nucleus is larger in pollinated ovaries than in non-pollinated ones. The zygote and primary endosperm nuclei rest at unicellular stage during dormancy period. At the end of dormancy break, the first division of primary endosperm takes place in the end of July and during first half of August. The primary endosperm nucleus goes into several successive divisions and covers the embryo sac, while zygote is still uninucleate. The divisions in zygote occur 15-20 days after it starts in endosperm nucleus. Thus, dormant fruitlets start growing following the enlargement of nucellar tissues and the division in primary endosperm nucleus occurs sometime in the end of July or first half of August, depending upon the location and climatic conditions.

Fruit growth of aonla after breaking its dormancy is slow in the first 15 days. In general, fruit attains 5-6 mm diameter by the end of August. The pointed and tapering fruitlets attain spherical shape. The fruits grow rapidly from middle of August to first week of October and complete almost 75 per cent of growth during this period. Thereafter, the growth rate again slows between second week of October and first week of November and increases slightly thereafter and ceases after November. Thus, fruit growth in aonla is double sigmoid. The maturity of cultivars varies depending upon the quantitative changes in fruit in November and December (Ram, 1971).

Rapid development of seeds starts when fruit attains a size of 1 cm in Banarsi. The length and fresh weight of seed increases rapidly until the fruit attains a size of 3.5 cm. The seed completes its development during the first rapid growth period of fruit and accumulates major part of the dry matter. There is no appreciable change in the length and fresh weight of the seed, although the fruit continues to grow. The dry weight of fruit increases until its maturity.

No appreciable changes occur in N and sugar content in the fruit of aonla during dormancy. Nitrogen content drops in the early stages followed by slight rise during the period of its first rapid growth and remains more or less constant during maturity period. However, sugars accumulate after dormancy break. A perceptible rise in reducing sugar content occurs in the fruit during August. The reducing and non reducing sugars content increases until the fruit attains diameter of about 3 cm and decreases thereafter. The fruits contain higher amount of reducing than non reducing sugar at all the stages of fruit growth (Nizamuddin *et al.*, 1982).

The dormant fruit contained a very low amount of cytokinin, which increased only after free nuclear division in endosperm and after the start of the division of zygote and thereafter increased rapidly during the early stages of seed growth. The cytokinin content was low in the fruit after the seed completed its growth indicating that the developing seed was the centre of cytokinin synthesis (Ram and Rao, 1976). The total gibberellins in aonla fruit attained a minor peak in August and a major peak in the beginning of October (1.3μ/g dry weight) and a sharp fall occurred thereafter. During fruit maturity, gibberellins content remained low without any appreciable change (Ram and Rao, 1978). During the dormancy break, the level of all the auxins in the fruit was very low. Levels of all the auxins increased during the first rapid growth period with a minor peak in second week of August and major peak in the beginning of October. Levels of all the auxins decreased between second week of October and third week of November. Similarly, total auxins content was low in mid April, which gradually increased to a high level by the end of May (Ram and Rao, 1981). After dormancy break, levels of inhibitors increased with a minor peak in the third week of August and major peak in the end of September and thereafter decreased to a low level by the end of October without further noticeable changes.

Both indeterminate and determinate shoots grow, while fruits are dormant and their dormancy break corresponds to the cessation of shoot growth. Auxins levels in the fruits increased with the onset of dormancy in April, gradually reaching a high level by the end of May and then decreased in June-July prior to dormancy break. Therefore, changes in auxins levels appear to regulate the dormancy of aonla fruit and that its supra optimal levels arrest the growth. This was further confirmed since externally applied IAA to the shoot tip moved to the fruits (Alleemullah and Ram, 1990). Absence of fruit growth suggests that auxins do not appear to be synthesized in its dormant tissues. After dormancy break, the levels of auxins, gibberellins, cytokinins and inhibitors increased remaining at high levels during the first rapid growth period of the fruit (Ram and Rao, 1978; 1981). The high levels of auxins during this growth phase were not inhibitory, possibly due to its interactions with gibberellins, cytokinins and inhibitors. This growth phase in aonla fruit appears to be controlled by the hormonal substances produced by the developing seed and the fruit. However, the factors regulating the second period of rapid fruit growth are not clearly understood since it did not match with rise in the levels of auxins, gibberellins and cytokinins (Ram and Rao, 1976, 1978 and 1981).

Seasonal Availability of Aonla

In North India, aonla fruits are available from November to February while, in South India from April to October and in Western India, from August to December. Besides popularization of aonla varieties with different maturity groups (early, mid and late maturity) and its expansion at comparatively higher elevation will further extend the availability of fresh aonla fruit in the country as summarized below:

Table 12.6: Availability of Aonla Fruits in India

Sl.No.	Region	Harvesting season
1.	Northern India	November to February
2.	Southern India	April to September
3.	Western India	August to November
4.	Eastern India (Wild aonla)	October to January

Maturity

Aonla fruits are harvested at full maturity. The maturity can be judged by one of the following criteria *i.e.*, specific gravity (1.07 to 1.10), TSS/acid ratio (5 to 6); change in fruit colour (green to shining whitish green or yellowish green); change in seed colour (from creamy white to brown). Apart from these indices, maturity of the commercial cultivars has been worked out *e.g.* Banarasi and Krishna mature 17-18 weeks after fruit set, while Kanchan and Francis take about 20 weeks. Chakaiya matures 23 weeks after fruit set.

Harvesting

The aonla fruits are usually hand picked or plucked with a locally made harvester by climbing the tree with the help of a bamboo or ladder. The fruits should be harvested in the morning hours and be collected in plastic crates. The fruits should not be allowed to fall on the ground as the injured fruits cause spoilage to other healthy fruits during packaging and storage.

Yield

A budded/grafted aonla tree starts bearing from third year after planting whereas, a seedling tree may take 6-8 years. Vegetatively propagated plants attain full bearing within 10-12 years. The yield varies greatly in different aonla cultivars. Banarasi is a shy bearer, Francis and NA-6 are average bearers, while Kanchan and NA-7 are prolific bearer. Number of factors *viz.*, genetic potential, environment and management contribute to the production. An aonla tree produces 100 to 300 kg fruits tree, thus gives the fruit yield of 15-20 t./ha.

Packaging

Poor handling and packaging cause up to 20 per cent losses in aonla fruits (Pathak, 2003). Considerable care should be taken to pack aonla fruits. Gunny bags and baskets are being used for packing of aonla fruits, though they have poor dimensional stability and stacking strength. Approximately 40-45 kg baskets made of pigeon pea stems, are used for packing aonla fruits with news paper as lining and aonla leaves as cushioning material. The Corrugated Fibre Board (CFB) boxes for packing aonla fruits may be used particularly for long distance transport.

Storage

The purpose of storage of aonla fruits is to extend the span of their availability. The mature fruits can be kept at room temperature for about 6 to 9 days depending upon the variety. Shelf life of fruits could be extended up to 12-18 days, if they are stored in low energy cool chambers. Aonla fruits could be stored up to 2 months at low temperature (5-7°C). The fruits may be kept up to 75 days in 15 per cent brine solution at room temperature (Singh and Kumar, 1997).

Processing

Aonla fruits, because of high acidity and astringent taste, are not commonly used for fresh consumption. It is consumed mainly in processed form. The excellent nutritive and therapeutic values of fruit offer great potentiality for processing into several quality products. In general, aonla fruits are utilized for three purposes as indicated below:

As Food Item

Aonla can be made into Ready to Serve (RTS), Nectar, Squash, Syrup, Jam, Preserve, Candy, Pickle, Sauce, Chutney, Dehydrated shreds, Dehydrated aonla segments (Supari), aonla powder and freeze dried powder.

Ayurvedic Medicinal Preparations

Aonla is used as one of the ingradients in several medicinal preparation like Chavanprash, Trifla, Trifla Prash, Trifla Churna, Amalki Grith, Brahed Dhatri, Madhumeh Churn, Brahed Dhatri Ghrit, Bavasir Nashak Mahausidhi.

Cosmetic Preparation

Aonla is also used in a number of cosmetic preparations like hair oil, shampoos and tooth paste.

Export Potential

Aonla has good demand in domestic market. With the changing scenario, there is growing demand for aonla and aonla based herbal/natural products in the international market as well. In Europian

FLOW DIAGRAM OF AONLA USES

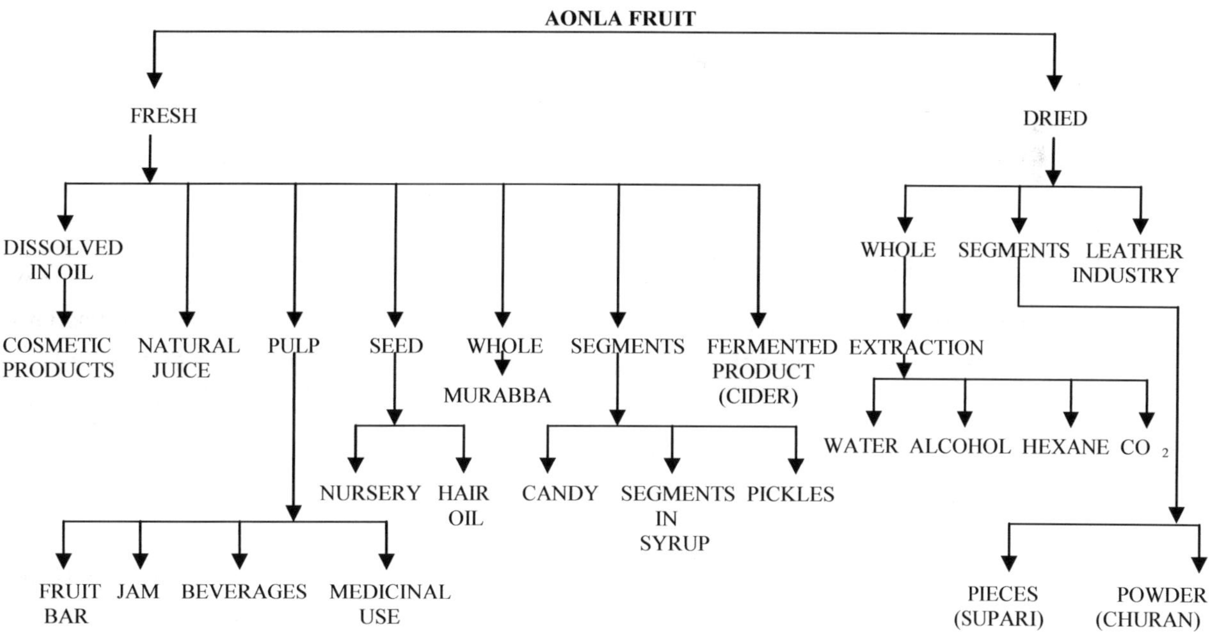

Figure 12.4

countries and USA, aonla fruits are processed into following products and sold in the market (Waizimizin, 2002).

1. Beverages (tea, juice, mixed with aonla).
2. Tonic, tinctures, capsules and tablets.
3. Cosmetics (face packs, anti-aging creams, shampoos and conditioner).

Production Constraints

Self Incompatibility

Most of the aonla cultivars are self incompatible and need pollinizer for proper fruit setting (Singh *et al.,* 1998). Bajpai (1963) and Mohammed and Ram (1990) reported that poor fruit set and retention are due to lack of pollinizer or pollinator or both. Higher fruit set and retention under open and cross-pollination may be due to compatible mating and stimulus provided by pollen. Singh *et al.* (2001) reported that the highest fruit set and retention were recorded in NA7 x NA6 combination followed by NA7 x Krishna among the following six combinations tried. The pollen parent also influences the fruit quality parameters.

Table 12.7: Effect of Different Pollen Parents on Fruit Set and Quality in Aronla

Crosses	*Fruit Set (%)*	*Fruit Retention (%)*	*Fruit Size (cm^3)*	*Fruit Weight (g)*	*TSS (ºB)*	*Ascorbic Acid (mg/100g)*
NA-7 (bagged)	2.35	–	–	–	–	–
NA-7 (open pollination)	61.43	18.54	14.50	45.00	12.00	5.90
NA-7 X Chakaiya	39.98	6.87	14/23	40/00	9.50	5.25
NA-7 X Francis	31.24	2.42	15.01	43.00	8.00	4.90
NA-7 X Krishna	46.86	9.43	15.91	50.00	11.00	5.12
NA7 X Banarasi	22.02	2.77	16.11	52.00	12.50	5.86
NA-7 X NA-6	58.42	10.42	15.00	41.00	10.50	5.68
NA-7 X NA-10	38.69	6.39	16.28	53.00	13.00	6.00
Mean	36.76	6.47	13.55	40.80	9.90	4.96
C.D. at 5 per cent	4.81	2.10	NS	1.03	2.30	0.16

Frost Susceptibility

Frost during winter causes injury to young plants. Sometimes, during December-January when temperature goes below 5°C mortality of young plants, shoots and branches of older plants is observed giving the plant a burnt appearance. The problem of frost damage can be minimized through provision of wind breaks, maintenance of optimum soil moisture, mulching, thatching of young plants, smudging and use of growth retarding chemicals.

Problem of Unfruitfulness

Many factors are responsible for unfruitful condition of plant *i.e.,* plantation of single cultivar or unknown seedling plantation, poor orchard health, lack of pollination, attack of insect/pests and diseases, occurrence of frost and abrupt climatic changes during flowering and fruit set. This problem can be minimized by planting of two cultivars in alternate rows and with provision of suitable pollinizer, manipulations in micro climate with plantation of wind breaks, irrigation and spray of chemicals like calcium carbonate and borax.

Off-season Flowering

In aonla, flower bud differentiation starts in first week of March and flowering takes place from last week of March to mid April. Removal of determinate shoots during July-October induces flowering in Northern India, but does not produce economic yield (Ram, 1971). In South India, aonla trees flower twice *i.e.*, February-March and June-July but later crop gives poor yield (Naik, 1949). Contrary to this, flowering in aonla was observed during October-November particularly in Pratapgarh district of U.P. This 'Off' season flowering has become a major constraint particularly in Chakaiya cultivar. Very a few such flowers bear fruits, which drop shortly due to prevailing low temperature afterwards in Northern India.

Insect Pests and Diseases

Insect Pests

Bark-eating Caterpillar [*Indarbela tetraonis* (Moore)]

The pest is widely distributed throughout the country, infesting a number of fruits, ornamentals and forest trees. This pest is very common in aonla orchards. Generally incidence of this pest is more common in neglected and poorly managed orchards. The infestation of this pest starts in April with the emergence of moths. The larvae of this moth are recorded feeding and boring on the bark of main trunk, stem and branches. In case of severe infestation, growth of the tree, flowering and fruiting are affected.

Symptoms

The attack of this pest may be identified by presence of irregular tunnels and patches covered with silken-web consisting of excreta and chewed up wood particles, on the shoots, branches, and trunk. Shelter holes may be seen particularly at the joints of shoots and branches. The young shoots dry and die, giving sickly look to the tree.

Management

1. Keep the orchards clean and healthy to prevent the infestation of this pest.
2. Detect early infestation by periodically looking out for drying young shoots.
3. Kill the caterpillars mechanically by inserting the iron spike in shelter holes made by these borers at early stage of infestation.
4. The larvae of this pest are parasitized by entomogenous fungus *Beauveria bassiana* in nature. This can be used as a potential bio-control agent.

Shoot Gall Maker [*Betousa stylophora* (Swinhoe)]

The pest attacks nursery plants and old bearing trees. The insect is active from June to December. The pest causes gall formation on stem and shoot. The larvae of this moth tunnel in the apical portion of the shoot and infested portion bulge into gall. When larva is active, reddish frass is extruded through a hole at one end. Fresh galls are generally formed from June to August. The full size galls measure 2.3 to 2.5 cm in length and 1 to 1.5 cm in width. With the beginning of hot weather, larva emerges out of the gall and pupates in a cocoon formed of webbed leaflets. This pest is important as its large-scale infestation may result in stunted growth of the trees affecting flowering and fruiting.

Symptoms

In the beginning of the infestation, terminal shoots swell, which increase in size with the passage of time. Full size galls can be seen in October- November.

Management

1. Over crowding of branches should be discouraged. Galled shoots should be pruned and destroyed along with the pest after harvest.

Pomegranate Butterfly [*Deudorix* (*Virachola*) *isocrates* (Fabr.)]

Deudorix isocrates, an important pest of pomegranate also attacks other fruits including aonla. Attack of this insect occurs during September-October, coinciding well with the fruiting season. The violet brown female butterfly lays shining white eggs, singly on young fruits. The larva of this pest bores the fruit and feeds on seeds, making this portion hollow from inside. The larva is flattened, stout built, covered with short hairs and measures about 2 cm long. After full-growth, larva comes out and pupates outside or remains inside. Infestation of this pest results in fruit loss. The entry and exit points of the larva also pave way for secondary infection by different pathogens.

Symptoms

Affected fruits are generally deformed at the point of entry of larvae. Frass may be seen exuding out of the borer hole. Such fruits weaken, rot and fell down before maturation. In case of severe attack, it may cause considerable loss.

Management

1. Cultivation of pomegranate and guava should be discouraged close to aonla plantation as these are major host plants of this pest.
2. Infested fruits should be identified, collected and destroyed to prevent further spread of infestation.

Mealy Bug [*Nipaecoccus viridis* (Newstead), *N. vestator* (Newstead)]

Incidence of these mealy bugs are noticed from March to July with the peak population in April-May. Among the nymphs, female adults are easily distinguished. Eggs are laid in ovisacs formed by the females. A female bug lays hundreds of eggs. Nymphs get settled on plant parts (leaf, shoot and inflorescence) and suck the sap. They attain maturity within 15 to 20 days. Excessive desaping by nymphs result in drying and dropping of leaves and flowers which affect growth of the tree, flowering and fruiting.

Symptoms

The attacked new shoots bend and twist with yellowing of leaves. In case of plants with severe infestation, twigs become leafless and dry. Excessive excretion of honeydew is noticed. Flowers dry up and drop.

Management

1. Clean cultivation and maintenance of health and vigour of the tree.
2. Prune affected parts and destroy them at early stage of infestation.

Aonla Aphids [*Cerciaphis emblica* (Patel and Kulkarny), *Schoutedonia emblica* (Patel and Kulkarny) and *Setaphis bougainvillae*. (Thunberg)]

The incidence of this pest is mainly seen from July to October with the peak period in September. The new shoots are infested at growing points. The nymphs and adult females suck the sap. Heavy attack of this pest affects the growth and vigour of the tree and ultimately affecting the flowering and fruiting.

Symptoms

The infested leaves turn yellow and dry up. Infested shoots appear bended and twisted at the growing points. Presence of ants also indicates the infestation of aphids.

Management

1. Clipping off and destruction of affected leaf and shoot.

Leaf Rollers [*Garcillaria acidula* and *Tonica* (*Psorosticha*) *ziziphy* Stainton]

The adults of these insects are miniature moths. Larvae of these moths bind the leaves together and feed there in. In case of heavy incidence leaves dry and drop leading to drying of twigs also.

Symptoms

The infestation may be identified by webbing of leaves, their withering and dropping.

Management

1. Avoid overcrowding of branches and maintain sanitation in the orchard.
2. Rolled leaves may be clipped off and destroyed along with the larvae in the beginning of infestation.

Stone Borer [*Curculio* sp.]

This pest is active from June to January. Tiny weevil emerges in June with the onset of rains. Emergence continues in July-August also, coinciding well with the fruiting season of aonla. The eggs are laid by excavating a cavity below the epicarp. Fruits of 1.5 to 2 cm in diameter are preferred for oviposition. Larva after hatching traverses through the mesocarp, reaches to stone, enters inside it and feeds there on seeds, destroying them completely. Full-grown larva comes out of the fruit by cutting small circular hole. Larvae drop down and go into the soil and diapause there till the next season. The infestation of this weevil is observed in Banarsi and Desi varieties of aonla.

Symptoms

Externally, the ovipositional site appears as small brown patch on the fruit. The infested fruits may also be identified by presence of exit hole.

Management

1. Deep ploughing of orchards after harvesting exposes diapausing larvae and is effective in bringing down pest population.

Fruit Midge [*Clinodiplosis* sp.]

Incidence of this pest occurs in the fruiting season in aonla from September to January. Adults are seen on wings in September-October and are miniature flies. The eggs are laid inside the fruits. Larvae after hatching feed on the content of the fruits. Immature larvae are creamish-white, while mature are pinkish-orange. The mature larva is about 3-3.5 mm in length and 1 mm in width. Fully matured larvae jump out of the fruits, drop down in the soil and pupate there. Full life cycle is completed in 20 to 28 days, with the last larval stage diapausing within the soil.

Symptoms

In the beginning, small grey-black spots appear at the site of infestation. It darkens to brown at later stage. Emergence hole is minute but visible easily. The affected parts of the fruit rot because of the damage done by the midge larvae and also fruits become susceptible to secondary infection by different pathogens. Desi varieties of aonla are more susceptible.

Management

1. Deep ploughing of the orchards after harvesting exposes the diapausing larvae and is effective in bringing down pest population.

Diseases

It is a hardy crop but there are a few important diseases, which are very dangerous for aonla cultivation. Rust is the most important disease of aonla. It is a serious problem especially in Rajasthan. It causes fruit and leaf infection and causes significant loss. Other important problem is different types of fruit rots. As fruits are perishable in nature, even slight injury causes storage and transit diseases. Some of the diseases of aonla are described below:

Rust [*Ravenelia emblicae* Styd.]

Rust of anola is an important problem especially in Rajasthan and reported very serious problem in Udaipur district. Recently, it is recorded in desi varieties of aonla in Lucknow and Pratapgarh (U.P.). Fruit and leaf infections are recorded.

Symptoms

On fruits initially a few black pustules appear which later develop in a ring. The pustules join together and cover big area of the fruit. The black spores are exposed after rupturing a papery covering. Fruits give dirty look and lose its market value. On leaves, pinkish brown pustules develop which may be arranged in groups or scattered as isolated pustule. It is considered that infection of fruit does not go on leaves and vice-versa Teleospores of *Ravenelia emblicae* cause the fruit and leaf infection.

Management

1. Three sprays of wettable sulphur (0.4 per cent) at an interval of one month from July are useful in controlling the disease.

Wilt

Recently wilting of aonla plants is recorded from Rajasthan.

Symptoms

Large number of plants in Rajasthan show bark cracking, bark splitting, defoliation and wilting symptoms. The main reason for wilting was attributed to frost injury, but association of *Fusarium sp.* was also found.

Management

1. Young plants should be covered during frost period and sufficient irrigation should be provided which protect plants from frost.
2. Mulching with available organic waste/black polyethene and cow dung pasting on tree trunk are quite effective in minimizing frost injury. As frost injury predisposes plants to fungal infection, proper protection from frost is required.

Sooty Mould

The aonla trees, which are infested by scale insect, are generally affected by sooty mould. In sooty mould infection several fungi are associated.

Symptoms

Sooty mould causes velvety covering of black fungal growth on the surface of leaves, twigs and flowers. These are restricted only to the surface and do not penetrate into leaves.

Management

1. Sooty mould can be managed by spaying starch @2 per cent.

Linchen [*Strigula elegans* (Fee.) Mull. Arg.]

Symptoms

Lichens are found on surface of trunk of grown up trees of aonla. It is seen in the form of whitish, pinkish, superficial patches of different shapes on main trunk and branches of tree.

Management

It can be managed by gunny rubbing, followed by spraying of trunk and branches with commercial caustic soda (1 per cent).

Blue Mould [*Penicillium citrinum*]

Occurrence of blue mould was first reported from Varanasi but it is a disease of very common wherever aonla is grown.

Symptoms

It causes brown patches and water-soaked areas on the fruit surface. As the disease progresses, three different types of colours develop in succession, *i.e.*, first bright yellow, then purple-brown and finally bluish green. There is exudation of drops of yellowish liquid on the fruit surface. The fruits emit a bad odour. The entire fruit finally gives a bluish-green pustulated or beaded appearance.

Management

The rot can be checked by:

1. Careful handling of fruits. Any injury on the fruit surface during harvesting and storage make the aonla fruits prone to blue mould.
2. Sanitary conditions in storage should be maintained and gas treatment with NCl_3 or Ozone can be given.
3. Treatment of fruits with borax or sodium chloride checks the blue mould infection.

4. Treatment with bavistin or topsin (0.1 per cent) control the blue mould.
5. Fruits can be smeared with mentha oil, which checks the rot.

Anthracnose [*Colletotrichum* state of *Glomerella cingulata* (Stoenm.) Spauld and Schrenk.]

Anthracnose on leaves and fruits of anola is common during August- September. It was first reported from Udaipur.

Symptoms

Initial symptom of the disease is in the form of minute, circular, brown to grey spots with yellowish margin on leaflets. The central area of the spot remains grayish raised with dot like fruiting bodies. On fruits, the depressed lesions are formed, which later turn dark in the center forming acervuli often arranged in rings. The lesion may vary in size and shape with spore masses appearing on fruiting bodies at high humidity. Consequently the fruits become shriveled and rot.

Soft Rot [*Phomopsis phyllanthi* Punith.]

The soft rot of aonla is more prevalent during December to February and it was first reported in orchards and markets of Allahabad and Fatehpur.

Symptoms

Smoke brown to black round lesions develop within 2-3 days of infection. The diseased parts later show olive brown discolouration with water soaked areas and cover the whole fruit within 8 days. The shape of the fruit is also deformed. Although fungus causes infection both in young and mature fruits, but mature fruits are more susceptible. Injury is essential for the development of disease.

Management

1. Treatment of fruits with difolatan (0.15 per cent), dithane M-45 and bavistin (0.1 per cent) just after harvest are effective in controlling the rot.

Dry Fruit Rot [*Phoma emblicae* Jamaluddin, Tandon and Tandon]

A post harvest dry rot on stored aonla fruit is caused by a species of *Phoma* and was first reported from Allahabad fruit markets. About 3 to 7 per cent of stored fruits are infected with this pathogen. Some times, freshly harvested fruits are also affected with this organism.

Symptoms

The infected areas are light brown and measure from 1.0 to 2.0 cm in diameter but subsequently they appear depressed and turn completely black. Occasionally pycnidia are observed on infected fruits.

Fruit Rot [*Phoma putaminum* Speg.]

The disease was first reported from Allahabad. It is of common occurrence during December-January.

Symptoms

The rot starts as a small pinkish brown necrotic spot which extends towards both ends of fruits forming eye shaped appearance. In severe cases lesions coalesce forming a bigger pustule. The mature lesions are dark brown in colour and severely infected fruits show wrinkling. The underlying tissues in the rotted fruits become soft. Fruits completely rot within 15 days. Fungus is seen to be present in the dropped fruits of local variety as well as on fruits attached to the lower branches of high yielding varieties *viz.* Chakaiya and Banarasi.

Fruit Rot [*Nigrospora sphaerica* (Sacc.) Massan]

Fruit rot caused by *Nigrospora sphaerica* was first recorded from Kanpur. It is common during the month of December. Loss due to this rot is estimated upto 5-10 per cent.

Dry Fruit Rot [*Cladosporium tenuissimum* Cooke, *C. cladospori-oidies* (Fries) de Vries.]

Dry rot was first reported from Allahabad. Loss due to this disease is estimated 2-5 per cent. Infection due to *C. tenuissimum* is noticed from November to February.

Symptoms

The disease starts as colourless area, slightly soft and subsequently progresses in a circular manner. Light brown mycelial growth of the fungus is evident on infected areas. The diameter of the lesions vary from 1.0 to 1.5 cm. Injury is necessary for manifesting the infection. In *C. cladosporioidies* the rot shows dark brown necrotic lesions. Slight growth of the organism appears in the necrotic cavity. The severity of infection is seen only in matured and ripe fruits but sometimes freshly harvested fruits also show infection.

Fruit Rot [*Pestalotia cruenta* Syd.]

Fruit rot caused by *Pestalotia cruenta* was observed in the local market of Allahabad. It is common during November.

Symptoms

The spots on the fruits are mostly irregular and brown. The disease usually starts as a brownish discolouration on the fruit surface, which develops slowly. Later the spots become mummy brown and skin around them develops light brown. At a relatively later stage the infected region becomes covered with fluffy aerial growth of the fungus. The internal parts of the diseased fruit show a dry dark brown area.

Fruit Rot [*Alternaria alternata* (Fr.) Keissler]

Fruit rot due to *Alternaria alternata* is recorded in dropped fruits of anola from Allahabad.

Symptoms

The rot starts as a small brownish spherical necrotic spot which increases in circular fashion with development of disease. In advance stage, the spots become dark brown to black and neighbouring spots coalesce. The central portion of the infected tissues becomes soft and pulpy.

General Methods of Controlling Fruit Rots

1. Harvesting should be done very carefully to avoid any injury to the fruits.
2. Fruits should be stored in clean containers.
3. Full sanitary measures should be adopted during storage and transit.
4. Sanitary conditions in storage should be maintained and gas treatment with NCl3 or Ozone should be given.
5. Treatment of fruits with borax or sodium chloride checks fruit rots.

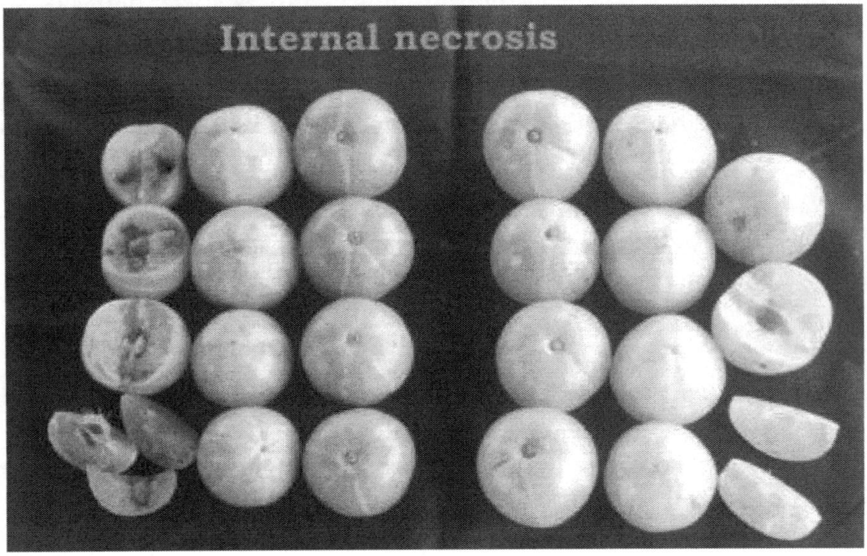

Figure 12.5

Physiological Disorder

Internal Necrosis

A physiological disorder known as internal fruit necrosis is observed in aonla fruits. Cultivar Francis appears to be highly susceptible followed by Banarasi.

Symptoms

The symptom starts with browning of innermost part of mesocarpic tissue at the time of endocarp hardening in the IInd or IIIrd week of September which later extends towards epicarp resulting into brownish black appearance of the flesh. In case of severe incidence, these black spots become corky and gummy pockets develop.

Infection has not been noticed on other aonla cultivars like Chakaiya, NA-6 and NA-7. Hence the cultivation of these varieties needs to be encouraged.

Management

1. Combined spray of zinc sulphate (0.4 per cent)+copper sulphate (0.4 per cent) and borax (0.4 per cent) during September- October is effective.

Future Strategies

1. Conservation of promising genotypes having high medicinal and processing attributes both in-situ and in field gene banks.
2. Breeding for development of frost tolerant, coloured varieties with high female flower production.
3. Standardization of canopy architecture and tree density for increasing productivity.

4. Aonla like many other trees has been in use for centuries in India. It is now important that traditional knowledge as well as new technologies/products of aonla are protected through patents.
5. Promotion for establishment of model nurseries and aonla orchards on problem soils for availability of genuine planting material and proper utilization of wastelands.
6. Encouraging Corporate sector to go for Integrated promotion of aonla plantation, processing and export promotion.
7. Advertising the attributes of aonla to the global community to make them aware for regular consumption of aonla in one or other form of value added products.
8. Scaling of a few-selected aonla products *viz.*, freeze drying of aonla powder, aonla supari, candy etc, which can attract international community.
9. Module for development of commercial processing units of aonla in the concentrated pockets.
10. Possibilities of extraction of alkaloid and formulation of antioxidant tablets from aonla powder.
11. Evaluating the properties of aonla based Ayurvedic preparations and their scaling.
12. Possibility of preparation of tooth paste and tooth powder from aonla.
13. Popularization of aonla based leather-tanning industry.

References

Alleemullah, M. and Ram, S. 1990. Causses of low fruit set and heavy fruit drop in Indian gooseberry (*Emblica officinalis* Gaertn). *Indian J. Hort.* 47: 270-277.

Amal, E.K.J. and Raghwan, R.S. 1957. Physiology and vitamin C in *Emblica officinalis. Proc. Indian Acad Sci.*, 47: 312-314.

Amin, R.S. 1978. Softwood grafting: A new technique for hard wood plants. *Current Sci.*, 47: 468-469.

Anonymous, 1999. Annual Report, *National Bureau of Plant Genetic Resources*, New Delhi, India.

Anonymous, 2002. Annual Report of Central Institute for Sub-tropical Horticulture, Lucknow, India.

Bajpai, P.N. 1957. Blossom biology and fruit set in *Phyllanthus emblica* L. *Indian J. Hort.* 14: 99-102.

Bajpai, P.N. 1963. Studies on bearing habit of aonla (*Emblica officianalis* Gaertn) syn. *Phyllanthus emblica* L. *Doctoral Thesis*, Agra University.Agra, India.

Bajpai, P.N. 1965. Studies on vegetative growth and development of male and female gametophyte in aonla (*Emblica officianalis* Gaertn). *Agra Univ. J. Res.* (*Sci*) 14: 167-86.

Bajpai, P.N. 1968. Studies on flowering and fruit development in aonla (*Emblica officinalis* Gaertn). *Hort. Adv.*, VII: 38-67.

Bajpai, P.N. 1969. Seasonal variation in vitamin C content of aonla fruit. *Hort. Adv.* VIII: 15-16.

Bajpai, P.N. and Shukla, H.S. 1985. Aonla, pp. 591-600. In: *Fruits of India: Tropical and Subtropical.* Naya Prokash, Calcutta, India.

Balasubramaniam, S. and Arulmozhiyan, R. 2003. Open pollination: A method for higher production in aonla (*Emblica officinalis* Gaertn) under rainfed vertisol. Paper presented in *Nat. Seminar on Production and Utilization of Aonla*, organized by All India Aonla Grower's Association, Salem, Tamil Nadu, India.

Barrett, M.F. 1956. Common Exotic Trees of South Florida. University of Florida Press, Gainesville, Florida.

Benthal, A.P. 1946. Trees of Calcutta and its Neighbourhood. Thacker Spink and Co. Ltd., Calcutta.

Chaundawat, B.S. 1990. *Arid Fruit Culture*, Oxford and IBH Publishing Co. Pvt. Ltd., New Delhi. Pp. 110-18.

Dhamkhar, D.S., P.M. Shah and Joshi, K.I. 1997. Seed germination and seedling growth in aonla (*Phyllanthus emblica* L.) as influenced by gibberellic acid and thiourea. *Journal of Applied Horticulture*, 3(1/2): 93-97.

Gopalan, I., and Mohanram, M. 1996. Fruit National Institute of Nutrition. ICMR, Hyderabad, Pp. 3-4.

Hooper, D. and Field, H. 1937. Useful plants and Drug, of Iran and Iraq. *Publ. 387. Bot Ser.* Vol 1, No.3. Field Museum of Natural History, Chicago.

Kadam, J.H., A.R. Karel, U.T. Desai and Chandan, P.M. 2003. In vivo micrografting- A new technique for vegetative propagation in aonla (*Emblica officinalis*). Paper presented in *Nat. Seminar on production and utilization of aonla*, organized by All India Aonla Growers Association, Salem, Tamilnadu, India.

Kalra, C.L. 1988. The chemistry and technology of amla (*Phyllanthus emblica* L.). A resume. *Indian Food Packer* 42(4): 67-82.

Karale, A.R. 2000. Improvement of aonla, wood apple and bael. p. 134. (In). *Improvement of Fruit Crops through Breeding* (Ed. More, T.A., Karale, A. R., Supe, V.S. and Desale, S.B.), Centre of Advance Studies in Horticulture (Fruit), Department of Horticulture, MPKV, Rahuri.

Mac Millan, H.F. 1943. Tropical planting and gardening with special reference to Ceylon (5th ed.). Mac Millan and Company.

Mehta, S.S. 2000. Introduction of commercial amla growing in South India. Paper presented in the Brainstorming session on the exploitation of the potential of arid fruit in Tamil Nadu, organised by Tamilnadu Fruit Grower Association of Tamil Nadu at Dindigul, Nov., pp. 51-55.

Mehta, S.S. 2001. A case study on development of aonla in Tamil Nadu. Proc. National Horticultural Conference held at New Delhi Nov.2001 (Ed. H.P. Singh, J.P. Negi, and J.C. Samuel. Pp 160-63.

Mishra, D., Lal, B., Pandey, D., Verma, A. and Pathak, R.K. 2007. Improving productivity of mango and aonla by top working. *Indian Horticulture*, 52(4): 32-33.

Mishra, M., Saxena, R.P., Srivastava, A.K. and Pathak, R.K. 1998. Effect of antioxidants. Phenol binding agents and explant waxing on in vitro browning of aonla (*E. officinalis* Gaertn). *Prog. Hort.*, 30(3-4): 128-134.

Mishra, M., Saxena, R.P., Pathak, R.K. and Srivastava, A.K. 1999. Studies on micropropagation of aonla. *Prog. Hort.*, 31(3-4): 116-122.

Mohammad, A. and Ram, S. 1990. Cause of low fruit set and heavy fruit drop in Indian gooseberry. *Indian J. Hort.*, 47(3): 270-277.

More, T.A., Karale, A.R., Supe, V.S. and Desale, S.B. 2000. Improvement in aonla through selection. Centre of Advance Studies in Horticulture.

Naik, K.C. 1949. *South Indian Fruits and their Culture*. P. Varadachary and Co., Madras.

Nath, V., Pathak, R.K., Singh, H.K. and Dwivedi, R. 2000. Farmers felt constraints and research need in aonla production. *Prog. Hort.*, 32(1): 10-19.

Nizamuddin, M., Hoffman, J. and Larm, O. 1982. Fractionation and characterization of carbohydrates from *Emblica officinalis* Gaertn. fruit. *Swedish J. Agric Res.*, 12: (1) 3-7.

Ojha, C.M. and Pathak, R.K. 1993. Changes in chemical composition of aonla cultivars during growth and development of fruits. *Prog. Hort.*, 25(1-2): 42-46.

Pandey, D., Pathak, R.K. and Tandon, D.K. 2005. Lakshmi-52: a new aonla selection. *Indian Horticulture*, 50(1): 20.

Pandey, D., Shukla, S.K., Nagar, A.K. and Yadav, R. C. 2006. Collection of aonla (*Emblica officinalis* Gaertn.) germplasm from Panna forest area of Madhya Pradesh. *National Symposium on Underutilized Horticultural Crops*, held at IIHR, Bangalore, 8-9th June, Abstract, pp. 37.

Pareek, O.P. 1998. Germplasm survey, collection and evaluation, Annual Report, AICRP on Arid Zone Fruits, Pp. 13-21 NRC for Arid Horticulture, Bikaner, India.

Pareek, O.P. 2001. Improvement of aonla and custard apple. Lecture delivered in winter school on Exploitation of underutilized fruits held at RCA, Udaipur, Rajasthan pp. 116-118.

Parmar, C. 2000. Amla and its wild Himalayan strain, New Crop Fact sheet. wysiwyg: www.hortpurd.amla/html.

Pathak, R.K. 1990. Salt tolerance studies in fruit crops. Paper presented at *XXIII Int. Hort. Cong.* held at Florence, Italy.

Pathak, R.K. 2003. Status report on genetic resources of Indian gooseberry–Aonla (*Emblica officinalis* Gaertn.) in south and south east Asia. *IPGRI*, New Delhi.

Pathak, R.K. 2003. Twenty five years of research and development of aonla. Lead paper presented in *National Seminar on Production and Utilization in Aonla* during August 2003 at Salem, Tamil Nadu, organized by Aonla Grower Association of India.

Pathak, R.K. and Srivastava, A.K. 1994. Flowering and fruit set studies in Indian gooseberry. *Proc. Ind. Hort. Cong.* Koyoto, Japan, p 90.

Pathak, R.K. and Pandey, S.D. 1985. Problems and Possibilities of aonla (*Emblica officinalis* Gaertn.) cultivation in wasteland. Abs. Pp 110. Paper presented in *National Symposium in plantation opportunities in India*, held at New Delhi.

Pathak, R.K. and Pathak, S. 2001. Fruit production in problematic soil. *Indian J. Hort.* 58 (122): 16-22.

Pathak, R.K., A.K. Srivastava, R. Dwivedi and H.K. Singh. 1993. Aonla Descriptor Res. Bull. Directorate of Res., NDUA&T, Faizabad.

Pathak, R.K., Pandey, D., Haseeb, M. and Tandon, D.K. 2003. The Aonla. Bull. CISH, Lucknow, India.

Pathak, R.K., Pandey, D., Singh, G. and Mishra, D. 2003. Approaches and strategies for precision farming in aonla. Pp. 176-190. In: *Proc. National Seminar-cum-workshop on Hi-Tech Horticulture and Precision Farming* (Ed) H.P. Singh, G. Singh, J.C. Singh, Samuel and R.K. Pathak). CISH, Lucknow.

Pathak, R.K., Singh, S. and Saroj, P.L. 2006. Aonla (*Emblica officinalis* Gaertn.), pp. 1-20. (In) *Advances in Arid Horticulture* (Ed) Saroj, P.L. and Awasthi, O.P. International Book Distributing Company, Lucknow.

Rajamanickam, C., Anbu, S., Balakrishnan, K. and Rajangum, J. 2002. Studies on propagation techniques in aonla. *South Indian Horticulture*, 50(1/3): 220-222.

Rajan, S. and Ram, S. 1982. A new approach towards vegetative propagation of aonla through cutting. *Progressive Hort.*, 14: 190-191.

Rajput V., Dhake, A.V. and Balasubramanyam, V.R. 2000. Aonla in marginal soils-yield rich returns. *Plant Hortitech.*, 1(5): 49-52.

Ram, S. 1971. Studies on the physiology of fruit growth in aonla (*Emblica officinalis* Garetn). *Ph.D. Thesis*, Kanpur University, Kanpur, India

Ram, S. 1982. Induction of off-season flowering in aonla (*Emblica officinalis* Gaertn) tree. Prog. Hort. 14(2-3): 180-183.

Ram, S. 1993. Aonla (*Emblica officinalis* Garetn) uses, botany and cultivation, *Res. Bull. 110*. Directorate of Experiment Station, G. B. Pantnagar. Pp. 1-26.

Ram, S. and Rao, T.R. 1976. Naturally occurring cytokinins in aonla (*Emblica officinalis* Gaertn) fruit. *New Phytol.*, 76: 441-448.

Ram, S. and Rao, T.R. 1978. Studies on naturally occurring gibberellins in aonla (*Emblica officinalis* Gaertn) fruit. *New Phytol.* 81: 513-519.

Ram, S. and Rao, T.R. 1981. Naturally occurring auxins and inhibitors and their role in fruit growth of aonla (*Emblica officinalis* Gaertn) fruit. *New Phytol.* 88: 53-66.

Rao, V.K. and Pathak, R.K. 1996. Studies on the effects of mulches in aonla orchards established on sodic land. I. Studies on soil properties. Indian J. Hort. 53(4): 251-354.

Rao, V.K. and Pathak, R.K. 1998. Effects of mulches on aonla (*Emblica officinalis* Gaertn) orchard in sodic soil. *Indian J. Hort.*, 55(1): 27-32.

Rawat, R.B.S. and Uniyal, R.C. 2003. Amla: An important species for Indian system of medicine. Paper presented in *National Seminar on Production and Utilization in Aonla* during August 2003 at Salem, Tamilnadu, organized by Aonla Grower Association of India.

Saroj, P.L., V. Nath and Vashistha, B.B. 2001. Effect of polycontainers on germination, seedling vigour, root charactes and budding success in aonla. *Indian J. Hort.*, 57(4): 300-304.

Sehgal, C.B. and Khurana, S. 1985. Morphogensis and plant regeneration from cultural endosperm of *Emblica officinalis* Gaertn. *Plant Cell Rep.* 45: 263-66.

Shankar, G. 1969. Aonla for your daily requirement of vitamin C. *Indian Hort.*, 13: 9-11-15.

Sharma, A.K., Sharma, K. and Baduala, S.D., 2000. Effect of transplanting time of rootstock and budding techniques on rapid propagation of aonla (*Emblica officinalis* Gaertn.). *Indian Journal of Soil and Conservation*, 28(3): 235-237.

Shukla, A.K., Shukla, A.K. and Vashishtha, B.B., 2004. Fruit breeding approaches and achievements. International Book Distributing Co., Lucknow.

Shukla, A.K. and Solanki, K.R., 2000. Plant aonla in situ in rakar soils. *Indian Hort.*, Jan-March, p. 12-13.

Singh, A.K., Singh, H.K., Verma, O. P. and Singh, K. 2002. Seed longevity in aonla (*Emblica officinalis* Gaertn.). *The Hort. J.*, 15(1): 33-39.

Singh, H.K. and Pandey, C.S. 2003. Effect of transplanting treatments and time of transplanting on few rooted budding of aonla (*Emblica officinalis* Gaertn.). *Haryana J. Hort. Sci.*, 32(3/4): 176-179.

Singh, H.K., Srivastava, A.K. and Dwivedi, R. 2001. Effect of pollinizer on fruit set and fruit quality of Narendra Aonla (*Emblica officinalis* Gaertn). *Ind. J. Agri,. Sci.*, 71(1): 65-6.

Singh, H.K., Srivastava, A.K. and Dwivedi, R. 1998. Pollination and fruit set behaviour of aonla (*Emblica officinalis* Gaertn). *Ind. J. Agric. Sci.*, 68(4): 204-5.

Singh, I.S., Pathak, R.K., Singh, H.K. and Dwivedi, R. 1993. Production and Post-harvest Technology of Aonla, Res. Bull. Directorate of Res. NDUA and T, Kumarganj, Faizabad.

Singh, R. and Kumar, S. 1997. Effect of different storage conditions on the shelf life of aonla (*Emblica offiinalis* G.) cv.Chakaiya. *Haryana J. Hort. Sci.*, 26(1-2) : 12-15.

Sturrock, D. 1959. Fruits of southern Florida. South Eastern Printing Co. Stuart, Florida, USA.

Tiwari, R.K. and Bajpai, C.K. 2002a. Propagation of aonla (*Emblica officinalis* Gaertn.) through grafting in polybags. *Indian J. Agric. Sci.*, 72(6): 353-354.

Tiwari, R.K. and Bajpai, C.K. 2002b. Studies on quality parameters of scion and rootstock on success of grafting in aonla. *Indian Journal of Agroforestry*, 4(2): 138-140.

Tiwari, R.K., Bajpai, C.K. and Dwivedi, R.P. 2005. Bench grafting in aonla (*Emblica officinalis* Gaertn.). *Indian J. Agric. Sci.*, 75(5): 296-297.

Tiwari, R.K., Bajpai, C.K. and Kumar, S. 2004. Standardization of bench grafting in aonla (*Emblica officinalis* Gaertn.). *Ramge Management and Agroforestry*, 25(2): 120-123.

Wagh, A.P., Chaudhary, M.F., Jadhaw, B.J., Damke, M.M. and Joshi, P.S. 2001. Studies on propagation in aonla (*Emblica officinalis* Gaertn.). *Journal of Soils and Crops*, 11(1): 74-77.

Webster, G.L. 1957. A monographic study of West Indian species of Phyllanthus. *J. Arnold Arbor.* XXXVIII (2): 57-79.

Williams, R.O. and Williams, R.O. Jr. 1951. Useful and ornamental plants of Trinidad and Tobago. 4th edn. Gaurdian Commercial Printery. Port of Spain, Trinidad.

Biodiversity in Horticultural Crops Vol. 2 *Pages* ***284–301***
Editor: **Professor K.V. Peter**
Published by: **DAYA PUBLISHING HOUSE**

Chapter 13

Orchids of Medicinal Value

***Ajay Kr. Singh*[1] *and T.S. Mehra*[2]**
[1]*Assistant Professor (Hort), KVK, Jakhdhar, Rudraprayag (GBPUA&T, Pantnagar) Uttarakhand, India*
[2]*Assistant Professor and Head Department of Forestry, DIBNS Manduwala, Dehradun, Uttarakhand, India*

Use of plants for treating ailments is as old as human civilization itself. Man has always relied on his surrounding vegetation for disease amelioration and through gradual use of several plant species. Many such species have stood the test of time for their therapeutic properties under various systems of medicine. Although traditional system of medicine in India has been using medicinal plants since Vedic period, but only a few species have been domesticated till date, and larger part remaining underutilized. Orchids represent the most diverse group among these underutilized plants. They are epiphytic, terrestrial and saprophytic with specific niche requirement. This exacting nature of the species is the main hurdle in domestication and cultivation of these species.

Orchids belong to one of the most highly evolved family of the entire plant kingdom-Orchidaceae. The family comprises over 30,000 individual species, and new discoveries being made and described every year. About 1220 species are reported from India, while 750 species are found in North East India and about 550 species (45 per cent of the country) are reported from Arunachal Pradesh alone (Rao,2004). Many of these orchids are rare, endangered and highly ornamental with long lasting flower qualities. Today economic importance of orchids is mainly because of their ornamental value. Around 150 species of orchids are ornamentally and commercially important. The rich diversity of Indian orchids includes many species with proven floricultural and/or herbal traits and their utility is being increasingly realized in the Ayurvedic system of medicine. About 40 orchid species are known to be used in various Ayurvedic formulations such as Ashtavarga, Jivinyavarga, Madhuravarga and Chayvanprash. Orchids also are of ethnobotanical interest linking aboriginal man plants for medicine. There are number of orchids which have been used for medicine alone or in combination with other plants through out the world. However, with the advent of modern civilization, wealth of knowledge available amongst the ethnic communities and traditional practitioner are losing ground. Numbers of medicinal orchids have become rare in their habitat because of unscientific harvesting and overexploitation from their natural habitats. It is therefore essential to conserve and propagate these plants and allow the ethnobotanical knowledge to perpetuate by systemically inventorying, studying and refining their use on modern scientific lines for the benefit of the mankind.

History Account

The word 'Orchid' originated from the Greek word *'Orchis'* which literally means 'Testicle'. It was Theophrastus who first used the Greek word *'Orchis'* to indicate the particular group of plants whose dried and chopped roots were used in the traditional pharmacopoeia of Greece and neighbouring Asia minor as antidepressants and stimulants, and even today in some rural areas of these countries "*Salep*", a nutritious drink prepared from the dried tubers of certain orchids, is used commonly. It was not until the mid eighteenth century that the name was applied to the whole family of orchids in Linnaeus's Species plantarum, which marked the beginning of modern plant taxonomy, where Linnaeus described orchids as two semi-globular underground tubers, placed side-by-side, resembling human testicles. Besides Theophrastus and Linnaeus, many other people were interested in orchids principally as medicinal plants, and this is testified by references in the medical books and herbals of men such as Dioscorides (first century A.D.), Otto Brunfels of Strasbourg (C.1488-1534), Leonhard Fuchs (1501-66) and John Gerard (1542-1612). As early as 200 BC the Chinese pharmacopoeia–"the Sang Nung Pen Tsao Ching", mentions *Dendrobium* as a source of tonic, astringent, analgesic, anti-inflammatory substances. In Europe, in the medieval belief of the doctrine of signatures, tubers of some of these plants were thought to confer virility and potency. Utility of orchids in Indian indigenous medicine system is known from Vedic period (Handa, 1986; Kaushik, 1985, 1988 and Kirtikar and Basu, 1935).

Orchids have tremendous ornamental value since ancient times. Perusal of historical records revealed that some of them also bear exotic names such as *Sita-Pushpa* and *Draupadi-Pushpa*. Legend goes around that Sita and Draupadi wore these orchids for ornamentation. Besides ornamental value, orchids are long known for their use as a source of medicine. Many such species are also mentioned in ancient Indian literature for their curative and aphrodisiac properties, and as symbols of sanctity (Hegde, 1984; Kaushik, 1985; Manilal and Sathish Kumar, 1986). In Ayurvedic system of medicines, a group of 8 herbs, *Ashtavarga*, consists of four orchid species namely Jeevak (*Microstylis wallichii*), Rishbhak (*Microstylis muscifera*), Ridhi (*Habenaria intermedia*) and Vridhi (*Habenaria edgeworthii*) forms the basis of a number of rejuvenating formulations and drugs (Abraham and Vatsala, 1981; Kaushik, 1985; Handa, 1986; Rao, 1979 and Uniyal, 1975). Other orchids commonly used as medicine in the Ayurvedic system are *Salem* (*Eulophia campestris*), *Jiwanti* (*Dendrobium alpestre, Eria muscicola, Pholidota* and *Ephemerantha spp*), *Rasna* (*Acampe papillosa, Vanda cristata*), *Salam Panja* (*Dactylorhiza hatagirea*) and *Shwethuli* (*Zeuxine strateumatica*) (Handa, 1986 and Kaushik, 1988; Chauhan, 1990).The antiviral and anticancerous properties of *Vanda parviflora* are positively tested (Rastogi and Dhawan, 1990) and the leaves of *Cymbidium giganteum* help blood clotting. Some orchids are also fed to milch cattle in North-Eastern India with the belief that *Dendrobs* enhance their milk yield and the *Cymbidium* improves their health (Pampahishey, 1974).

The correct identity of different drugs used in Ayurveda has long been debated and discussed by different authors. Handa (1986) has referred *Microstylis wallichii* (*Malaxis wallichii* or *Malaxis acuminata*) as *Jeevak* whereas the same is referred as Rishbak (Uniyal, 1975). Handa (1986) also referred *Habenaria acuminata* as *Kakoli*, whereas, in Ayurveda *Roscoea procera* is considered as Kakoli. Similar confusion is also there in respect to the identity of *Ridhi* and *Vridhi*. Handa (1986) mentioned *Habenaria intermedia* as *Ridhi* and *Vridhi*, but in Ayurveda these come from two separate species. *H. intermedia* D.Don and *H. edgeworthii* Hook. f., respectively (Dey, 1980). The correct identity of different drugs used in Ayurveda is discussed by the following authors–Chunekar, 1969; Nadkarni, 1972 and Uniyal, 1975; Honda, 1986; Chauhan, 1990.

In this chapter the Indian Orchid species that are being used in different indigenous systems of medicine are briefly described with respect to their geographical distribution, botanical features, chemistry and uses by following the work of Hooker, 1884; Collett, 1902; Kirtikar and Basu, 1935; Chopra *et al.*, 1958; Uniyal, and Issar, 1969 and Uniyal, 1975, 1989. Deva and Naithani, 1986; Handa, 1986; Kaushik and Dhiman, 2000; Kapoor, 2001.

Acampe papillosa (Lindl.) Lindl.

Epiphyte, often form large clumps on trees in tropical zone. Genus is named so because of its brittle flowers (Derived from Greek word *akampes* meaning rigid). Distributed in Eastern Himalayas and North West India, at altitudes below 1000 m, plants grow up to 90cm tall. Leaves fleshy obliquely notched at the apex. Inflorescence very short, 4cm long to the maximum, rarely branched, densely many flowered. Flower up to 1.8cm across, fragrant and long lasting. Blooms in August-September. Sepals and petals are yellow with brown blotches and three lobed lips are whitish.

Use

Roots are used in rheumatism. Commonly known as Rasna.

Acampe praemorsa (Roxb.) Blatt.

Epiphyte with woody stem, up to 60 cm tall. Distributed in Southern India, West Bengal, Orissa, Chota Nagpur, Konkan and Western Ghats. Inflorescence about 6 cm long. Flowers in May-August and about 1cm across. Petals and sepals are creamy yellow with brown stripes. Lip is also creamy yellow with red stripes.

Use

Known as 'Taliyamravada'. Roots are used in rheumatism.

Bulbophyllum neilgherrense Wight.

Epiphytic, distributed in Malabar and Nilgiri hills in Southern India. The genus name is derived from Greek *bulbos* meaning bulb and *phyllon* meaning leaf, in reference to the leafy pseudobulbs in most of the species. Slightly compressed, smooth and ovoid conical pseudobulbs are placed distantly on a creeping rhizome. Leaves 10-15cm long, 3.5-4cm broad, oblong-obtuse and fleshy. Inflorescence up to 18cm long, many flowered. Flowers about 1cm across. Petals and sepals brownish yellow to purplish green. Lip greenish brown to purple and has entire or toothed auricles.

Use

Pseudobulbs are used as tonic for restoration of adolescence.

Coelogyne corymbosa Lindl.

The genus name is derived from Greek words *Koilos* (hollow) and *gyne* (female), most likely in reference to the deeply set stigmatic cavity found in the genus. Epiphytic with creeping habit. Distributed in North Eastern Himalayas at altitudes 1600-2500m amsl, Pseudobulbs clustered, leaves 2, lance shaped, leathery; inflorescence 2-4 flowered, loosely arranged; flowers sweet scented, about 7cm across, white with 4 broad yellow eyes bordered in orange red on the lip.

Use

Used in burn. Juice from the fresh pseudobulbs is applied on the wounds to relieve pain.

Coelogyne punctulata Lindl.

Genus name is derived from Greek *koilos* (hollow) and *gyne* (Female) suggestive to the deeply excavated stigma. Epiphytic, distributed in Meghalaya at altitudes 1000-2000m amsl Pseudobulbs turbinately ovoid, two leaved. Flowers in March-April.

Use

Used in burn injuries and for healing of wounds. Pseudobulbs are dried and powdered and applied externally

Calanthe triplicata (Willem) Ames

Genus name is derived from *Kalos* meaning beautiful and *anthe* meaning flower, in allusion to strikingly beautiful flowers in most of the species. This is the type species of the genus. Terrestrial grows up to 1m long. Leaves large, plicate, strongly ribbed, up to 60cm long and 10-12cm wide. Inflorescence erect, up to 1m long, densely-many flowered towards the apex. Flowers in May-August, white, up to 5cm across. Lip three lobed, side lobes spreading, mid lobe deeply divided into two segments. Distributed in North-Eastern Himalayas at altitudes 1000-2000m amsl. Grows on moist, humus and shaded forest floor.

Use

Extract of roots is used in treating diarrhoea and tooth decay.

Cymbidium aloifolium (L.) Sw.

Epiphytic, distributed in North Eastern Himalayas, Assam, West Bengal and Andaman. Pseudobulbs short and stout. Leaves thick, upto 50cm long and 3cm broad. Raceme long, pendulous, many flowered. Flowers 4-5cm across, sepals and petals are yellow, having broad medial purple stripe. Lip is purple with darker lines of the same colour and a whitish patch at the base of the apical lobe. Flowers in April-May.

Use

Salep prepared from the rhizomes is used as nutrient and demulcent. Rhizomes are also used as emetic and purgative.

Cymbidium ensifolium (L.) Sw.

The genus name is derived from Greek word *Kymbos* meaning boat or punt and probably refers to petals and sepals of Cymbidium flower, which are boat shaped or to lip which curls up like the prow of a boat. Epiphytic, distributed in Sikkim, Meghalaya and Arunachal Pradesh at altitudes 700-1200m amsl. Leaves sword shaped, up to 90cm long. Flowers in May-June. Sepals and petals green, streaked with red. Lip glabrous, white, spotted with brown red.

Use

Decoction of rhizome is used in gonorrhea. Decoction of flowers is also used in eyesores.

***Cypripedium elegans* Reichb.f.**

Epiphytic or lithophytic. Distributed in North Eastern Himalayas at altitudes 1200-2500m. amsl. Pseudobulbs ovoid, bilaterally flattened. Leaves narrow up to 70cm long and unequally bilobed at the apex. Inflorescence pendulous, closely many flowered. Flowers bell shaped, straw yellow or cream yellow. Flowers in September-December.

Use

Used in nervous irritability, hysteria, spasms, fits and in derangements of functioning of brain such as madness, because of its sedative and antispasmodic properties. It is also used in rheumatism. Plant part used is root.

Cymbidium longifolium D. Don

Epiphytic. Distributed in the Himalayas at altitudes 1500-2200m amsl. Leaves linear, channeled below. Inflorescence sub erect, shorter than leaves. Flowers in September-October. Sepals and petals greenish, streaked with purple. Lip is white with purple spots.

Use

Taken as emetic. Powder from the dried pseudobulbs is taken as aquatic solution in empty stomach. Salep from the plant is taken as demulcent.

Dactylorhiza hatagirea (D.Don)Soo

The genus name is derived from Greek *dactylos* (finger) and *rhiza* (root), in allusion to the digitate tuberoids of the genus. Terrestrial, distributed in Western and North East Himalayas. Tubers palmate. Inflorescence densely many flowered, cylindrical. Flowers purplish-lilac or pinkish, rarely white. Flowers in June-September. Grows at altitudes 2500-3500m, amsl in moist and humus conditions.

Use

Known as *Salam Panja*. Used as expectorant, aphrodisiac and astringent. Paste prepared from the bulb is applied externally in bone fractures, severe cuts and wounds. Same is taken orally as cure for cough. Extract from the tuber is used to relieve hoarseness. Salep made from tuber is also used in seminal debility, diarrhea, dysentery, and fever for its nutritive value.

Dendrobium crumenatum Sw.

The genus name is derived from *dendros* meaning tree and *bios* meaning life, in allusion to aerial epiphytic existence of most of the species. Epiphytic, distributed in North East India and Andaman Islands. Stem up to 100cm long, branched. Flowers, on the leafless ends of the branches, large, white, sepals and petals nerved. Lip has a primrose disc and pink veins.

Use

Used in skin eruptions. Leaves of the plant are used as poultice in boil and pimples.

Dendrobium monticola P. F. Hunt and Summerh

Epiphytic, distributed in North Eastern Himalayas at altitudes 500-1000m. amsl. Grows on tree trunks in primary forests. Pseudobulbous, up to 15cm long, 2-3 leaved. Inflorescence leaf opposed.

Raceme erect, 3-7 flowered. Flowers greenish or whitish, with purple lines on the lip. Lip triangular at the apex. Flowers in April-June.

Use

Commonly known as *Jiwanti,* applied as poultice in boils, pimples and other skin eruptions.

Dendrobium nobile Lindl.

Probably the most commonly cultivated *Dendrobium* species world wide. State flower of Sikkim. Epiphytic, distributed in North Eastern India, prefers sunny condition, up to 1500m. Stems up to 50cm long, erect, compressed and becomes ground with age. Inflorescence from each node of stem. Flowers very showy, very variable. Petals and sepals white at the base and gradually rosy or mauve towards tip. Lip rich maroon at base, then whitish or yellowish with mauve or purplish margin. Flowers in March-May.

Use

Used for speedy healing of wounds. Seeds of the species are applied externally on the fresh wounds.

Dendrobium ovatum (L.) Kraenzl.

Epiphytic, distributed in Western Ghats and Karnataka, up to 1700m. amsl. Pseudobulbs up to 50cm long, leafless at the time of flowering. Inflorescence in racemes, from the apical part of the stem, about 15cm long, many flowered. Flowers cream coloured, lip greenish. Flowering in September-January.

Use

Extract taken from entire plant is taken in stomach ache. It helps in improving bile secretion. Plant extract is also used as laxative.

Desmotrichum fimbriatum Bl. Bijdr

Distributed throughout the Western Ghats, parts of Sikkim and Assam. Rhizome creeping annulate stems 60-90cm long. Leaf terminal, solitary, 10-20cm long, linear, oblong, obtuse with numerous parallel slender veins. Flowers remaining open for a few hours only, 1-3 from near the base of a leaf, 2-2.5cm across, white. Sepals oblong-lanceolate, subacute, spreading. Petals narrower than the sepals, linear-oblong, obtuse.

Use

The fruit is sweet and aphrodisiac. The plant is stimulant and tonic.

Epipactis helleborine (L.) Crantz.

Genus name derived from ancient Greek *epipaktis* meaning a plant that curdled milk. This is the type species. Terrestrial, distributed in Himalayas at altitudes 2000-3500m. amsl. Characterized by fibrous roots, leafy stem and terminal raceme of drooping flowers. Flowers greenish, variously marked with yellow, white or purple. Flowers in July.

Use

Roots are used for treating insanity.

Eria pannea Lindl.

Genus name is derived from Greek *erion* meaning wool in reference to the wooly indumentum of the perianth. Epiphytic, distributed in North Eastern Himalayas at 500-1000m. amsl. Usually grows on tree trunks in dense tropical primary forests. Rhizomatous, 20-25cm high, creepy, wooly, leaves terete. Flowers sweet scented, sepals and petals covered with wool from outside, inner surface golden brown, lip dark brown.

Use

Decoction prepared from roots and leaves is used in boneache. Also used for ague, paste of whole plant is mixed in bathing water.

Eria spicata (D.Don) Hand.

Epiphytic, distributed in North Eastern Himalayas at 500-1300m amsl. Stems very stout, up to 20cm long. Inflorescence one sided, densely many flowered in an ovoid spike. Flowers white to straw coloured, side of lip streaked with yellow. Flowering in November to February.

Use

Used in headache and stomach ailments. Paste prepared from shoot is applied externally on forehead for relief from headache, same is taken orally in stomach ailments.

Eulophia dabia (Don) Hochr.

Genus name derived from Greek *eu* (Well, true) and *lophos* (plume) in reference to the crests on the lip. Terrestrial, distributed in lower altitudes (200-300m amsl) in Andamans, North Bengal and Deccan. Pseudobulbs irregular, deformed. Leaves two, shed before flowering. Inflorescence about 50cm long, many flowered. Flowers pale pink, has darker linings. Flowers in March-May.

Use

Commonly known as *Salem* or *Amrits*. Used as aphrodisiac. Pills are prepared from bulb paste. Usually two pills in a glass full of milk are consumed about three hours prior to bedtime. It is also used as appetizer, in cough, in stomatis and heart trouble. In the Yunani system, it is also used as astringent.

Eulophia compestris Wall.

The species is distributed in sub-Himalayan tract of India Nepal. Tubers irregularly oblong, often lobed. Leaves two, 25-40cm long, linear, acuminate, and plicate. Flowering scape 15-30cm. long, sheathed at interval by loose membranous bracts. Flowers dropping, subsecund, yellowish or green with pink or purple markings, about 2.5cm across. Sepals slightly attached to the base of the lip, linear-lanceolate, acute or acuminate, 5-7 nerved. Petals spreading, narrower than the sepals, oblanceolate, 3-5 nerved. Ovary slender and stalked. Fruit capsule 2 cm long ellipsoid.

Use

Tuber is an appetizer; stomachic, tonic, aphrodisiac, alterative and blood purifier.

Eulophia epidendraea C.E.C. Fischer

Genus name derived from Greek *eu* (True) and *Lophos* (plume), descriptive of beautiful crest on the lip. Terrestrial, distributed in West Bengal and South India. Pseudobulbs large, leaves grass-like with a stout mid-rib. Flower scape up to a meter long. Sepals and petals green with reddish nerves, lip white with red crested nerves.

Use

Tubers are used as vermifuge.

Eulophia herbacea Lindl.

Terrestrial, distributed in eastern Himalayas, Orissa, West Bengal and Karnataka. Tuberous, stem up to 20cm high, slender and sheathed. Leaves many nerved. Scape up to 1m long, raceme short. Flowers 7-10, white, nerved in purple. Flowering in June-July.

Use

Tubers are used as vermifuge

Eulophia nuda Lindl.

Terrestrial, distributed in Sikkim, Meghalaya, Manipur, Assam, Orissa, Chota Nagpur and Vizagapatnam hill, in the tropical valleys at 500-1500m amsl. Tubers spherical. Inflorescence up to 60cm tall, un-branched, 7-8 flowered. Sepals, petals pale green, sepals having brown markings on the outer surface. Lip pale green at base, white in the middle and flushed with pink at the apex. Flowering in April-July.

Use

Commonly known as *Manya* or *Goruma*. Tubers are used for treating tumors, in scrofulous glands of the neck and in bronchitis. Also used as blood purifier, appetizer and vermifuge.

Eulophia ramentacea Lindl.ex Wight

Terrestrial, distributed in Western Ghats. Tuberous, leaves elliptic-lanceolate. Scape up to 90 cm long. Flowers yellow, about 1.5cm across, sepals 5 nerved petals subsimilar. Lip large, rounded.

Use

Tubers are used in scrofulous glands of the neck.

Flickingeria macraei (Lindl) Seidenf

Genus name is after Edward A. Flickinger. Epiphytic, distributed in North East Himalayas, Nilgiri hills at altitudes 1500-2500m amsl. Characterized by creeping and rooting rhizomes, single flowered inflorescence on top of pseudobulbs. Flowers whitish, lip is peculiarly shaped with two long wavy keels, apical lobe pentagonal and side lobes running backward. Flowers in April-May.

Use

Commonly known as *Jeevanti*. Pseudobulbs are used as aphrodisiac, astringent and expectorant. Also used as cure for asthma, bronchitis and night blindness.

Gymnadenia orchidis Lindl.

Genus name derived from Greek *gymnos* (naked) and *aden* (gland) in reference to the naked viscidium in the genus. Terrestrial, distributed in Sikkim and Kashmir in the temperate zone at altitudes around 3000m amsl. Characterized by parallel and contiguous bases of anther loculi and plicature of rostellum. Grows up to 50cm tall. Digitately tuberous. Scape narrowly-cylindrical, densely many flowered. Flowers 5-6cm across, pink with violet tinge. Flowers in August.

Use

Used as adaptogen. Roots are directly chewed for remedy from gastric, liver and urine disorder. Also taken in diarrhoea, dysentery and fever.

Habenaria commelinifolia Wall.

Genus name derived from Greek *habena* (reins) in allusion to the long, strap-like division of the petals and lip. Terrestrial, distributed in Kumaon, Central India, Chota Nagpur, Orissa and Karnataka. Plants up to 1m tall. Inflorescence 20-25cm long, many flowered. Flowers white, 1.5-2cm across. Flowering in August-September.

Use

Used as a source of Salep.

Habenaria edgeworthii Hook.f.ex Collett.

Terrestrial, distributed in Western Himalayas in the temperate zone at altitudes 2500-3000m amsl. Plant grows up to 90cm tall. Tubers white in colour, oblong, in pairs, 2cm long and 1cm in diameter. Inflorescence up to 30cm long, many flowered. Flowers greenish yellow, about 1.5cm across, lip is longer than lateral sepals, spur is upwardly curved. Flowering in July-August (Figure 13.1).

Use

Commonly known as *Vridhi*. One of the constituents of *Ashtavarga* in Ayurvedic system, which is used for preparation of a number of rejuvenating tonics, such as *Chyavanprash*. Acts as blood purifier.

Habenaria intermedia D.Don

Terrestrial, distributed in Eastern and North Eastern Himalayas at altitudes 2000-2300m amsl. Tuberous, roots white, oblong or elliptic ovoid. Stem teret, 20-25cm tall, inflorescence 4-6 flowered. Flowers greenish white, about 5cm across. Petals prominently 5 nerved. Flowering in July-August (Figure 13.2).

Use

Known as *Ridhi*. It is also one of the ingredients in *Ashtavaraga*. Acts as rejuvenator and expectorant.

Liparis rostrata Reichb.f.

Genus name derived from Greek *liparos* (Shiny or greasy), in reference to the glossy leaves. Minutely pseudobulbous, stem 2-7cm, leaves 2, opposite. Distributed in temperate Himalayas.

Use

Tubers are used in stomach problem.

Luisia trichorrhiza (Hook).Blume

Genus name is in honour of Don Luis de Torres, a Spanish botanist of 19th century. Commonly known as Bee orchids. Characterized by terete stems and leaves. Epiphytic, distributed in Sikkim, Garhwal (Uttarakhand), Orissa and Chota Nagpur (MS), at lower elevations. Stem grows up to 25cm, very stout. Flowers pale green with pale purple lining, greenish at the ridges of apical lobe. Flowers in March-April.

Use

Used in muscular pain. Ointment prepared from tuber is applied externally.

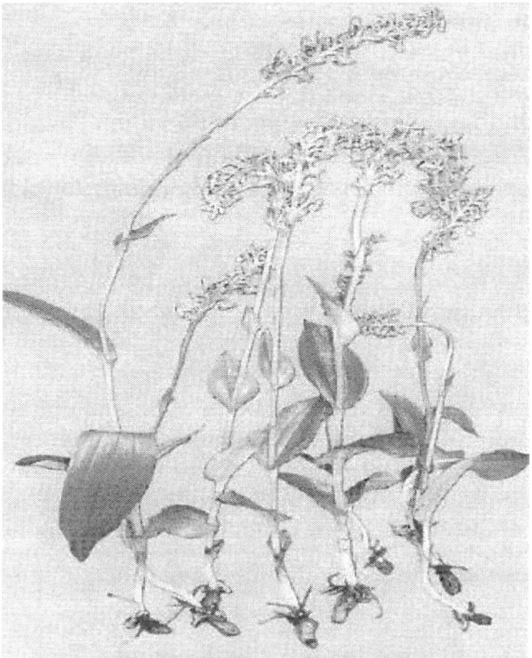

Figure 13.1: *Habenaria edgeworthii* **Hook.f.ex (Vridhi)**

Figure 13.2: *Habenaria intermedia* **D.Don (Ridhi)**

Figure 13.3: *Microstylis wallichii* **Lindl. (Jeevak)**

Figure 13.4: *Microstylis muscifera* **Lindl. (Rishbhak)**

Figure 13.1–13.4: Medicinal Orchids of 'Ashtavarga' (8 Wonder Drugs of Vedic Period)

Malaxis acuminata D.Don/*Microstylis wallichii* Lindl.

Genus name is derived from the Greek *malaxis* (softening), in allusion to the tenderness of the leaves. Terrestrial, distributed in Himachal to North Eastern Himalayas, Western Ghats and Nilgiri hills at altitudes around 1850 to 2300 m amsl. Stems up to 10cm, sheathed below. Leaves often shaded in purple with undulated margins. Inflorescence 25-30cm long, ribbed, laxly many flowered. Flowers yellowish green in colour. Lip with indistinct anterior lobe. Flowers in June-September (Figure 13.3).

Use

Known as *Jeevak*. One of the constituents of *Ashtavarga*. Besides being used as tonic, also used in treatment of tuberculosis and to increase the sperm production. Paste prepared from tuber is applied on abscess and sores as poultice.

Malaxis cylindrostachya (Lindl.) O.Ktze

Terrestrial, distributed in Eastern Himalayas at altitudes 2500-3500m amsl. Plants differ from *M. muscifera* (Lindl.)O.Ktze in having single leaf and slender stem and scape. Stem up to 15cm long. Inflorescence about 15cm, densely many flowered. Flowers pale yellow to pale green. Flowering in July-August.

Use

Used as physical tonic under the name *Salep*.

Malaxis muscifera (Lindl.) O.Ktze/*Microstylis muscifera* Lindl

Terrestrial, distributed in Himalayas in temperate zone, from Kashmir to Sikkim at altitudes up to 3200m amsl. Pseudobulbs fleshy, two leaved. Plants grow up to 45cm tall. Inflorescence 20-25cm, raceme laxly flowered. Flowers yellowish green, about 0.5cm across. Flowering in July-September.(Figure 13.4)

Use

Known as *Rishbhak*. One of the drugs in *Ashtavarga*. Used as rejuvenator of tissues.

Oberonia caulescens Lindl.

Genus name is after the mythical king of the Fairies: Oberon. Epiphytic, distributed in NorthEast Himalayas at altitudes 1000-2000m amsl. Plants up to 60cm tall. Leaves fleshy, 4-6 in number. Inflorescence slender. Flowers pale red brown, minute. Flowering in July-August.

Use

Used for curing liver ailments. Paste from tuber mixed with Black Pepper powder (usually in 3:1 ratio) is dissolved in a glass full of cold water. The mixture is consumed in empty stomach.

Pecteilis susannae (L.) Rafin

Genus name derived from the Greek *pectein* (comb), in reference to the pectinate side lobes of the lip. Terrestrial, distributed in Himalayas from Garhwal to Arunachal Pradesh, Deccan peninsula and Kerala. Plants up to 80cm tall. Inflorescence 4-8cm long, a few flowered. Flowers large, about 10cm across, white, showy, fragrant during night. Flowers in August.

Use

Tubers are used in treatment of boil on palm.

Pholidota articulata Lindl.

Genus name derived from the Greek *pholidotos* meaning scaly, probably in reference to the large inflorescence bracts. Epiphytic, distributed in Western and North Eastern Himalayas at altitudes 600-1000m amsl. Stems branching, jointed, leaves in pair. Inflorescence up to 15 cm long densely flowered, drooping. Flowers off white or yellowish white, bracts yellowish with green shades. Flowering in July-August.

Use

Pseudobulbs are used as tonic.

Pholidota imbricata (Roxb.) Lindl.

Type species of the genus. Epiphytic, distributed in Andaman, Orissa, Chota Nagpur and North Eastern Himalayas. Pseudobulbs furrowed, tufted, bearing single leaf. Inflorescence, drooping, up to 25cm long, densely many flowered. Flowers pale white, have brownish tinge floral bracts pale brown with dark brown dots. Flowering in June-August.

Use

Used in abdominal pain and rheumatism. Juice extracted by crushing the bulbs is taken twice a day until cure of the ailment.

Pleione maculata (Lindl.) Lindl. and Paxt.

Epiphytic, distributed in North Eastern Himalayas at altitudes 1000-1700m amsl. Pseudobulbs, clustered, conical or shortly cylindrical, 2 leaved. Leaves deciduous. Inflorescence 5-6cm long, from the base of leafless pseudobulbs. Flowers very showy, 6cm across. Sepals and petals uniformly white or creamy white. Lip purple lined at the base and side lobes, mid lobe white and having purple blotches in the margin.

Use

Used in stomach ache and liver problems.

Rhynchostylis retusa (L.) Bl.

Genus name is derived from Greek *rhynchos* meaning beak and *stylis* meaning column, in reference to beaked column in *R. retusa* Bl., the type species. Epiphytic, distributed in Western and North Eastern Himalayas at lower elevations, West Bengal, Tripura, Assam Orissa, Western Ghats, Malabar Coast and Chota Nagpur. Commonly known as 'Fox Tail Orchids' due to their brush like flower spike. Highly variable, stem very stout, leaves fleshy, deeply channeled. Inflorescence drooping, compactly many flowered. Sepals and petals white spotted with pink, lip white at the base and purple at tip. Flowering in April-June. A pure white form is also there.

Use

Commonly known as *Banda* or *Rasna*. *Rasnaditaralsar* used in treatment of rheumatism is prepared from it. Fresh plants are used as emollient.

Saccolabium papillosum Lindl.

The species is distributed in foot hills of North East Himalayas and parts of West Bengal. Stem 60-90 cm stout, erect, clustered, bearing many nearly horizontal leaves. Leaves coriaceous, 7.5-10 cm.

long, apex truncate or deeply notched. Inflorescence recemose, 4-8 flowered. Flowers 8mm in diameter, bracts triangular, sepals and petals yellow, blotched with brown. Sepals subequal, oblong, sub acute, spreading. Petals narrower than sepals, sub-spathulate. Anthers broadly conical. Fruit capsule 3.2 cm long, fusiform and rigid.

Use

Used against rheumatism

Satyrium nepalense D.Don

Genus name derived from the Greek *Satyrion*. Satyri were the lascivious demigods in Greek mythology. The name is so applied because of the presumed aphrodisiac properties of the tubers. Terrestrial, distributed in Himalayas at altitudes 2200–3000 m amsl. Plant robust, erect, 15-50cm tall. Inflorescence spicate, up to 15cm tall, densely many flowered. Flowers pink or white, fragrant. Flowering in September-October.

Use

Known as *Salam Misri*. Dried tubers are used as tonic. Also for treating malaria and dysentery.

Spiranthes sinensis (Pers.) Ames

Genus name is derived from the Greek *speira* meaning coil and *anthe* meaning flower, in allusion to the spiral inflorescence. Terrestrial distributed in Himalayas and Nilgiri hills, usually in grassy lands at altitudes up to 2000m amsl. Plants upto 50cm tall, have cylindrical hairy roots. Inflorescence spiral, densely many flowered. Flowers white or pink with white lip. Flowering in April-May.

Use

Paste prepared from shoots and bulbs is applied on sores as poultice.

Tropidia curculigoides Lindl.

Genus name is probably in allusion to the boat shaped lip in many species of the genus. Terrestrial, distributed in North Eastern Himalayas at lower elevations. Plants up to 60cm tall, stem erect, slender. Inflorescence few flowered, terminal or leaf opposite. Flowers 1cm across, white, lightly tinged with yellow. Flowering in October-November.

Use

Used in diarrhea. Decoction of root is taken directly.

Vanda tessellata (Roxb.)Hook.f.ex Don

Genus name is derived from the Sanskrit word referring this species. Epiphytic, widely distributed in India. Plants climbing, leaves narrow, tridentate at apex. Inflorescence sub erect, 8-10 flowered. Flowers 5-6cm across. Sepals-petals yellowish blue or green, tessellated with brown. Flowering in April-June.

Use

Known as *'Bandanika'*. Roots of the species are used in dyspepsia, bronchitis, rheumatism and fever. Juice extracted from leaf is used as ear drop in otitis.

Table 13.1: Lesser Known Orchids of Medicinal Value

Sl.No.	*Botanical Name*	*Parts Used*	*Therapeutic Uses*
1.	*Aerides odorata* Lour.	Whole plant	As anti tuberculosis, for healing wounds. Juice of leaves to heal boils in ears and nose.
2.	*Aerides crispum* Lindl.	Whole Plant	As oil for Earache
3.	*Ansellia humilis*	Root	As antidote to bed dreams
4.	*Anthogonium gracile* Lindl.	Root	As gummy substance in medicines
5.	*Calanthe sylvatica*	Flowers	Flowers used to stop nose bleeding
6.	*Cephalanthera longifolia*	Rhizome	Roots and Rhizome as a tonic
7.	*Chusua roborowskyi* (Maxim.) Hunt	Pseudobulb	In rheumatism
8.	*Cleisostoma williamsonii* (Reichb.F.) Garay	Whole plant	As an astringent. In bone fracture
9.	*Coelogyne ovalis* Lindl.	Whole plant	As aphrodisiac, used for cough, urinary infections and eye disorders.
10.	*Cremastra appendiculata*	Root	Roots paste used for toothache and as emollient and as an antidote to snakebite.
11.	*Cypripedium maculosum*	Whole plant	To enhance longevity
12.	*Cypripedium pubscenes* Wild	Whole plant	In nervous disorder
13.	*Dendrobium densiflorum* Wall.	Stem, Bark	Leaves crushed to paste with salt and applied on fractured area to set bone.
14.	*Epipactis helleborine* (L) *Crantz.*	Root	In insanity
15.	*Goodyera oblongifolia*	Whole plant	In gout
16.	*Goodyera pubescens*	Root	Antidote for mad dog bite
17.	*Goodyera repens* (L.)R.Br.	Whole plant	In snake bites and eye disorder, stomach
18.	*Habenaria clavigera* (Lindl.)	Whole plant	As anti AIDS agent
19.	*Habenaria crinifera* Lindl.	Whole plant	In headache
20.	*Habenaria latilabris*	Root, tuber	As blood purifier and rejuvenating tonic
21.	*Habenaria longicorniculata*	Whole plant	In internal pain and swelling
22.	*Luisia tenuifolia* Bl.	Rhizomes	As emollient, in boils, abscesses and tumours
23.	*Microstylis cylindrostachya*	Tuber	As tonic
24.	*Peristylus stocksii* (Hook. f.) Kranz	Whole plant	In debility
25.	*Phaius tancarvilleae*	Pseudobulb	Paste used to heal swellings of hands and legs, poultice to soothe pain of abscess.
26.	*Pholidota chinensis*	Pseudobulbs	Aqueous extract for scrofula, feverish stomachache, and toothache
27.	*Pholidota imbricata* Hook.	Pseudobulb	In abdominal pain and rheumatism
28.	*Pholidota pallida*	Pseudobulb	In rheumatism
29.	*Pleione maculata* (Lindl.) Lindl. and Paxt.	Pseudobulb	In stomachache and liver ailments
30.	*Vanda roxburghii* Br.	Leaves	In rheumatism, disorder of nervous system
31.	*Vanda teres* Lindl.	Whole plant	In general debility

Table is prepared from following : Bhattacharjee,1998; Chandra,1989; Chopra *et al.*,1956,1969; Chauhan, 1984,1990; Chunekar,1969; Dey,1980; Handa,1986; Hegde,1984,1988; Jain *et. al.*, 2006; Kapoor, 2001; Kaushik,1985,1988; Mukesh Kumar *et al.*,1987; Naithani,1984, 1985; Rao, 2004. Samant *et al*, 1998; Uniyal, 1975, 1989; and Uniyal and Issar, 1969.

Vanda cristata Lindl.

Epiphytic, distributed in Kumaon Hills, Assam and North Eastern Himalayas at altitudes 1000-2000m amsl. Stem twisted in appearance, leaves narrowly oblong, bilobed at apex, horizontally arranged. Inflorescence axillary, 3-6 flowered. Flowers variable, yellowish to olive green, 5-6cm across. Flowering in May-September.

Use

Known as *Rasna*. Used as expectorant.

Vanda spathulata Spreng.

Epiphytic, distributed in Kerala. Stem up to 1cm, leaves flat, about 10cm long and 3cm wide. Leaves and scapes often marked with blood red spots. Inflorescence up to 45cm long, few to many flowered. Flowers golden yellow, about 3cm across. Flowering almost throughout the year.

Use

Known as *Ponnamponmariva*. Used in asthma and mania, and also to temper bile and abate frenzy.

Vanda testacea (Lindl.) Reichb.f.

Epiphytic, distributed in Western Himalayas, North Eastern Himalayas, Assam, Kerala and Karnataka. Stems thick, sheathed, 15-20cm long. Inflorescence erect, axillary from existing or fallen leaves, few flowered. Flower about 1.5cm across. Flowering in May-July.

Use

Known as *Rasna*. Used in rheumatism, nervous problems, scorpion stings, in cuts, wounds and in ear-ache.

Zeuxine strateumatica Lindl.

Naturally distributed throughout the outer Himalayan range of India. Stem 5-25cm long. Leaves 2.5-5.0cm. long, linear, acuminate, claspine, sessile on the hyaline truncate sheaths, gradually passing into linear bracts upwards. Inflorescence densely flowered recemose1.3-5.0cm long, bracts 8mm long. Sepals unequal, greenish white. Petals greenish white, slightly shorter than the dorsal sepal, 1.6mm broad, obtuse. Anthers depressed.

Use

Tubers are used as salep and as a tonic in combination with the roots of *Cymbidium aloifolium*.

Chemical Properties

Importance of orchids as a source of medicine is mainly due to presence of secondary metabolites like alkaloids, glycosides, steroids, flavanoids etc. Many orchid species have potential chemical properties and there is need for chemical evaluation for whether they can be exploited in the form of drugs or not. Bose and Bhattacharjee (1980) gave a detailed interesting account of the uses of orchids in small scale village industry, their uses for medicinal, food and family planning purposes, existence of glycosides and alkaloids in plants, and also the religious influence it has over the nature inhabitants in some parts of the world.

Several heterocyclic compounds, phenolic, esters, natural fluorenone and benzyle derivatives were also reported from the extracts of Indian orchids. Some of these compounds like Batasasin-III found common in many species were reported in 26 species of nine genera of four subtribes namely Coelogyninae, Dendrobinae, Bulbophyllenae etc. (Veerraju *et al.*, 1989). The first natural fluorenone derivatives, namely dengibsin and dengibsinin were isolated from the chloroform extracts of *Dendrobium gibsonii* (Talapatra *et al.*, 1985). Friedalin, sitosterol and new compound cymbinodin A- a phenanthroquinone were isolated from *Cymbidium aloifoilium* (Barua *et al.*, 1990). Majumdar and Maity (1988) isolated a new phenolic compound, flaccidin, a 9, 10-dihydrophenanthropyrone derivatives from the extracts of *Coelogyne flaccida*. Two phenanthropyrone derivatives, designated as isoflaccidin and iso-oxoflaccidin, were also isolated from the methanolic extracts of *Coelogyne flaccida* (Majumdar and Maity, 1991). A new steroidal compound of biogenetic importance was isolated from the plant of *Coelogyne uniflora* (Majumdar and Pal., 1990) and two bibenzyl derivatives cumulatin and tristin were also isolated from *Dendrobium cumulatum* and *Bulbophyllum trist*, respectively (Majumdar and Pal., 1993). Roots of *V. tessellata* contain an alkyl perulate. The dried whole herb also contains long chain alkanes and alkanol sistosterol, resin, saponin, tannins, fatty acids, colouring agents, etc (Das *et al*, 1967; Prasad *et al.*, 1968; Chawla *et al.*,1992). Lawler (1984) reported that several Ayurvedic type preparations containing this plant (root or whole plant) were used as aphrodisiac and given for impotence and barrenness.

References

Abraham, A. and Vatsala, P., 1981. *Introduction to Orchids.* Tropical Botanical Garden and Research Institute, Trivandrum.

Barua, A.K., Ghosh, B.B., Roy, S. and Patra, A., 1990. Cymbinodin-A, a phenanthroquinone from *Cymbidium aloifolium. Phytochemistry*, 29(9): 3046-47.

Bhattacharjee, S.K., 1998. *Handbook of Medicinal Plants.* Pointer Publisher, Jaipur. India.

Bose, T.K. and Bhattacharjee, S.K., 1980. Orchids of India. Noya Prokash, Calcutta.

Chandra, V., 1989. Medicinal plants used by the Tribal of Arunachal Pradesh. A preliminary study. *J. Econ. Tax. Bot.*13 (2): 391-394.

Chauhan, N.S., 1984. Medicinal wealth of Pabbar Valley in Himachal Pradesh, P.G. Thesis. HPKV, Palampur.

Chauhan, N.S., 1990. Medicinal Orchids of Himachal Pradesh. *J. Orchids Soc. India* 4: 99-105.

Chawla, A.S., Sharma, A.K., Handa, S.S. and Dhar, K.L., 1992. Chemical studies and anti inflammatory activity of *Vanda roxburghii. Indian J. Pharmaceutical Sci*, 54: 159-61.

Chopra, R.N., Chopra, I.C., Handa, K.L. and Kapoor, L.D., 1958. *Indigenous drugs of India.* Academic Publisher, Calcutta.

Chopra, R.N.; Chopra, I.C. and Verma, B.S., 1969. *Supplement to the Glossary of Indian Medicinal Plants*, Publication and Information Directorate, New Delhi.

Chopra, R.N., Nayar, S.L. and Chopra, I.C. 1956. *Glossary of Indian Medicinal Plants*. C.S.I.R, New Delhi.

Chunekar, K.C., 1969. *Vanaspatika Anusandhan Darshika.* Vidya Bhawan, Varanasi.

Collett, Henery., 1902. *Flora Simelensis.* Reprint 1971. Bishen Singh Mahendra Pal Singh, Dehradun.

Das, S., Bhattacharya, A. and Bhattacharya, A.K., 1967. Active constituents of *Vanda roxburghii. J Indian Chem Soc*, 44: 804-5.

Deva, Som and H.B.Naithani., 1986. *The Orchid flora of North-West Himalaya.* Print and Media Associates, New Delhi.

Dey, A.C., 1980. *Indian Medicinal Plants- used in Ayurvedic preparation.* Bishen Singh. Mahendra Pal Singh. Dehradun. India. pp 13-15.

Handa, S.S., 1986. Orchids for drugs and chemicals. In: *Biology, Conservation, and Culture of Orchids* (Vij, S.P., Ed.). Affiliated East-West Press, New Delhi.

Hegde, S.N., 1984. *Orchids of Arunachal Pradesh*, Forest Department, Arunachal Pradesh, Itanager.

Hegde, S.N. and Ingalhalli, R.S., 1988. A note on the medicinal usage of some orchids, *Arunachal Pradesh Forest News*, 6(1): 11-18.

Hooker, J.D., 1884. *Flora of British India*, 7th Vols. London. Reprinted by Bishen Singh Mahendra Pal Singh, Dehradun. (Reprint, 1978).

Issar, R.K., 1981.Traditionally important medicinal plants and folklore of Uttarakhand Himalaya for animal treatment. *J. Scientific Res. Plants and Medicine*, 2: 61-66.

Jain, J.B., Kumana, Sheetal, C. and Bhattacharjee, B., 2006. Medicinal flora of Madhya Pradesh and Chattishgarh: A Review. *Indian Journal of Traditional Knowledge.* 5(2): 227-242.

Kapoor, L.D., 2001. *CRC Handbook of Ayurvedic Medicinal Plants*. CRC Press, Boca Raton.

Kaushik, P., 1985. *Glimpses of medical botany in Atharveda* (*Kand IV*). The Vedic Path, 48: 64-67.

Kaushik, P., 1988. *Indigenous Medicinal Plants.* Today and Tomorrow's Printers and Publishers. New Delhi. pp157-199.

Kaushik, P. and Dhiman, A. K., 2000. *Medicinal plants and raw drugs of India.* Bishen Singh Mahendra Pal Singh, Dehradun.

Kirtikar, K.R. and Basu, B.D., 1935. *Indian Medicinal Plants,* 2nd Ed. (Reprint Ed.1976.) Bishen Singh Mahendra Pal Singh, Dehradun.

Lawler, L.J., 1984. Ethnobotany of the Orchidaceace. *In*: Arditti, J., Ed, *Orchid Biology: Review and Perspectives*-3. Ithaca, Cornell University Press, 27-149.

Majumdar, P.L. and Maity, D.C.1988. Flaccidin, a 9, 10-dihydrophenanthropyrone derivatives from the orchid *Coelogyne flaccida*. *Phytochemistry.* 27(3): 899-901.

Majumdar, P.L. and Maity, D.C.1991. Isoflaccidin and iso-oxoflaccidin, stilbenoids from *Coelogyne flaccida*. *Phytochemistry.* 30(3): 971-974.

Majumdar, P.L. and Pal, S. 1990. A steroidal ester from *Coelogyne uniflora. Phytochemistry*; 29(8): 2717-20.

Majumdar, P.L., and Pal, S. 1993. Cumulatin and tristin, two bibenzyl derivatives from the orchids *Dendrobium cumulatin* and bulbophyllum triste. *Phytochemistry.* 32: 1561-6.

Manilal, K.S. and Satish Kumar, C. 1986. *Researches in Indian Orchids-* A review in biology, Conservation and Culture of orchids (Vij, S.P.Ed.)Affiliated East West Press, New Delhi.

Muktesh Kumar, Sasidharan, N. and Renuka, C. 1987.Medicinal orchids of Kerala forests. *Indian J Forestry*. 10: 216-9.

Nadkarni, K. M. 1972. *The Indian Materia Medica*, Revised and enlarged by Nadkarni, A.K., 2 Vols.1954.

Naithani, B.D., 1984 and 1985. *Flora of Chamoli.* Vols I and II. Botanical Survey of India. Howrah. New Delhi.

Pempahishey, K.T. 1974. Orchid eaters of 'Shangri-La'. *Amer. Orchid Soc. Bull.*, 43: 716-25.

Prasad, D.N, Satyawati, G.V., Das Gupta, B.and Das, P.K., 1968. Anti-inflammatory activity of the steroidal fraction obtained from *Vanda roxburghii.* Abstract of paper. 1st Congress.

Rao, A.S., 1979. *Orchids of India*. National Book Trust, New Delhi.

Rao, A.N., 2004. Medicinal Orchid Wealth of Arunachal Pradesh. *ENVIS Newsletter* on ENVIS NODE on Indian medicinal plants. Bangalore, 1 (2): 1-5. April-June.

Rastogi, R.P. and Dhawan, B.N., 1990. Anticancer and antiviral activities in Indian medicinal plants: a review. *Drug Dev. Res.*, 30: 215-20.

Samant, S.S., Dhar, U. and Palni,L.M.S.,1998. *Medicinal plants of Indian Himalaya. Diversity, Distribution Potential Values.* Gyanodaya Prakhashan, Nainital.G.B.Pant Institute of Himalayan Environment and Development, Koshi- Katarmal, Almora.India.

Talapatra, S.K., Bose, S., Malik, A.K. and Talapatra, B. 1985. On the chemistry of Indian Orchidaceae plants-II. Dengibsin and dengibsinin, the first natural fluorenone derivatives from *Dendrobium gibsonii* Lindl. *Tetrahedron*, 41(13): 2765-69.

Uniyal, M.R., 1975. *Astavarga.* In "Sandigdha Vanaushadhi Anka of Dhanwantri Patrika". Shri Jwala Ayurved Bhawan, Aligarh.

Uniyal, M.R., 1989. *Medicinal flora of Garhwal Himalaya*, Shri Baidyanath Ayurved Bhawan Pvt. Ltd., Nagpur.

Uniyal, M.R and Issar, R.K., 1969. Commercially and traditionally important medicinal plants in Mandakini valley of Uttarakhand Himalayas. *J. Res. Indian Med.*, 4(1): 83.

Veerraju, P., Rao, Rao, N.S.P. and Rao, L. J., 1989. Amoenumin, a 9, 10-dihydro-5H phenanthro-(4, 5-b, c, d)-pyran from *Dendrobium amoenum. Phytochemistry*, 28(3): 950-51.

Part IV

Biodiversity of Ornamentals

Biodiversity in Horticultural Crops Vol. 2 *Pages* ***305–320***
Editor: **Professor K.V. Peter**
Published by: **DAYA PUBLISHING HOUSE**

Chapter 14

Plumeria: Temple Trees

Jacob Varghese
Department of Botany, S.H.College, Thevara, Cochin – 13

Plumeria, a member of Apocynaceae flowering plant family, is native to tropical and sub-tropical America. The ornamental varieties of this tree are grown in tropics in gardens and parks and for landscaping around the world. This tree is a common feature of the Hawaii islands, that many people misconstrue to be indigenous to Hawaii. Known as Frangipani or Temple tree or 'Champa'; it is frequently associated with temples and graveyards. Buddhists, Muslims and Christians alike consider the tree as a symbol of immortality because of its unusual power of producing leaves and flowers even after it has been uprooted from the soil. *Plumeria* is conceivably the loveliest of all tropical American ornamental trees grown in the tropics.

Plant Characteristics

A large deciduous shrub or small tree, *Plumeria* grows to a maximum height of 10 m. It's broad and usually round–headed canopy spreads almost equal to the height of the tree itself. The bole is short and the branching starts close to the ground level with an artistic pattern. The succulent, thick branches fork in a regular manner, and copious latex oozes out when they are cut. The branches are round and with blunt ends and unappealing when the tree is bare and leafless. Leaves are simple with alternate phyllotaxy and are deciduous during the winter season. They are seen crowded in spirals near the end of the branches. Flowers bloom in upright clusters in terminal cymes at the tips of branches. The species and hybrids vary in tree size, compactness, branching character, leaf size, deciduous nature, flower size and colour etc. Based on plant habit and leaf morphology, the ornamental varieties belong to the following 4 groups:

Plumeria rubra

Red; pink and red tinctured flowers; some having a yellow patch at the mouth region are seen among its cultivars. The species is native to Mexico, Central America and Venezuela. Most of the cultivated varieties like *Butter cup, Gerald, P S P Boagar, Pink Selection, Purple Star, Slaughter Pink, and Golden Twist* all belong to *P. rubra* category.

Plumeria alba

White, white and yellow tinctured and deep yellow flowers. It is the national flower of Nicaragua and Laos. The leaves of *P. alba* are quite narrow and corrugated, unlike other species. *Paul Weissich, Curcusiana, Pudica* (with saddle shaped leaves) are two important varieties of this group.

Plumeria obtusa

White and yellow tinctured flowers; leaves deep green colored with leaf blade having a round tip. The ornamental varieties in *P. obtusa* species are short–statured with profuse branching, non–deciduous leaves which are well suited to be grown in grass lawns and meadows and also widely used in landscaping purpose. *Pink Bud, Wind Meal, Stenopetella and White Singapore* are all varieties of this group.

Plumeria acuminata

White and yellow tinctured flowers; leaves medium green colored with leaf tip having a pointed tip. The wild *Plumeria* trees seen across India belong to this species.

Other than the above mentioned groups, dwarf varieties are gaining importance as garden plants. These varieties, with their short stature (maximum height of 1.7m), with smaller leaves and flowers and profuse branching, are suited to be grown in large pots also. Dwarf varieties with their umbrella like plant habit, requires no pruning in any stage of growth. The flower shades range from pure white to pink. *Singapore Beauty, Singapore Yellow, Indonesian White, White Dwarf, Mini Wind Meal* are the chief members of this group. *'Bali curly white and yellow'* is a newly introduced cultivar from Indonesia with flowers having double petals.

Economic Value

The flowers are fragrant all through the day and night especially at night when they bloom. The fragrance is wafted to a great distance when the tree is in blossom. The sweet scented flowers of *Plumeria* have a mixed of fragrance of jasmine, citrus, gardenia and other strange odors. In Java it is customary to present flowers of *Plumeria* while welcoming distinguished guests. The volatile oil extracted from the flowers is used widely in perfumery industry, cosmetics and toiletries. *P.obtusa* varieties with their white flowers are conventionally used in the extraction of flower essence. Steam distillation is adopted for this extraction procedure.

The 5 petals with their unique and perfect symmetry play a key role in psychiatric treatment for easing mental tension. In the psychiatric field *Plumeria* flowers are called the 'flower of perfection'. They are also significant in the alternate system of medical treatment–Aromatherapy–where essential oils extracted from the flowers and other plant parts are put to extensive uses. Mature wood of the tree is widely used in the production musical instruments.

Planting Materials and Propagation

Stem cuttings selected from old trees,–at least two years–are generally used for propagation. Tip cuttings of 30 to 60cm in length, selected from healthy growing, nonbearing branches are allowed to 'cure' in a dry, shady place for 2–3 weeks before planting. It is advisable to remove the mature leaves, retaining only the young ones at the tip portion.

Figure 14.1

Figure 14.2

Figure 14.3

Figure 14.4

Figure 14.5

Figure 14.6

Figure 14.7

Figure 14.8

Figure 14.9

Figure 14.10

Figure 14.11

Figure 14.12

Figure 14.13

Figure 14.14

Figure 14.15

Figure 14.16

Figure 14.17

Figure 14.18

Figure 14.19

Figure 14.20

Figure 14.21

Figure 14.22

Figure 14.23

Figure 14.24

New hybrid varieties can be propagated only by air-layering. For this, mature, healthy branch is selected. A ring of 3cm size bark is removed from the branch 45cm from the tip. The skin-removed portion is kept dry for about 2–3 weeks. In this period the bark above the girdled portion begin to bulge. After 2–3 weeks, the wounded portion is covered with a mixture of moist coco-pith and red soil in equal quantity and wrapped with plastic strap. With in a month, adventitious roots appear on the surface of the plastic covered portion and the twig is ready for propagation. The stem is cut below the rooted region and is used for planting.

Hybrid varieties are also propagated by hormone-treated stem cuttings. For this, mature healthy vegetative stem cutting is selected. After 'curing' in shade for 1–2 weeks, the lower, 6cm portion of the cutting is dipped in a solution containing 20 ppm of Indole Butyric Acid or any other rooting hormone and 2 per cent of Carbendazim (fungicide) until adventitious roots appear at the cut-end portion.

Plumeria seldom produces fruits and seeds. The architecture of the flower with a lower long tubular portion of the corolla makes it inaccessible for the insect pollinators to carry out entamophilous pollination. Mature flower buds or young flowers can be artificially pollinated by carefully collecting the pollens situated in the tubular corolla and transferring to the stigma deeply located inside the corolla tube. This can be achieved by using a clean brush with fine bristles. A successful pollination results in fruit setting within a few weeks. The fruits seen in pairs are large-sized which dehisce on maturity to release the seeds. The seeds are winged with a dry and broad wing on one side. Seeds from trees with white flowers usually produce white flowered plants. Similarly dark red and yellow varieties produce their respective colours. Usually multicoloured plants give rise to a range of flower colours which are unpredictable, although it will tend to reflect parent plant's flower quality. In *Plumeria* hybridization is usually done between *P. obtusa* and *P.accuminata*. The hybrid plants like the *'Lolani'* variety, produces flowers with a display of shades of yellow and white. Hybridization among *P.rubra* varieties will yield only hybrids with red, pink or yellow flower colours.

Seeds are collected when the pods are split open. They are planted in a shallow vessel containing moistened potting mixture with the winged portion in upright position. Seeds will germinate in about 2 weeks. Seedlings can be used for transplantation after they have produced 4-5 true leaves.

Plants propagated through stem-cuttings generally produce flowers with in one year of growth, depending on the maturity of the stem and age of the mother tree from which the cutting was selected. Seedlings take about three years for flower production. Even though the cuttings maintain the quality of the cultivar, they produce only shallow roots and with limited life span for the saplings. Seed–propagated plants produce deeper roots and the tree has a typical symmetrical growth pattern with a much longer life. New hybrid varieties with riotous colours are produced mainly through seed propagation.

Cultural Practice

Plumeria is grown as an ornamental in spacious gardens and for landscaping. Areas with ambient light conditions and well drained soil are the best suited for growing of this tree. The Singapore and Indonesian dwarf varieties of *Plumeria* are also grown in large containers with the pot-mixture having a minimum depth of 60 cm. For healthy growth and profuse flowering, the tree requires at least 6-7 hours of direct light or else the branches will become leggy with very little flower production.

For planting on the ground, a pit of 45 x 45x 45 cm is prepared. The pit is filled with a planting mixture with red soil and garden compost in 3:1 ratio. The 'cured' stem cutting is planted with one third of the lower end inserted in the mixture. A strong support is provided for easy and upright establishment. The same mixture is used for growing the plant in containers with proper drainage facilities. The pH of the medium should be around neutral value (7.8). It takes about one month for the cutting to produce new buds and leaves. In the early stage of the growth, *Plumeria* requires only urea, which is meant to promote better leaf production and healthy development. After the tree is fully established, 2 per cent 18:18:18 (N.P.K) can be given in biweekly interval. At the time of flower induction, N.P.K with lesser nitrogen content is recommended which will facilitate profuse flowering. Along with chemical fertilizers, bio-manures like compost, dried, powdered cow dung are also good. The fallen leaves of the tree can be used as a leaf mold, which is a good bio-manure similar to partly decomposed cow-dung.

As the tree is adapted to semi-dry conditions, watering is required only during severe summer in alternate days. Over watering will lead to excessive leaf formation and reduction in blooming. Watering and water-logging during monsoon period should be avoided. Water-logging will lead to unusual leaf fall affecting the growth of the tree. Pot grown plants can be watered when the pot mixture is almost fully dried up.

Many cultivars of *Plumeria* have free flowering nature. Others bloom 2 or 3 times annually. Flowering is more prevalent in March–April. The flower bunch remains on the tree for about 2 weeks. The profuse branching of the tree is directly related to the free flowering nature. *P.obtusa* hybrids show excessive branching, leading to small size with a shrubby habit and copious flowering. A newly introduced *P.rubra* variety *'Brancher'* also shows numerous branches with almost 15 branches naturally produced from the main stem. This variety has beautiful yellowish-white large-sized flowers adding to its adorable nature.

After the abscission of the older flower bunch, spent-away branches sprout new branches, leading to the production of new bunches. *Plumeria* is a deciduous tree, shedding leaves during winter, *P.obtusa* varieties, however, show leaves on the tree almost all through the year.

Disease and Pest Manangement

Water-logging in the root zone of the tree can lead to fungal rot of roots. Rotting of roots may result in abnormal shedding of leaves and death of young branches. Cultural practice avoiding excessive watering and keeping the soil almost dry can protect *Plumeria* from root rot. Systemic fungicides containing Carbendazin or Hexaconazole are recommended to get rid off the symptoms of decay. When young plant is affected with rot, the plant is uprooted from the soil and the rotted portion and stem are removed and remaining healthy portion is treated with a fungicide. After fungicide treatment, it is kept in shade for 1-2 weeks and used for planting.

A white 'powder like' coating formed on the under side of the leaves indicates presence of white flies as a colony, sucking away the plant sap, leading to yellowing and dropping of immature leaves. Insecticides containing Imidachloropid(1 per cent) is recommended to overcome the insect attack.

References

Alfred byrd. Graf 1986. Tropica–Colour cyclopedia of exotic plants and trees. Roehrs Company, East Ruther ford, N.J. USA.

Bailey, S.H. 1953. The standard cyclopedia of Horticulture Vol. III Macmillan Co. New York USA.

Gopalaswamiyengaar, K.S. 1991. Complete Gandering in India. Gopalaswamy Parthasarthy, 'Srinivas' 237/46, 5th Road, Chamarajapet, Bangalore.

Swaminathan, M.S. and Kochar, S.L. 2003. An Atlas of Major flowering trees in India, Macmillan India Ltd, New Delhi.